An Approach to the Extension of a Theorem Prover by Advanced Structuring Mechanisms

Vom Fachbereich Mathematik und Informatik
der Universität Bremen
genehmigte Dissertation von **Maksym Bortin**

Gutachter: PD Dr. Christoph Lüth
Prof. Dr. Hans-Jörg Kreowski

Datum des Promotionskolloquiums: 13.04.2010

An Approach to the Extension of a Theorem Prover by Advanced Structuring Mechanisms

Maksym Bortin

Bibliografische Information der Deutschen Nationalbibliothek

Die Deutsche Nationalbibliothek verzeichnet diese Publikation in der Deutschen Nationalbibliografie; detaillierte bibliografische Daten sind im Internet über http://dnb.d-nb.de abrufbar.

ISBN 978-3-8325-2502-6

Logos Verlag Berlin GmbH
Comeniushof, Gubener Str. 47,
10243 Berlin
Tel.: +49 (0)30 42 85 10 90
Fax: +49 (0)30 42 85 10 92
INTERNET: http://www.logos-verlag.de

Contents

Chapter 1

Introduction

The presented work gives a formal foundation for an extension of the theorem prover Isabelle by morphisms in order to provide mechanisms within the framework for structuring and reuse of abstract developments. The approach has been also implemented in Isabelle, such that a case study, demonstrating how it impacts theory development in the practice, is presented here as well. The approach and its implementation have been mainly developed within the scope of the project '*Abstraktion und Wiederverwendung*' ('*Abstraction for Reuse*'), supported by DFG (German Research Council), and the software package, including a manual, is available at the project web-site `http://www.informatik.uni-bremen.de/~cxl/awe`

The aim of this chapter is to give an overview of the Isabelle system, emphasising aspects like underlying logic and system architecture, which are central for this work. Further, this introduction also sketches related work as well as already available structuring concepts in Isabelle and gives motivations for integration of more advanced structuring by morphisms.

Although the presented approach is tailored to Isabelle, the formal foundation focuses on its general principles and techniques. Based on these, it could be adapted and applied also to other formal systems satisfying some essential properties like the presence of proof terms.

1.1 The theorem prover Isabelle

Isabelle [Nipkow *et al.* 2002] is a logical framework in form of an interactive generic LCF-style theorem prover implemented in Standard ML. ML is a functional programming language which has been designed and developed with the motivation to encode logical inference steps as functions in a programming language [Milner 1985], which also explains the name: ML stands for *meta language*. One of the predecessors of Isabelle is Edinburgh LCF (abbreviating *Logic for Computable Functions*), a programmable proof-checker [Paulson 1990]. The key idea of the LCF-approach is to use an abstract ML data type (usually called `thm`, a shortening for *theorem*) to represent the set of derivable sentences in a logic. This means that the proof checker provides only few basic functions which can construct or manipulate values of the type `thm`, forming the logical kernel. So, the kernel encodes basic axioms and inference rules of a logic, such that any value having the type `thm` must have been *proved*, or in other words constructed via these inference rules. Such a construction can be considered as a finite sequence of applications of the basic kernel functions, and hence can be separately represented by so-called *proof terms*. Using these, to any value of the type `thm` a proof term representing its deduction can be recorded and assigned. On the other hand, any (well-formed) proof term determines some derivable proposition, interpreting the recorded deduction steps by the logical kernel. The support of formal proof terms by a proof system turns out to be important in the context of structuring, abstraction and reuse of formal developments.

In particular, Isabelle's logical kernel implements the so-called *meta logic*, which is an intuitionistic fragment of higher order logic extended with Hindley-Milner polymorphism and type classes. Moreover, proof terms corresponding to the meta-logical inference steps under the Curry-Howard isomorphism, are formalised as an algebraic data type in ML and recorded by the framework for any deduction. The idea behind the choice of the meta logic is to give a possibility to encode other (usually stronger) logics in terms of the meta logic. Such encoded logic is then called *object logic*. This approach also explains the name *logical framework*, which may be understood as a meta-level inference system within which other deductive systems can be specified [Pfenning 2001].

Formalisations of object logics as well as further developments inside these are organised via hierarchical *theories*. Any Isabelle theory is formed by the basic elements: type classes, type constructors, constants and axioms, and can *extend* (import) a finite number of another theories, thus importing all the basic elements from there. Importing gives a basic structuring mechanism in form of a theory hierarchy. At the top-level, Isabelle has been extended by Isar [Wenzel 2001], an environment for interactive theory development with a powerful high-level human-readable proof language, theory presentation, etc. In Isabelle/Isar theories are represented using the following form

theory $\mathcal{T}$
imports $\mathcal{T}_1 \ldots \mathcal{T}_n$
begin
<declarations and proofs>
end

Isar can be seen as a prime example of an extension of the framework by new powerful concepts without changing the meta logic. In the same manner, numerous further derived concepts and tools have been integrated into Isabelle. The implementation of the approach, presented in this thesis, can be basically seen as one of these.

On the theory level, Isabelle's meta logic is represented by a fixed special theory forming the root element in the hierarchy. Roughly, a theory, encoding an object logic, imports the meta-logical theory and introduces basic object-logical connectives together with the axioms specifying these by means of the meta-logical operations. Then, derived object-logical connectives can be specified by *definitions*, i.e. by a special sort of axioms identifying a fresh operation symbol with a closed term. Different object logics have been already formalised in Isabelle, e.g. intuitionistic and classical first order logics, set theory, and classical higher order logic (HOL, for short). HOL is the most developed object logic in Isabelle, providing such derived concepts as algebraic (freely generated) data types, extensible records, (co)inductive and recursive definitions, and containing theory developments in various areas of mathematics and computer science, *inter alia* in number theory, algebra, program semantics and verification.

1.2 Consistency and correctness

According to the LCF-architecture of Isabelle, the question of the correctness of derived results reduces to the following problems.

- Consistency of the meta logic. As mentioned above, Isabelle's meta logic is a weak intuitionistic fragment of the higher order logic, and the rules are standard and well-understood, such that they can be assumed to be consistent.

- Correct implementation of the rules. Due to the weak meta logic, the primitive inference rules are implemented by relatively small number (15) of functions. The validation of these is also facilitated by the fact that they are written in a functional programming language. Some of the implementations are more involved, while some are fairly easy to validate. However, Isabelle has been intensively used in research already for two decades, and indeed, in the early versions some problems in the implementation of the logical kernel have been discovered and fixed. This fact should emphasise, that the implementation is not just trusted, but is in a continuous process of validation, pursued by the developers and users. In the meantime, it can be seen as a reasonable hypothesis that the inference rules of the meta logic are correctly implemented by the kernel functions.

- Sound axiomatisations of the object logics. For such well-explored logics like HOL or the first order logic there are standard axiomatisations for which soundness has been already proved by other means.

- Validation of axioms introduced apart of the object-logical axiomatisation. This is the actual problem, since this way inconsistencies may be introduced, inconspicuously making further deductions worthless. Hence, the standard development strategy in Isabelle is to introduce new axioms only in the case when these form a *conservative extension* to the considered object logic. As already mentioned, definitions form such extensions. Let us call a theory development involving only conservative extensions to an object logic a *conservative development.* Then the question of correctness of propositions, derived in a conservative development, can be seamlessly reduced to the correctness assumptions from the previous three points.

Altogether, these points justify that propositions, derived in a conservative development, are reasonably correct, i.e. correspond to valid sentences in the considered object logic. This property makes Isabelle a valuable tool, e.g. for reasoning about safety critical systems. At this point the natural question arises how such systems can be specified and structured in Isabelle. Treatment of this problem is one of the main subjects of the presented work.

1.3 Specification, parametrisation and structuring in Isabelle

Polymorphism and type classes

As already mentioned, the meta logic of Isabelle is based on the polymorphic simply typed lambda calculus, such that the type system is similar to the type system of the underlying programming language ML. This Hindley-Milner polymorphism provides a basic possibility for abstraction and reuse within the meta logic: the standard example function $reverse :: \alpha\ list \Rightarrow \alpha\ list$, reverting a list of elements, abstracts over the type of these elements, such that propositions like $reverse(reverse(x)) = x$ are proven once for all possible type instantiations. On the other hand, this type system is also simple enough, such that type inference is decidable in polynomial time.

The type system has been further refined by the concept of type classes, providing order-sorted polymorphism. Initially, type classes have been introduced in Isabelle as a tool to support formulation of different object logics [Nipkow 1993]. So, two type classes where one is declared to be a subclass of the other allow, e.g. to distinguish between first and second order quantifications in a formalisation of the second order logic, restricting the particular polymorphic type parameters. For example, we can declare two type classes c_1, c_2, and some polymorphic predicate P, and restrict its type parameter to the class c_1. This declaration prevents that some proposition $P(t)$, where the type of the term t is of class c_2 (and hence not necessarily of class c_1) can be derived, because it is not correctly typed. The only possibility to change this is given by an additional specification of c_2 as a subclass of c_1, in the case when c_1 is not already specified to be a subclass of c_2.

Local proof contexts

The basic concept for structured specifications in Isabelle are so-called *local proof contexts* [Kammüller *et al.* 1999], or locales for short. It is based on the fact that in presence of a higher order meta logic, predicates having arbitrary functions as parameters can be *defined*. So, this approach fits seamlessly into the conservative development approach in Isabelle, mentioned above. Taking into account the polymorphism, such higher order predicates can capture axiomatisations of, e.g. algebraic structures parameterised over the type of domain elements:

locale *group* =
fixes $e :: \alpha$
$\circ :: \alpha \Rightarrow \alpha \Rightarrow \alpha$
$inv :: \alpha \Rightarrow \alpha$
assumes `<group axioms>`

This declaration is then internally translated into a definition of a higher order polymorphic predicate for groups, where the type variable α and the constants $e, \circ, inv$ form the parameters of this locale. Such a locale can be further used as a proof context, which essentially means that the corresponding predicate is implicitly added to the meta-logical assumptions of a conjecture using fixed (type) variables for parameters. Moreover, instantiations of a locale can be specified as well: an instance comprises a structure satisfying the underlying predicate. So, in the group locale above one needs to give a type for α together with interpretations for the group operations on this type, which produces so-called *proof obligations*, i.e. in this case that the given interpretations satisfy the group axioms. If these can be proved, all properties shown with the group locale as a proof context are automatically instantiated by the given interpretation: this step is essentially an application of the *modus ponens* rule provided by the kernel. Locales can be also structured as follows: a locale can either extend another locales, which is specified by its declaration, or it can be stated that a locale L_1 is a *sublocale* of L_2 [Ballarin 2009]. In the latter case, the locale L_2 is interpreted in the context L_1, which can produce proof obligations as well. If these can be proved, the properties shown in the context L_2 are automatically propagated to L_1.

Specifying type classes

Further, Isabelle has been also extended with a possibility to specify type classes, similarly to the approach in the programming language Haskell, e.g. [Haftmann 2009]. This built-in mechanism is based on the concept of locales, and thus essentially provides types qualified by predicates. For example, the class of pre-orders can be specified in Isabelle/HOL as follows:

class *preorder* =
fixes $\leq \ :: \ \alpha \Rightarrow \alpha \Rightarrow bool$
assumes
`reflexive :` $\forall a.\ a \leq a$
`transitive :` $\forall a\ b\ c.\ a \leq b \wedge b \leq c \longrightarrow a \leq c$

That is, *preorder* qualifies those types α for which an operation $\leq \ :: \ \alpha \Rightarrow \alpha \Rightarrow$ *bool*, satisfying the reflexive and transitive laws, exists.

Now, the class of linear orderings can be specified as a subclass of pre-orders:

class *linorder* = *preorder* +
...

Using *linorder*, e.g. a polymorphic sorting function on lists *sort* with the type $(\alpha :: linorder)\, list \Rightarrow (\alpha :: linorder)\, list$ can be defined. The mechanism for building instances of such type classes is provided by the **instantiation**-command. So we can use the type *nat*, which represents the natural numbers in Isabelle/HOL, as an instance of *linorder*:

instantiation *nat* :: *linorder*

begin
defining $\leq$ *as the less-or-equal order on the natural numbers and proving the laws*
end
This way the function *sort* can be applied to any list of natural numbers.

1.4 Axiomatic theories and morphisms

Altogether, locale proof contexts provide a light-weight extension of the framework with structured parameterised specifications inside the meta logic. One of the basic problems of this approach is that it lacks the possibility to specify dependencies between the type variables occurring in the types of parameters, apart of the equality. Let us sketch the problem, formalising Kleene algebras using a locale and then trying to extend this to so-called *typed Kleene algebras* [Kozen 1998]. The latter concept will be further seized in the case study (Section 6.1.1)

Kleene algebras can be formalised by means of a local proof context in Isabelle, e.g. as follows:

locale *Kleene-Algebra* =
fixes 0 :: α
1 :: α
$\_ + \_$ :: $\alpha \Rightarrow \alpha \Rightarrow \alpha$
$\_ \cdot \_$ :: $\alpha \Rightarrow \alpha \Rightarrow \alpha$
$\_^{\star}$ :: $\alpha \Rightarrow \alpha$
assumes `<axioms of the Kleene algebra>`

So, the type parameter α represents the type of values in the domain of a Kleene algebra. The notion of typed Kleene algebra generalises this: now domain elements are considered as arrows, thus having some (possibly different) source and target types. This allows us to specify that, e.g. multiplication has not necessarily be defined for any two domain elements ([Kozen 1998] as well as the presented case study give enough motivation for this treatment). The natural way to express this in Isabelle is to introduce a new type constructor, say K, with the rank 2. That is, $(\alpha, \beta)\, K$ constructs a type for any two argument types α, β. Using this, we can take those types, which are constructed via K, as the possible types of domain values, and then change the type of multiplication to $\_ \cdot \_ \;::\; (\alpha, \beta)\, K \Rightarrow (\beta, \gamma)\, K \Rightarrow (\alpha, \gamma)\, K$. The actual problem now is, that the type constructor K can only be declared outside the locale above. The reason is, of course, that a locale basically denotes a predicate, and predicates cannot have type constructors as parameters, due to the rank-1 polymorphism.

The actual solution for this problem is a specification of K as a *type constructor class* with the typed Kleene algebra operations as its class operations. The concept of type constructor classes is also known from the functional programming language Haskell [Peyton Jones 2003]. In contrast to programming languages, in Isabelle type constructor classes can be also specified by arbitrary conditions on the class operations. From the discussion above follows that this can only be done in form of additional axioms, since type constructors cannot be qualified by predicates. For the particular typed Kleene algebra specification this means that the rules placed in the **assumes**-clause of the locale above have to be restated in form of proper axioms outside the locale.

There are the following important observations about this approach:

- any specification of a type constructor class can only be done by means of a theory, not a local proof context;
- such a theory will in general form a non-conservative extension to the object logic;
- even if we could derive some properties of the specified type constructor class in Isabelle, we could not propagate these results anywhere, as we can do by instantiation of locale proof contexts, because there is no notion of *what* is an instance of such a class or more generally, *what* is an instance of the theory specifying this class.

The approach to the extension of the framework by signature and theory morphisms, initially introduced in [Bortin, Johnsen and Lüth 2006] and explicitly described in this thesis, provides exact answers to these questions within Isabelle. More precisely, in accordance with locale-sublocale relation we can say that an *instance* of a theory is any other theory (also called *subtheory*) such that a theory morphism from the former to the latter exists. In particular, regarding the example above, any theory morphism from the source theory, axiomatising typed Kleene algebras, to some target theory establishes that this target theory provides some structure forming a typed Kleene algebra. However, this structure could comprise an axiomatic specification as well. Generally, let us call a subtheory *proper* if it as part of a conservative development. Then the existence of a proper subtheory of some axiomatic specification can be considered as a proof that the original specification is consistent. That the presented notions of morphisms indeed provide these properties is founded in a proof-theoretical way, which is in turn provided by the fact that our environment is a formal proof system.

The morphism approach moreover gives advanced structuring of theories extending the theory hierarchy, i.e. the structuring by import mentioned above. This property allows to model parametrisation in a quite simple way: a theory is considered to be a parameter of another theory if a path from the former to the latter in the hierarchy graph exists. An application of such a parametrisation is then provided by a theory morphism having the parameter theory as its source. This gives us the basic device for reuse of abstract developments *in-the-large*.

Furthermore, those parametrisations where a theory extends its parameter theory in a conservative way will be called *well-formed*. This sort of parametrisations plays a special rôle for axiomatic developments, because instantiations of well-formed parametrisations give conservative extensions as well. This is sketched in Figure 1.1 where the left sub-tree represents some axiomatic, specificational development, while the small triangle inside represents some well-formed parametrisation within it. Of course, axiomatic developments can and

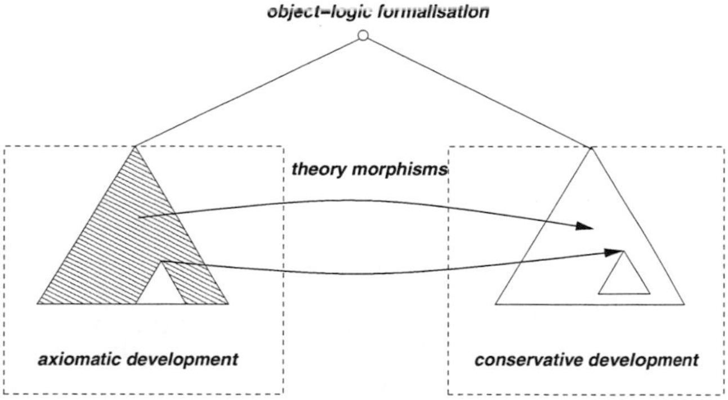

Figure 1.1: The different development approaches related by morphisms

should be structured by morphisms as much as possible, in order to achieve a high level of abstraction and also in order to emphasise interrelations between different structures/theories, which might even stem from different areas of mathematics or computer science.

Polymorphism and type classes revisited

Now, type classes play a similar rôle in axiomatic developments as intended for formalisations of object logics, as mentioned above. This is not very surprising, since object logics are axiomatic specifications as well. Let us for example

consider a declaration of a polymorphic function in Isabelle/HOL:
consts $f :: (\alpha :: \mathit{linorder})\ T \Rightarrow \alpha\ T$
which can be a part of a specification of a type constructor class T.

Alternatively, we can take the same specification, but now replace *linorder* by some fixed parameter type class c without super classes:
consts $f :: (\alpha :: c)\ T \Rightarrow \alpha\ T$
Taking into account theory morphisms, the latter version is more general, since we can obtain the former by an instantiation of c by *linorder*. Moreover, using a theory morphism the type constructor T could be instantiated by *list*, such that the constant f could be in turn instantiated by the sorting function *sort*, mentioned above.

This small example should motivate the treatment of type classes by morphisms: it allows us to use the native polymorphism of the prover in axiomatic specifications, which is then maintained by interpretations given by morphisms. In other words, considering declarations like $f :: (\alpha :: c)\ T \Rightarrow \alpha\ T$ as type parametrisations *inside* the meta logic, morphisms allow to interpret such parametrisations by other parametrisations.

1.5 Related work

The basic idea of structuring of developments by morphisms is well-known in the area of specification languages. So, in [Burstall and Goguen 1980] the category with equational theories as objects and theory morphisms as arrows forms the basis for a denotational semantics of the algebraic specification language *Clear*. Further, the notion of a theory in [Burstall and Goguen 1980] differs from the notion, used here, as follows. Any *theory presentation* $\langle \Sigma, E \rangle$ in Clear, where E is a finite set of Σ-identities, corresponds to a *theory* $\mathcal{T}_E$ here, representing Σ by constant declarations and E by axioms. Then the *theory* in Clear, *presented by* $\langle \Sigma, E \rangle$, is $\langle \Sigma, \overline{E} \rangle$ where $\overline{E}$ denotes the model-theoretically defined closure of E. In contrast, the corresponding notion here will be expressed closing E under the derivability relation on the syntactic level, and denoted by $\mathcal{T}_E^*$. Then $\overline{E}$ can be obtained from $\mathcal{T}_E^*$ by taking only first order identities. This relies on the fact that $\mathcal{T}_E$ formalises $\langle \Sigma, E \rangle$, i.e. a presentation of a first order equational theory, which are known to be complete. In general, for an arbitrary theory $\mathcal{T}$, consistency questions will be 'syntactically' treated here, using the concepts of theory morphisms and conservative extensions. For completeness questions on $\mathcal{T}$, beside the syntactic closure $\mathcal{T}^*$ we would need some model-theoretical counterpart $\overline{\mathcal{T}}$. This complex of problems will not be considered here.

CASL

Let us also consider the *Common Algebraic Specification Language* (CASL) [Mosses 2004] as another example of a specification language. CASL is a first order language and a further development of Clear, not restricted to equational specifications. Let us compare parameterised specifications in CASL to parameterised theories, introduced here. In CASL
spec SP[P **with** *required symbols and properties*]
declares a specification SP as parameterised over a specification P, which provides some required symbols and properties. Then the specification SP[SP′] denotes the instantiation of SP by SP′ if SP′ is an instance of P, i.e. there is a specification morphism from P to SP′. Morphisms in CASL are usually left implicit if clear from the context, but can also be made explicit by a declaration of a *view*, e.g.
view SP′_as_P : P **to** SP′ = < *instantiations* >
Hence, instantiation is actually an example of a specification-building operation. Any specification, built by instantiation like SP[SP′], can be *flattened*, i.e. expressed by the union of the particular specifications, renaming additionally the symbols in accordance with morphisms. This is essentially the semantics of the instantiation. In Isabelle extended by morphisms we can use the integrated framework command for instantiation as follows
instantiate_theory SP **by_thymorph** SP′_as_P
which proceeds as the following diagram shows:

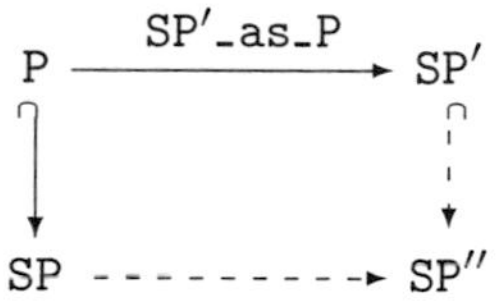

That is, P becomes a separate theory representing the parameters, such that SP imports P, and the resulting theory SP″ corresponds to the *flattened* specification SP[SP′]. The reason for this treatment is that theories in a logical framework like Isabelle are not logical entities, but merely a meta-structuring mechanism. Consequently, specification-building operations, like instantiation in this case, cannot be syntactically abstracted within an Isabelle theory by a construct similar to SP[SP′], such that only their semantics can be addressed.

Furthermore, there are also several systems based on specification languages where morphisms appear in different forms. Let us closely consider two of these: Specware [Srinivas and Jullig 1995] and PVS (*Prototype Verification System*) [Owre at al. 1999].

Specware

Specware is a framework for the composition and refinement founded on the category of specifications and, in this sense related to Clear. In particular, this means that the special notions of parametrisation and instantiation presented here, are covered in Specware generally by diagrams and pushouts. Following the *correct-by-construction* principle, Specware is used in program construction by composition and refinement of specifications. In order to verify emerging proof obligations Specware employs external theorem provers, such that the correctness of derived results depends on different factors like correctness of involved theorem prover, correctness of translations between prover and the system, etc. Even though these problems are of technical and not conceptual nature, they cannot be completely neglected. In this sense, the Specware approach can be seen as dual to the presented: Specware focuses on structuring and refinements and relies on external theorem provers, whereas here formally founded structuring mechanisms are integrated into an LCF-style framework with an expressive logic, powerful proof support and rich library of theories.

PVS

Another approach, which is closely related to the presented has been implemented in PVS, a higher order specification language. In contrast to Specware, which involves external provers, PVS is integrated with a theorem prover and support tools. There are various aspects in which PVS differs from Isabelle. Some of these (see also [Griffioen and Huisman 1998] for further aspects) should be highlighted in this context:

- the underlying logic of PVS is the higher order logic, which is much stronger than Isabelle's meta logic and corresponds to Isabelle/HOL, i.e. to an object logic in Isabelle;
- expressive, but hardly manageable type system: any nonempty predicate is considered as a type, such that dependent types, equivalence on types, sub-typing, etc. are possible;
- the PVS-prover is not LCF-style, based on several primitive concepts like primitive types `int`, `real`, or definitions of recursive functions, etc.

whereas in Isabelle/HOL all these concepts are rigorously derived by conservative extensions and are merely supported by integrated tools, which strongly increases the reliability of derived results.

Also parametrisation of theories is a basic concept in PVS, providing parametric polymorphism. So, a formalisation of groups in PVS may look as follows:

```
group  [G : TYPE,
        e : G,
        o : [G,G->G],
        inv : [G->G] ] : THEORY
BEGIN
 ASSUMING <group axioms> ENDASSUMING
 <derived types, operations and properties>
END groups
```

The formal type parameter G in this formalisation corresponds to the type variable α in the group locale in Isabelle. Then the correspondence between the formal parameters for the group operations and group axioms is clear. Further, any property derived in the local proof context for groups in Isabelle, corresponds to a derived property in the group theory above. However, the main difference is that in the group theory in PVS we can also construct new types and operations, like the type of the subgroup generated by an element x of type G. This cannot be done by means of local proof contexts in Isabelle. A possibility to treat this kind of problems in Isabelle by means of morphisms is shown by the stack example below. Furthermore, actual parameters can be supplied for formal parameters of a parameterised PVS-theory if it is imported, e.g.

```
IMPORTING group[int, 0, +, - ]
```

This step generates the corresponding instances of `<group axioms>` as proof obligations, and allows access to all the derivations in group.

Moreover, PVS also provides the mechanism of *theory interpretations* [Owre and Shankar 2001], which largely subsumes parametrisation. This approach is similar to the presented. So, one can specify

```
group : THEORY
BEGIN
 G : TYPE
 e : G
 o : [G,G->G]
```

```
 inv : [G->G]
 ASSUMING <group axioms> ENDASSUMING
 <derived types, operations and properties>
END groups
```

and then achieve the same instantiation effect using theory interpretation based on *mappings* of types and constants:

```
IMPORTING group{{ G := int, e := 0, o := +, inv := - }}
```

The main difference between parametrisation and mappings is that in the former case actual parameters have to be supplied either for all or for none of formal parameters, while in the latter case the mappings can be arbitrarily partial, such that a part of symbols can remain uninterpreted. Led by this principle, axioms containing uninterpreted symbols do not appear in proof obligations and are asserted again as axioms.

In the presented approach, theory interpretations are covered by morphisms and instantiation of parameterised theories. Moreover, mappings constituting a theory interpretation are here abstracted by separate notions of signature and theory morphisms, providing the possibility to manipulate, e.g. to compose, these. In contrast to PVS, the mappings of morphisms are constrained to be total here, which is not a proper restriction: the distinction between interpreted and uninterpreted symbols is achieved in form of separation between the symbols in the parameter theory and the symbols in the body theory of a parametrisation. Such treatment seems to be more clear: due to the mix of partial mappings and supplies of actual parameters, a PVS theory having several `IMPORTING` declarations, is intuitively not easy to grasp. Furthermore, the mechanism of theory interpretations in PVS is not constructive: `IMPORTING` above instantiates all derived properties of groups asserting thus their validity, which then relies on a correct implementation of different involved mechanisms, whereas here we manipulate proof terms and rely only on the correctness of the logical kernel.

On the other hand, since PVS is based on a specification language, we can refer to instantiations and theory interpretations syntactically in the same manner as in CASL. Moreover, several instances of a structure can be considered in the usual way, which is again not possible in this form in Isabelle. So, for example in order to specify group homomorphisms in PVS, we can consider two instances of the group specification above just by

```
G1, G2 : THEORY group
```

Here G1 and G2 refer to distinguishable copies of the theory group, provided by implicit renaming of type and operation symbols. In the presented approach similar things have to be done 'manually': two theories as copies of the group axiomatisation have to be created, which is merely supported by theory instantiation with explicit renaming.

Example: correctness of a stack implementation

Next, in order to demonstrate how theory interpretations in PVS correspond to morphisms and instantiation in Isabelle/HOL extended by morphisms, let us consider the example in [Owre and Shankar 2001], showing interpretation of a well-known stack specification by the quotient type and operations on it, induced by an equivalence relation on a concrete implementation of a stack. This example basically shows how the correctness relation between an abstract specification and a concrete implementation can be established, following the general principles described in [Ehrig and Kreowski 1999].

The diagram in Figure 1.2 sketches how it works and an outline of the involved Isabelle theories can be found in Appendix A. So, the theory *StackSpec*

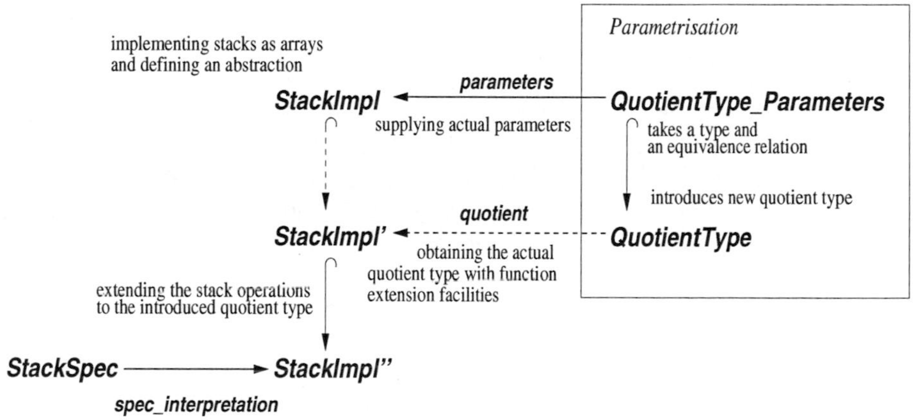

Figure 1.2: Stack-example in Isabelle with morphisms

specifies an abstract data type α *stack*, drawn from the elements of a fixed arbitrary type α, in the usual way. In other words, *StackSpec* specifies the type constructor class $\alpha\,stack$ with the polymorphic operations *empty*, *is_empty*, *top*, *pop*, *push* satisfying well-known properties.

Further, the theory *StackImpl* contains an implementation of a stack by a tuple, comprising an infinite array (represented by a mapping $f :: nat \Rightarrow \alpha$) and a natural number $size :: nat$, such that $size - 1$ points to the top element of a non-empty stack, which is in turn characterised by $size \neq 0$. Then, in *StackImpl* the equivalence relation *eqR* on such tuples is defined, identifying those stacks which sizes are equal and the elements in the particular arrays are equal up to the size. Hence, the stack operations, which are defined in such a way that they never access the elements in array above its size, can be shown to be congruences on *eqR*.

Now, the core of this example forms the theory *QuotientType* parameterised over the theory *QuotientType_Parameters*, where the former simply imports the latter. This parametrisation represents the following general design pattern *in-the-large*: for a given type $\alpha\ T$ and an equivalence relation $\simeq$ on it, the induced quotient type $\alpha\ TQ$ is constructed. Moreover, the 'lifting' facilities, i.e. higher order functions assigning to any $\simeq$-congruence f the corresponding extension $f^\sharp$ to the quotient type $\alpha\ TQ$, are introduced in *QuotientType* as well.

Applying this parametrisation in the stack-example means that the actual parameter $\alpha\ stack$ in *StackImpl* is supplied for $\alpha\ T$ as well as *eqR* for $\simeq$. This has been done by the construction of the theory morphism *parameters*. Consequently, the instantiation of *QuotientType* by *parameters* together with the renaming of the quotient type constructor *TQ* to *stackQ* extends *StackImpl* to *StackImpl'*, introducing the new type constructor *stackQ* which represents the type of 'abstracted' stacks. Moreover, extension functions together with their properties are translated to *StackImpl'* along the theory morphism *quotient* as well. Using these facilities, the concrete stack operations are extended to the operations on the quotient type, which then satisfy the specification of a stack in *StackSpec*. This is emphasised by the construction of the theory morphism *spec_interpretation* from *StackSpec* to *StackImpl''*.

Altogether, the correspondence to the stack-example in [Owre and Shankar 2001] is the following:

- `stack` – *StackSpec*;
- `cstack` – *StackImpl*;
- `equivalence_class[T : TYPE, == : (equivalence?[T])]` – corresponds to the ⟨*QuotientType_Parameters*, *QuotientType*⟩ parametrisation such that `T` corresponds to $\alpha\ T$ and `==` to $\simeq$;
- the last step here, constructing the theory morphism *spec_interpretation*,

corresponds to the last step in the theory `cstack` in PVS, namely to the theory interpretation

```
IMPORTING stack[t]{{ stack := estack, ...}}
```

but is more explicit, since Isabelle does not provide type conversion mechanisms used in PVS.

It should be noted, that the introduction of the quotient type constructor TQ is quite different from the corresponding step in PVS, because of the different type treatments of PVS and Isabelle, as mentioned above. So, in *QuotientType* the Isabelle/HOL **typedef** concept is used, which basically introduces $\alpha\ TQ$ as the type of equivalence classes on $\alpha\ T$ induced by $\simeq$ merely in terms of an injective mapping, assigning to any element of type $\alpha\ TQ$ an $\simeq$-equivalence class, which is consequently a predicate on elements of type $\alpha\ T$. Generally, this construction requires non-emptiness proof, and also introduces a so-called *type definition* axiom – another example of axioms forming conservative extensions beside definitions.

Finally, as mentioned in [Owre and Shankar 2001], operations in Specware can largely be simulated by means of theory interpretations in PVS. Thus, a similar statement holds for Isabelle extended by theory morphisms and instantiation of parameterised theories.

Summary

The next points summarise the most important properties and benefits of the presented approach.

- Light-weight extension of a powerful theorem prover by advanced structuring mechanisms. Light-weight means here that no changes to the framework, let alone the logical kernel, are necessary.

- Allows interpretations and refinements of abstract developments. This property is interesting, e.g. in the context of formal program development, such that lot of tedious proof work can be saved by careful planning and design. Furthermore, such program development methodologies like PROSPECTRA [Krieg-Brückner *et al.*1991], could be directly applied in a logical framework like Isabelle.

- The approach is constructive: any proposition, obtained via translation along a theory morphism, is not just asserted – it has been replayed by the logical kernel from a constructed proof term. In particular, this means that the validity of such propositions does not depend on the implementation of the morphism extension, but only on the correctness of the logical kernel and adequacy of the axioms on which the proofs are based. On the other hand, what depends on the implementation is the correctness of translation of proof terms, i.e. that translated proof terms will always be accepted by the kernel. This aspect is founded by the presented work.

1.6 The structure of the thesis

This thesis is structured as follows. In the first part, the aspects of the framework which are crucial for the morphism extension are formalised. The central notion is that of theories, which are formed by type classes, type constructors, operations, and axioms. Based on this, the sets of sorts, types, terms, and proof terms as well as relations between these are then defined. Further, the notions of signature morphism, their normalisation and composition are introduced. Moreover, the way of construction of signature morphisms in Isabelle is described. The notions of signature morphism and their composition are then extended to theory morphisms, allowing a categorical view on theories and theory morphisms. Furthermore, application of parameterised theories will be declared by means of the presented instantiation mechanism for theories by theory morphisms. This completes the first part.

The second part is devoted to a case study in axiomatic development supported by morphisms in the higher order logic of Isabelle. A formalisation of the notion of an allegory [Bird and de Moor 1997] as a type constructor class plays the central rôle for the case study. Further, the connections between allegories and other algebraic structures, e.g. lattices and Kleene algebras are investigated via morphisms. It will be shown that special kinds of lattices model allegories, while a special kind of allegories, called *locally complete*, model Kleene algebras. Some examples of how morphisms can support abstraction steps, generalising a development, are presented here as well.

Furthermore, in the context of locally complete allegories, an abstract design tactic *in-the-large* for the construction of divide-and-conquer algorithms satisfying some given specification, is formalised by means of *hylomorphisms*. Using this design tactic and the fact that relations together with the usual operations like composition and intersection form a locally complete allegory, we obtain a more concrete Divide-and-Conquer tactic in the relational context. Finally, its applicability is demonstrated by the examples of QuickSort and insertion/update of entries in balanced binary search trees, deriving respective polymorphic functions in the 'correct-by-construction' manner.

Acknowledgement

I would like to thank Prof. Dr. Bernd Krieg-Brückner for providing an excellent environment for young researchers in his working group. I also thank all my colleagues at the University of Bremen and the research group Safe Cognitive Systems at German Research Center for Artificial Intelligence (DFKI) for many interesting talks and discussions. Very special thanks go to my parents for so much support.

Part I

Extending a Logical Framework by Morphisms

Chapter 2

Foundations

As mentioned in the introduction, theories allow us to structure formal developments in a logical framework in form of a theory hierarchy. Furthermore, in order to introduce the notions of signature and theory morphisms, we also need to specify the setup of a theory, and the signature assigned to a theory. So, this chapter gives a formal foundation for these concepts with emphasis on the essential notions for the approach like α-conversion, definitions, and derivability of propositions.

2.1 The set of theories

Elements of the set $\mathcal{T}h$ are called *theories*. There is the special theory $\mathsf{Pure} \in \mathcal{T}h$, representing the meta logic, which will be called the *meta-logical* theory.

The parent relation. The relation $Par \subset \mathcal{T}h \times \mathcal{T}h$ is called the *parent* relation, such that elements of the image of a set $\{\mathcal{T}\}$ under Par, denoted by $Par(\mathcal{T})$, are called *parent* theories of $\mathcal{T}$. The parent relation satisfies the following properties:

1. Par is an acyclic relation;

2. $Par(\mathcal{T}) \in \mathcal{P}_{fin}(\mathcal{T}h)$ holds for any $\mathcal{T} \in \mathcal{T}h$, i.e. the set of parents of a theory is finite,

3. $Par(\mathsf{Pure}) = \emptyset$;

4. *Pure* $\in Anc(\mathcal{T})$ for any $\mathcal{T} \in \mathcal{Th}$;

where Anc denotes the reflexive transitive closure of Par. The elements of $Anc(\mathcal{T})$ are called *ancestors* of $\mathcal{T}$, while any $\mathcal{T}' \in Anc(\mathcal{T})$ with $\mathcal{T}' \neq \mathcal{T}$ is also called a *proper* ancestor of $\mathcal{T}$. Since Par is finitely branching, any theory has a finite number of ancestors as well. In other words, the set $\mathcal{Th}$ is structured hierarchically by Par with *Pure* on the top.

Further, to any two theories $\mathcal{T}_1, \mathcal{T}_2$ we can assign the following subsets of $\mathcal{Th}$:

$$\begin{array}{lcl} Dom_{thy}(\mathcal{T}_1, \mathcal{T}_2) & \stackrel{def}{=} & Anc(\mathcal{T}_1) \setminus Anc(\mathcal{T}_2) \\ Cod_{thy}(\mathcal{T}_1, \mathcal{T}_2) & \stackrel{def}{=} & Anc(\mathcal{T}_2) \setminus Anc(\mathcal{T}_1) \\ Global_{thy}(\mathcal{T}_1, \mathcal{T}_2) & \stackrel{def}{=} & Anc(\mathcal{T}_1) \cap Anc(\mathcal{T}_2) \end{array}$$

called *domain*, *codomain* and *global* theories of $\mathcal{T}_1$ and $\mathcal{T}_2$, respectively. Obviously, by definition these sets are pairwise disjoint.

Theory developments. A finite set $\mathcal{D} \subset \mathcal{Th}$ is called a (*theory*) *development* if $\mathcal{D} = \bigcup_{\mathcal{T} \in \mathcal{D}} Anc(\mathcal{T})$ holds, i.e. if $\mathcal{D}$ is closed under the parent relation. It follows immediately that *Pure* $\in \mathcal{D}$ holds for any development $\mathcal{D}$.

Setup of a theory. The structure of theories in general and of the meta-logical theory *Pure* in particular, will be specified stepwise in the following sections introducing the elements of which a theory consists, starting with type classes.

2.2 Class signatures

Any theory $\mathcal{T} \in \mathcal{Th}$ has a finite set of *type classes* (these will be also called *classes* for short, if no confusion arises), such that the *class signature* $\Sigma_{class}(\mathcal{T})$ of $\mathcal{T}$ is the union of the type classes from all $\mathcal{T}' \in Anc(\mathcal{T})$, together with a *class relation* $\prec_{\Sigma_{class}(\mathcal{T})} \subseteq \Sigma_{class}(\mathcal{T}) \times \Sigma_{class}(\mathcal{T})$. Class relations are constrained to be *acyclic*, i.e. there is no class c such that $c \prec^{+}_{\Sigma_{class}(\mathcal{T})} c$, where $\prec^{+}_{\Sigma_{class}(\mathcal{T})}$ denotes the transitive closure of $\prec_{\Sigma_{class}(\mathcal{T})}$.

Further, let $\prec^{*}_{\Sigma_{class}(\mathcal{T})}$ denote the reflexive transitive closure of $\prec_{\Sigma_{class}(\mathcal{T})}$. If $c' \prec^{*}_{\Sigma_{class}(\mathcal{T})} c$ holds then c is said to be a *super class* of c', and c' a *subclass* of c. The set of super classes of $c \in \Sigma_{class}(\mathcal{T})$ is denoted by $Super_{\Sigma_{class}(\mathcal{T})}(c)$. In case when $c' \in Super_{\Sigma_{class}(\mathcal{T})}(c)$ with $c \neq c'$, c' is also said to be a *proper* super class of c.

Let $\mathcal{T}_1$ and $\mathcal{T}_2$ be two theories with $\mathcal{T}_1 \in Anc(\mathcal{T}_2)$. That $\Sigma_{class}(\mathcal{T}_1) \subseteq \Sigma_{class}(\mathcal{T}_2)$ holds, follows from the definition of a class signature, since $\mathcal{T} \in Anc(\mathcal{T}_1)$ implies $\mathcal{T} \in Anc(\mathcal{T}_2)$ for any theory $\mathcal{T}$. Moreover, in this case the particular class relations have to respect the ancestor relation, i.e. have to satisfy the following conditions:

1. for any $c, c' \in \Sigma_{class}(\mathcal{T}_1)$, $c \prec_{\Sigma_{class}(\mathcal{T}_1)} c'$ holds iff $c \prec_{\Sigma_{class}(\mathcal{T}_2)} c'$ holds;
2. $c \prec_{\Sigma_{class}(\mathcal{T}_2)} c'$ implies $c' \in \Sigma_{class}(\mathcal{T}_1)$ for any $c \in \Sigma_{class}(\mathcal{T}_1)$ and $c' \in \Sigma_{class}(\mathcal{T}_2)$.

This conditions essentially require $Super_{\Sigma_{class}(\mathcal{T}_2)}(c) = Super_{\Sigma_{class}(\mathcal{T}_1)}(c)$ for any $c \in \Sigma_{class}(\mathcal{T}_1)$. Consequently, $Super_{\Sigma_{class}(\mathcal{T}_2)}(c) \subseteq \Sigma_{class}(\mathcal{T}_1)$ holds for any $c \in \Sigma_{class}(\mathcal{T}_1)$.

Further, for any two theories $\mathcal{T}_1$ and $\mathcal{T}_2$, domain, codomain, and global classes can be defined as follows:

$$
\begin{aligned}
Dom_{class}(\mathcal{T}_1, \mathcal{T}_2) &\stackrel{def}{=} \bigcup_{\mathcal{T} \in Dom_{thy}(\mathcal{T}_1, \mathcal{T}_2)} \Sigma_{class}(\mathcal{T}) \\
Cod_{class}(\mathcal{T}_1, \mathcal{T}_2) &\stackrel{def}{=} \bigcup_{\mathcal{T} \in Cod_{thy}(\mathcal{T}_1, \mathcal{T}_2)} \Sigma_{class}(\mathcal{T}) \\
Global_{class}(\mathcal{T}_1, \mathcal{T}_2) &\stackrel{def}{=} \bigcup_{\mathcal{T} \in Global_{thy}(\mathcal{T}_1, \mathcal{T}_2)} \Sigma_{class}(\mathcal{T})
\end{aligned}
$$

In particular, using that class relations respect the ancestor relation, we obtain that the super classes of a global class are global as well, i.e.

$$Super_{\Sigma_{class}(\mathcal{T}_i)}(c) \subseteq Global_{class}(\mathcal{T}_1, \mathcal{T}_2) \tag{2.1}$$

holds for any $\mathcal{T}_1, \mathcal{T}_2 \in \mathcal{T}h$, any global $c \in Global_{class}(\mathcal{T}_1, \mathcal{T}_2)$, and $i = 1, 2$.

Meta logic. The set of type classes of the theory *Pure* is empty, such that $\Sigma_{class}(Pure) = \emptyset$ holds.

2.3 Sorts

Let $\mathcal{T}$ be a theory. The set $Sorts(\Sigma_{class}(\mathcal{T}))$ of *sorts* over the class signature $\Sigma_{class}(\mathcal{T})$ is defined by $Sorts(\Sigma_{class}(\mathcal{T})) \stackrel{def}{=} \mathcal{P}(\Sigma_{class}(\mathcal{T}))$, i.e. any subset of $\Sigma_{class}(\mathcal{T})$ forms a sort over $\Sigma_{class}(\mathcal{T})$. In particular, super classes always form a nonempty sort, i.e. $Super_{\Sigma_{class}(\mathcal{T})}(c) \in Sorts(\Sigma_{class}(\mathcal{T}))$ holds for any $c \in$

$\Sigma_{class}(\mathcal{T})$.

Definition 1 (subsort relation, isomorphic sorts)
Let $\mathcal{T}$ be a theory. Then the class relation $\prec_{\Sigma_{class}(\mathcal{T})}$ induces the ***subsort*** relation $\preceq^{\mathcal{T}}_{sort} \subseteq Sorts(\Sigma_{class}(\mathcal{T})) \times Sorts(\Sigma_{class}(\mathcal{T}))$ defined by: $S \preceq^{\mathcal{T}}_{sort} S'$ holds iff for any class $c' \in S'$ there exists a class $c \in S$ satisfying $c \prec^{*}_{\Sigma_{class}(\mathcal{T})} c'$, i.e. a subclass of c'.

Furthermore, sorts $S, S' \in Sorts(\Sigma_{class}(\mathcal{T}))$ are called ***isomorphic***, denoted by $S \simeq^{\mathcal{T}}_{sort} S'$, if $S \preceq^{\mathcal{T}}_{sort} S'$ and $S' \preceq^{\mathcal{T}}_{sort} S$ holds. Since $\preceq^{\mathcal{T}}_{sort}$ is a pre-order on sorts over $\Sigma_{class}(\mathcal{T})$, $\simeq^{\mathcal{T}}_{sort}$ is an equivalence relation on these sorts.

If $S \preceq^{\mathcal{T}}_{sort} S'$ holds then S is also called a ***subsort*** of S' and S' a ***super sort*** of S. Obviously, $S' \subseteq S$ implies that S is a subsort of S'. In particular, for the empty sort we obtain:

1. $\emptyset \in Sorts(\Sigma_{class}(\mathcal{T}))$, for any theory $\mathcal{T}$;

2. $S \preceq^{\mathcal{T}}_{sort} \emptyset$, for any $\mathcal{T}$ and $S \in Sorts(\Sigma_{class}(\mathcal{T}))$, i.e. $\emptyset$ is a super sort for any sort;

3. $\emptyset \preceq^{\mathcal{T}}_{sort} S$ implies $S = \emptyset$, i.e. the empty sort has no proper super sorts.

Further, we define a reduction relation on sorts, which will allow us to decide whether two sorts are isomorphic.

Definition 2 (reduction on sorts)
Let S and S' be two sorts in $Sorts(\Sigma_{class}(\mathcal{T}))$. Then $S \rightsquigarrow^{\mathcal{T}}_{sort} S'$ holds iff there exist $c, c' \in S$ with $c \prec^{+}_{\Sigma_{class}(\mathcal{T})} c'$, and $S' = S \setminus \{c'\}$. The relation $\rightsquigarrow^{\mathcal{T}}_{sort}$ will be called the *reduction* on sorts and in the case $S \rightsquigarrow^{\mathcal{T}}_{sort} S'$ we say that S *reduces to* S'.

The following points summarise the properties of the reduction relation, which will be used in the sequel.

1. $\rightsquigarrow^{\mathcal{T}}_{sort} \subseteq \supset$;

2. $\rightsquigarrow^{\mathcal{T}}_{sort} \subseteq \simeq^{\mathcal{T}}_{sort}$. If $S \rightsquigarrow^{\mathcal{T}}_{sort} S'$ then $S \preceq^{\mathcal{T}}_{sort} S'$ holds obviously, since $S' \subset S$, by the previous point. On the other hand, $S' \preceq^{\mathcal{T}}_{sort} S$ holds as well, since we can assign to the class c', which is removed from S, some class $c \in S'$ with $c \prec^{+}_{\Sigma_{class}(\mathcal{T})} c'$;

3. $\rightsquigarrow^{\mathcal{T}}_{sort}$ is terminating. Furthermore, any sort $S \in Sorts(\Sigma_{class}(\mathcal{T}))$ has the unique normal form w.r.t. $\rightsquigarrow^{\mathcal{T}}_{sort}$ denoted by $S{\downarrow}^{\mathcal{T}}_{sort}$. Suppose S_1 and S_2 are two normal forms of S. Then, by the previous point, $S_1 \simeq^{\mathcal{T}}_{sort} S_2$. Hence, for any $c \in S_1$ there is $c' \in S_2$ with $c' \prec^*_{\Sigma_{class}(\mathcal{T})} c$. But then there is also some $c'' \in S_1$ with $c'' \prec^*_{\Sigma_{class}(\mathcal{T})} c'$. Since $c'' \prec^+_{\Sigma_{class}(\mathcal{T})} c$ would contradict that S_1 is a normal form, we have $c = c''$ and hence also $c = c'$, since $\prec_{\Sigma_{class}(\mathcal{T})}$ is acyclic. Thus, $c \in S_2$ holds.

Furthermore, for the normal form we can conclude

1. $S{\downarrow}^{\mathcal{T}}_{sort} \simeq^{\mathcal{T}}_{sort} S$;
2. $S{\downarrow}^{\mathcal{T}}_{sort} \subseteq S$;
3. if $c \in S$ then there exists $c{\downarrow} \in S{\downarrow}^{\mathcal{T}}_{sort}$ such that $c{\downarrow} \prec^*_{\Sigma_{class}(\mathcal{T})} c$;
4. $Super_{\Sigma_{class}(\mathcal{T})}(c){\downarrow}^{\mathcal{T}}_{sort} = \{c\}$ for all $c \in \Sigma_{class}(\mathcal{T})$.

For example, let $\mathcal{T}$ be a theory with $\Sigma_{class}(\mathcal{T}) = \{c_1, c_2, c_3\}$, $c_2 \prec_{\Sigma_{class}(\mathcal{T})} c_1$ and $c_3 \prec_{\Sigma_{class}(\mathcal{T})} c_1$. Let $S_1 \overset{def}{=} \{c_1, c_2\}$ and $S_2 \overset{def}{=} \{c_1, c_2, c_3\}$. Then the sort S_1 reduces to $\{c_2\}$, such that $S_1{\downarrow}^{\mathcal{T}}_{sort} = \{c_2\}$ holds, which also confirms the identity above, that the normal form of the super classes of c_2 is equal to $\{c_2\}$, since $Super_{\Sigma_{class}(\mathcal{T})}(c_2) = S_1$. Furthermore, the sort S_2 reduces to $\{c_2, c_3\}$, which also cannot be further reduced, such that $S_2{\downarrow}^{\mathcal{T}}_{sort} = \{c_2, c_3\}$ holds. For instance, we have $S_2{\downarrow}^{\mathcal{T}}_{sort} \preceq^{\mathcal{T}}_{sort} S_2$, since we can assign either c_2 or c_3 to c_1, and c_2, c_3 to themselves.

Finally, the following two properties of the normal form will be applied in the construction of signature morphisms.

Lemma 2.1 *If $S{\downarrow}^{\mathcal{T}}_{sort} = \{c\}$ then $S \subseteq Super_{\Sigma_{class}(\mathcal{T})}(c)$.*

Proof. (by contradiction) Assuming $S{\downarrow}^{\mathcal{T}}_{sort} = \{c\}$ and $S \not\subseteq Super_{\Sigma_{class}(\mathcal{T})}(c)$, we obtain some $c' \in S$ with $c' \notin Super_{\Sigma_{class}(\mathcal{T})}(c)$. Since $c' \in S$ there exists $c'{\downarrow} \in S{\downarrow}^{\mathcal{T}}_{sort}$ with $c'{\downarrow} \prec^*_{\Sigma_{class}(\mathcal{T})} c'$. This implies $c'{\downarrow} = c$, and hence $c \prec^*_{\Sigma_{class}(\mathcal{T})} c'$, which is equivalent to $c' \in Super_{\Sigma_{class}(\mathcal{T})}(c)$ giving a contradiction. □

The other direction is not always true: we could have, e.g. $Super_{\Sigma_{class}(\mathcal{T})}(c) = \{c, c'\}$ and $S = \{c'\}$, such that $S{\downarrow}^{\mathcal{T}}_{sort} = S \neq \{c\}$. So, $c \in S$ is additionally assumed.

Lemma 2.2 *If $S \subseteq Super_{\Sigma_{class}(\mathcal{T})}(c)$ and $c \in S$ then $S{\downarrow}^{\mathcal{T}}_{sort} = \{c\}$.*

Proof. Firstly, $c \in S{\downarrow}^{\mathcal{T}}_{sort}$ holds, because there cannot be some proper subsort of c in S such that c could be reduced. Secondly, if there is some $c' \in S{\downarrow}^{\mathcal{T}}_{sort}$ with $c' \neq c$ then $c' \in S$ holds as well. With $S \subseteq \mathit{Super}_{\Sigma_{class}(\mathcal{T})}(c)$ we can conclude: $c \prec^{+}_{\Sigma_{class}(\mathcal{T})} c'$. This gives the reduction $S{\downarrow}^{\mathcal{T}}_{sort} \rightsquigarrow^{\mathcal{T}}_{sort} S{\downarrow}^{\mathcal{T}}_{sort} \setminus \{c'\}$, contradicting to the definition of normal form. □

2.4 Type signatures

Any theory $\mathcal{T} \in \mathcal{T}h$ has a finite set of *type constructors*, such that the *type signature* $\Sigma_{type}(\mathcal{T})$ of $\mathcal{T}$ contains the type constructors from any $\mathcal{T}' \in \mathit{Anc}(\mathcal{T})$. Moreover, any type constructor $C \in \Sigma_{type}(\mathcal{T})$ has

1. a fixed rank $\mathit{rank}_{\mathcal{T}}(C) \in \mathbb{N}$;
2. a finite set of arities, denoted by $\mathit{arities\text{-}of}_{\mathcal{T}}(C)$. An *arity* of C is a sequence of sorts $(S_1, \dots, S_n, S_{n+1})$, where $n = \mathit{rank}_{\mathcal{T}}(C)$, and $S_i \in \mathit{Sorts}(\Sigma_{class}(\mathcal{T}))$, for $i = 1, \dots, n+1$.

Let $\mathcal{T}_1, \mathcal{T}_2$ be theories with $\mathcal{T}_1 \in \mathit{Anc}(\mathcal{T}_2)$. Then $\Sigma_{type}(\mathcal{T}_1) \subseteq \Sigma_{type}(\mathcal{T}_2)$ holds by the definition of type signatures. Moreover, ranks and arities respect the ancestor relation, i.e. satisfy the following conditions:

1. $\mathit{rank}_{\mathcal{T}_1}(C) = \mathit{rank}_{\mathcal{T}_2}(C)$, for any $C \in \Sigma_{type}(\mathcal{T}_1)$;
2. $\mathit{arities\text{-}of}_{\mathcal{T}_1}(C) \subseteq \mathit{arities\text{-}of}_{\mathcal{T}_2}(C)$, for any $C \in \Sigma_{type}(\mathcal{T}_1)$.

Further, for any two theories $\mathcal{T}_1$ and $\mathcal{T}_2$, domain, codomain, and global type constructors are defined as follows:

$$\begin{aligned}
\mathit{Dom}_{type}(\mathcal{T}_1, \mathcal{T}_2) &\stackrel{def}{=} \bigcup_{\mathcal{T} \in \mathit{Dom}_{thy}(\mathcal{T}_1, \mathcal{T}_2)} \Sigma_{type}(\mathcal{T}) \\
\mathit{Cod}_{type}(\mathcal{T}_1, \mathcal{T}_2) &\stackrel{def}{=} \bigcup_{\mathcal{T} \in \mathit{Cod}_{thy}(\mathcal{T}_1, \mathcal{T}_2)} \Sigma_{type}(\mathcal{T}) \\
\mathit{Global}_{type}(\mathcal{T}_1, \mathcal{T}_2) &\stackrel{def}{=} \bigcup_{\mathcal{T} \in \mathit{Global}_{thy}(\mathcal{T}_1, \mathcal{T}_2)} \Sigma_{type}(\mathcal{T})
\end{aligned}$$

From these definitions we obtain $\mathit{Global}_{type}(\mathcal{T}_1, \mathcal{T}_2) = \Sigma_{type}(\mathcal{T}_1) \cap \Sigma_{type}(\mathcal{T}_2)$, $\mathit{Dom}_{type}(\mathcal{T}_1, \mathcal{T}_2) = \Sigma_{type}(\mathcal{T}_1) \setminus \mathit{Global}_{type}(\mathcal{T}_1, \mathcal{T}_2)$, $\mathit{Cod}_{type}(\mathcal{T}_1, \mathcal{T}_2) = \Sigma_{type}(\mathcal{T}_2) \setminus \mathit{Global}_{type}(\mathcal{T}_1, \mathcal{T}_2)$.

Meta logic. The set of type constructors of the meta-logical theory *Pure* is $\{\texttt{prop}, \Rightarrow\}$, such that $\Sigma_{type}(\mathit{Pure}) = \{\texttt{prop}, \Rightarrow\}$ holds. The rank of prop is 0 and of $\Rightarrow$ is 2. Moreover, $\Rightarrow$ will be used as a right-associative infix operator. Both prop and $\Rightarrow$ have no arities in *Pure*.

2.5 Types

Let $\mathcal{X}_{type}$ denote a fixed infinite countable set of *type variables*. Mappings from the set of type variables to the set $\mathit{Sorts}(\Sigma_{class}(\mathcal{T}))$ for a theory $\mathcal{T}$, will be called *sort assignments* over $\mathcal{T}$. Next, we define, how types are built over a type signature.

Definition 3 (set of types)
Let $\mathcal{T}$ be a theory and ϑ a sort assignment over $\mathcal{T}$. The set of *types* over ϑ and $\Sigma_{type}(\mathcal{T})$, denoted by $\mathit{Types}_\vartheta(\Sigma_{type}(\mathcal{T}))$, is inductively defined as follows;

(i) if $\alpha \in \mathcal{X}_{type}$ then $\alpha ::_{sort} \vartheta(\alpha) \subset \mathit{Types}_\vartheta(\Sigma_{type}(\mathcal{T}))$;

(ii) if $C \in \Sigma_{type}(\mathcal{T})$, $\kappa_1, \ldots, \kappa_n \in \mathit{Types}_\vartheta(\Sigma_{type}(\mathcal{T}))$ where $n = \mathit{rank}_\mathcal{T}(C)$, then the *application of the type constructor* $(\kappa_1, \ldots, \kappa_n)C$ is an element of $\mathit{Types}_\vartheta(\Sigma_{type}(\mathcal{T}))$, (the empty sequence will be omitted in applications, e.g. we write C instead of $()C$, while for sequences of the length 1 the parenthesis will be omitted, e.g. we write $\kappa\ C$ instead of $(\kappa)\ C$).

The set of *types over* $\Sigma_{type}(\mathcal{T})$, denoted by $\mathit{Types}(\Sigma_{type}(\mathcal{T}))$, is then the union of $\mathit{Types}_\vartheta(\Sigma_{type}(\mathcal{T}))$ for all sort assignments ϑ over $\mathcal{T}$, i.e.

$$\mathit{Types}(\Sigma_{type}(\mathcal{T})) \stackrel{def}{=} \bigcup_{\vartheta:\mathcal{X}_{type} \to \mathit{Sorts}(\Sigma_{class}(\mathcal{T}))} \mathit{Types}_\vartheta(\Sigma_{type}(\mathcal{T})).$$

Occurrence of type variables in a type. Any $\kappa \in \mathit{Types}(\Sigma_{type}(\mathcal{T}))$ determines the finite set $\mathit{typevars}(\kappa) \subset \mathcal{X}_{type}$ of type variables which occur in κ. Moreover, any $\kappa \in \mathit{Types}(\Sigma_{type}(\mathcal{T}))$ determines the infinite set of sort assignments ϑ over $\mathcal{T}$ with $\kappa \in \mathit{Types}_\vartheta(\Sigma_{type}(\mathcal{T}))$, since type variables are labelled by sorts. However, all of this sort assignments agree on the type variables occurring in κ. So, let ϑ_κ denote some sort assignment from this set, such that the value $\vartheta_\kappa(\alpha)$ does not depend on the actual choice of ϑ_κ for any $\alpha \in \mathit{typevars}(\kappa)$. So, e.g. if $\mathit{typevars}(\kappa) = \emptyset$ holds, i.e. κ is a non-polymorphic type, then ϑ_κ can be any sort assignment.

By the way, the set of types specified here, corresponds in Isabelle implementation to so-called *certified types*. That is, in an 'uncertified' type the sort labels are not restricted by a sort assignment, such that it can be 'certified' iff it constitutes a correct sort assignment.

of-sort relation. Next, a relation between types and sorts over the type and class signatures of a theory, respectively, will be defined.

Definition 4 (of-sort relation)
Let $\mathcal{T}$ be a theory and ϑ a sort assignment over $\mathcal{T}$. The relation $\textit{of-sort}_{\mathcal{T}}(\vartheta) \subseteq \textit{Types}_{\vartheta}(\Sigma_{type}(\mathcal{T})) \times \textit{Sorts}(\Sigma_{class}(\mathcal{T}))$ is inductively defined as follows:

(i) if $\kappa = \alpha ::_{sort} \vartheta(\alpha)$ with $\alpha \in \mathcal{X}_{type}$ and $\vartheta(\alpha) \preceq^{\mathcal{T}}_{sort} S$ then κ $\textit{of-sort}_{\mathcal{T}}(\vartheta)$ S;

(ii) if $\kappa = (\kappa_1, \ldots, \kappa_n)C$ with $\kappa_1, \ldots, \kappa_n \in \textit{Types}_{\vartheta}(\Sigma_{type}(\mathcal{T}))$, $C \in \Sigma_{type}(\mathcal{T})$, and there exists $(S_1, \ldots, S_n, S_{n+1}) \in \textit{arities-of}_{\mathcal{T}}(C)$ with $S_{n+1} \preceq^{\mathcal{T}}_{sort} S$ and κ_i $\textit{of-sort}_{\mathcal{T}}(\vartheta)$ S_i for all $1 \leq i \leq n$ then κ $\textit{of-sort}_{\mathcal{T}}(\vartheta)$ S.

Then, the relation $\textit{of-sort}_{\mathcal{T}} \subseteq \textit{Types}(\Sigma_{type}(\mathcal{T})) \times \textit{Sorts}(\Sigma_{class}(\mathcal{T}))$ is defined as the union of all $\textit{of-sort}_{\mathcal{T}}(\vartheta)$ relations for any sort assignment ϑ over $\mathcal{T}$, i.e.

$$\textit{of-sort}_{\mathcal{T}} \stackrel{def}{=} \bigcup_{\vartheta : \mathcal{X}_{type} \to \textit{Sorts}(\Sigma_{class}(\mathcal{T}))} \textit{of-sort}_{\mathcal{T}}(\vartheta).$$

By the definition, the of-sort relation respects the subsort relations in the following sense:

Lemma 2.3 *The rule*

$$\frac{\kappa \ \textit{of-sort}_{\mathcal{T}}(\vartheta)\ S \qquad S \preceq^{\mathcal{T}}_{sort} S'}{\kappa \ \textit{of-sort}_{\mathcal{T}}(\vartheta)\ S'}$$

holds for any ϑ, $\kappa \in \textit{Types}_{\vartheta}(\Sigma_{type}(\mathcal{T}))$ *and* $S, S' \in \textit{Sorts}(\Sigma_{class}(\mathcal{T}))$.

Proof. (by cases on κ)
(i) Let $\kappa = \alpha ::_{sort} \vartheta(\alpha)$ with $\alpha \in \mathcal{X}_{type}$. By Definition 4, from κ $\textit{of-sort}_{\mathcal{T}}(\vartheta)$ S we can then conclude $\vartheta(\alpha) \preceq^{\mathcal{T}}_{sort} S$ and, hence κ $\textit{of-sort}_{\mathcal{T}}(\vartheta)$ S', by transitivity of $\preceq^{\mathcal{T}}_{sort}$.
(ii) Let $\kappa = (\kappa_1, \ldots, \kappa_n)C$ with $C \in \Sigma_{type}(\mathcal{T})$, $\textit{rank}_{\mathcal{T}}(C) = n$, and $\kappa_1, \ldots, \kappa_n \in \textit{Types}_{\vartheta}(\Sigma_{type}(\mathcal{T}))$. Again, by Definition 4, from κ $\textit{of-sort}_{\mathcal{T}}(\vartheta)$ S we obtain some arity $(S_1, \ldots, S_n, S_{n+1}) \in \textit{arities-of}_{\mathcal{T}}(C)$ with κ_i $\textit{of-sort}_{\mathcal{T}}(\vartheta)$ S_i for $1 \leq i \leq n$ and $S_{n+1} \preceq^{\mathcal{T}}_{sort} S$. By transitivity of $\preceq^{\mathcal{T}}_{sort}$, we have $S_{n+1} \preceq^{\mathcal{T}}_{sort} S'$, and, thus κ $\textit{of-sort}_{\mathcal{T}}(\vartheta)$ S'. □

Substitution of type variables in a type. Let $\kappa \in \mathit{Types}(\Sigma_{type}(\mathcal{T}))$. A sequence $[\alpha_1 := \kappa_1, \ldots, \alpha_n := \kappa_n]$, where $\alpha_1, \ldots, \alpha_n \in \mathcal{X}_{type}$ and $\kappa_1, \ldots, \kappa_n \in \mathit{Types}_{\vartheta}(\Sigma_{type}(\mathcal{T}))$ for some sort assignment ϑ, is called a *substitution in* κ and denoted by $\overline{[\alpha_i := \kappa_i]}_n$, if the following conditions hold:

1. $\alpha_1, \ldots, \alpha_n$ are pairwise distinct type variables;
2. $\vartheta(\alpha) = \vartheta_{\kappa}(\alpha)$ for any $\alpha \in (\mathit{typevars}(\kappa) \setminus \{\alpha_1, \ldots, \alpha_n\})$;
3. κ_i *of-sort*$_{\mathcal{T}}$ $\vartheta_{\kappa}(\alpha_i)$ for all $1 \leq i \leq n$ and $\alpha_i \in \mathit{typevars}(\kappa)$.

So, $\kappa\overline{[\alpha_i := \kappa_i]}_n \in \mathit{Types}_{\vartheta}(\Sigma_{type}(\mathcal{T}))$ denotes the type, which is obtained from κ by simultaneous substitution of $\alpha_i ::_{sort} \vartheta_{\kappa}(\alpha_i)$ by κ_i for $1 \leq i \leq n$. An important property of the substitution of type variables is that it preserves *of-sort*$_{\mathcal{T}}$ in the following sense:

$$\frac{\kappa \ \textit{of-sort}_{\mathcal{T}}\ S}{\kappa\overline{[\alpha_i := \kappa_i]}_n \ \textit{of-sort}_{\mathcal{T}}(\vartheta)\ S} \tag{2.2}$$

with $\kappa_i \in \mathit{Types}_{\vartheta}(\Sigma_{type}(\mathcal{T}))$, because of the restrictions on the substitution.

α-conversion on types. Two types $\kappa_1, \kappa_2 \in \mathit{Types}(\Sigma_{type}(\mathcal{T}))$ are said to be *α-convertible* by a bijection $\theta : \mathit{typevars}(\kappa_1) \rightarrow \mathit{typevars}(\kappa_2)$, denoted by $\kappa_1 \stackrel{\alpha}{=}_{\theta} \kappa_2$, if the following conditions hold, where $\mathit{typevars}(\kappa_1) = \{\alpha_1, \ldots, \alpha_n\}$ is assumed.

1. $\vartheta_{\kappa_1}(\alpha) = \vartheta_{\kappa_2}(\theta(\alpha))$, for any $\alpha \in \{\alpha_1, \ldots, \alpha_n\}$;
2. $\kappa_2 = \kappa_1\overline{[\alpha_i := \theta(\alpha_i) ::_{sort} \vartheta_{\kappa_1}(\alpha_i)]}_n$.

It remains to justify that $\overline{[\alpha_i := \theta(\alpha_i) ::_{sort} \vartheta_{\kappa_1}(\alpha_i)]}_n$ is indeed a substitution in κ_1. To this end we note that for all $1 \leq i \leq n$ firstly $\theta(\alpha_i) ::_{sort} \vartheta_{\kappa_1}(\alpha_i) \in \mathit{Types}_{\vartheta_{\kappa_2}}(\Sigma_{type}(\mathcal{T}))$ and, secondly $\theta(\alpha_i) ::_{sort} \vartheta_{\kappa_1}(\alpha_i)$ *of-sort*$_{\mathcal{T}}$ $\vartheta_{\kappa_1}(\alpha_i)$ hold, since $\preceq^{\mathcal{T}}_{sort}$ is reflexive.

If a bijection θ with $\kappa_1 \stackrel{\alpha}{=}_{\theta} \kappa_2$ exists then it is uniquely determined by κ_1 and κ_2. Moreover, if $\kappa_1 \stackrel{\alpha}{=}_{\theta} \kappa_2$ holds then $\kappa_2 \stackrel{\alpha}{=}_{\theta^{-1}} \kappa_1$ holds as well. The notation $\kappa_1 \stackrel{\alpha}{=} \kappa_2$ is used to express that there exists some θ with $\kappa_1 \stackrel{\alpha}{=}_{\theta} \kappa_2$. So, $\stackrel{\alpha}{=}$ is an equivalence relation on the set $\mathit{Types}(\Sigma_{type}(\mathcal{T}))$ for any theory $\mathcal{T}$.

The following lemma emphasises that α-convertible types are equal up to a consistent renaming of their type variables.

Lemma 2.4 *Let $\kappa, \kappa' \in Types(\Sigma_{type}(\mathcal{T}))$ with $typevars(\kappa) \subseteq \{\alpha_1, \ldots, \alpha_n\}$ and $\kappa \stackrel{\alpha}{=}_\theta \kappa'$. Then $\kappa\overline{[\alpha_i := \kappa_i]_n} = \kappa'\overline{[\theta'(\alpha_i) := \kappa_i]_n}$ holds as well, where $\theta' : \{\alpha_1, \ldots, \alpha_n\} \rightarrow \mathcal{X}_{type}$ is a mapping, which extends θ as follows:*

$$\theta'(\alpha_i) = \begin{cases} \theta(\alpha_i) & \text{if } \alpha_i \in typevars(\kappa), \\ \text{some } \beta \notin typevars(\kappa') & \text{otherwise.} \end{cases}$$

Proof. Let $typevars(\kappa) = \{\gamma_1, \ldots, \gamma_m\}$. By the definition of α-conversion, from $\kappa \stackrel{\alpha}{=}_\theta \kappa'$ we have $\kappa' = \kappa\overline{[\gamma_j := \theta(\gamma_j) ::_{sort} \vartheta_\kappa(\gamma_j)]_m}$. Thus, we can reason:

$$\begin{array}{ll} \kappa'\overline{[\theta'(\alpha_i) := \kappa_i]_n} & = \\ (\kappa\overline{[\gamma_j := \theta(\gamma_j) ::_{sort} \vartheta_\kappa(\gamma_j)]_m})\overline{[\theta'(\alpha_i) := \kappa_i]_n} & = \\ & \text{since } typevars(\kappa) = \{\gamma_1, \ldots, \gamma_m\} \\ \kappa\overline{[\gamma_j := (\theta(\gamma_j) ::_{sort} \vartheta_\kappa(\gamma_j))\overline{[\theta'(\alpha_i) := \kappa_i]_n}]_m} & = \\ & \text{since } \theta \text{ is a bijection and} \\ & typevars(\kappa) \subseteq \{\alpha_1, \ldots, \alpha_n\} \\ \kappa\overline{[\alpha_i := \kappa_i]_n} & \end{array}$$

□

Furthermore, as a consequence from (2.2) we obtain that α-conversion respects the of-sort relation, i.e. the following rule holds:

$$\frac{\kappa_1 \stackrel{\alpha}{=} \kappa_2 \quad \kappa_1 \ \textit{of-sort}_{\mathcal{T}}\ S}{\kappa_2 \ \textit{of-sort}_{\mathcal{T}}\ S} \qquad (2.3)$$

2.6 Operation signature

Any theory $\mathcal{T} \in \mathcal{T}h$ has a finite set of *constants* (also called *operations*), such that the *operation signature* $\Sigma_{op}(\mathcal{T})$ of $\mathcal{T}$ contains the constants from any $\mathcal{T}' \in Anc(\mathcal{T})$. Moreover, any constant $f \in \Sigma_{op}(\mathcal{T})$ has a fixed type $\textit{type-of}_{\mathcal{T}}(f) \in Types(\Sigma_{type}(\mathcal{T}))$.

Let $\mathcal{T}_1, \mathcal{T}_2$ be theories with $\mathcal{T}_1 \in Anc(\mathcal{T}_2)$. Similarly to class and type signatures, $\Sigma_{op}(\mathcal{T}_1) \subseteq \Sigma_{op}(\mathcal{T}_2)$ follows immediately from the definition of operation signatures. Moreover, the function $\textit{type-of}_{\mathcal{T}}$ respects the ancestors relation as follows: $\textit{type-of}_{\mathcal{T}_1}(f) = \textit{type-of}_{\mathcal{T}_2}(f)$ holds for any $f \in \Sigma_{op}(\mathcal{T}_1)$.

For two theories $\mathcal{T}_1$ and $\mathcal{T}_2$, domain, codomain, and global constants are defined as follows:

$$\begin{array}{lcl} Dom_{op}(\mathcal{T}_1,\mathcal{T}_2) & \stackrel{def}{=} & \bigcup_{\mathcal{T}\in Dom_{thy}(\mathcal{T}_1,\mathcal{T}_2)} \Sigma_{op}(\mathcal{T}) \\ Cod_{op}(\mathcal{T}_1,\mathcal{T}_2) & \stackrel{def}{=} & \bigcup_{\mathcal{T}\in Cod_{thy}(\mathcal{T}_1,\mathcal{T}_2)} \Sigma_{op}(\mathcal{T}) \\ Global_{op}(\mathcal{T}_1,\mathcal{T}_2) & \stackrel{def}{=} & \bigcup_{\mathcal{T}\in Global_{thy}(\mathcal{T}_1,\mathcal{T}_2)} \Sigma_{op}(\mathcal{T}) \end{array}$$

Meta logic. The set of constants of *Pure* is $\{\equiv, \bigwedge, \Longrightarrow\}$, where $\bigwedge$ is called *meta-quantification* with $\textit{type-of}_{\mathsf{Pure}}(\bigwedge) = (\alpha \Rightarrow \texttt{prop}) \Rightarrow \texttt{prop}$, $\equiv$ is called *meta-equality* with $\textit{type-of}_{\mathsf{Pure}}(\equiv) = \alpha \Rightarrow \alpha \Rightarrow \texttt{prop}$, and $\Longrightarrow$ is called *meta-implication* with $\textit{type-of}_{\mathsf{Pure}}(\Longrightarrow) = \texttt{prop} \Rightarrow \texttt{prop} \Rightarrow \texttt{prop}$. The sort assignment information has been omitted in the types above, because $\Sigma_{class}(\mathsf{Pure}) = \emptyset$, allows only one sort assignment, namely that maps any type variable to $\emptyset$. So, e.g. $\alpha \Rightarrow \alpha \Rightarrow \texttt{prop}$ is a shortening of the type $\alpha ::_{sort} \emptyset \Rightarrow \alpha ::_{sort} \emptyset \Rightarrow \texttt{prop}$. Such shortenings, avoiding explicit sort information, will be used from now on as much as possible.

Moreover, $\equiv$ and $\Longrightarrow$ are used as infix operators, $\Longrightarrow$ is right-associative, and $\bigwedge$ is used as a binder.

2.7 Terms

Let $\mathcal{X}_{term}$ denote a fixed infinite countable set of *term variables*. Further, let $\mathcal{T}$ be a theory and ϑ a sort assignment over $\mathcal{T}$. Mappings, assigning to any term variable a type $\kappa \in \textit{Types}_{\vartheta}(\Sigma_{type}(\mathcal{T}))$, will be called *type assignments* over ϑ. Next we define how terms are built over a subset of an operation signature.

Definition 5 (set of terms)
Let $\mathcal{T}$ be a theory, ϑ a sort assignment over $\mathcal{T}$, and ϱ_ϑ a type assignment over ϑ. Further, let $\Sigma \subseteq \Sigma_{op}(\mathcal{T})$. The set of *terms* over ϱ_ϑ and Σ, denoted by $\textit{Terms}_{\varrho_\vartheta}(\Sigma)$, is inductively defined as follows:

(i) if $x \in \mathcal{X}_{term}$ then $x ::_{type} \varrho_\vartheta(x) \in \textit{Terms}_{\varrho_\vartheta}(\Sigma)$;

(ii) if $f \in \Sigma$, $\kappa = \textit{type-of}_{\mathcal{T}}(f)$, $\textit{typevars}(\kappa) \subseteq \{\alpha_1, \ldots, \alpha_n\}$, and there exist $\kappa_1, \ldots, \kappa_n \in \textit{Types}_{\vartheta}(\Sigma_{type}(\mathcal{T}))$ such that $[\alpha_i := \kappa_i]_n$ is a substitution in κ, then $f ::_{type} \kappa[\alpha_i := \kappa_i]_n \in \textit{Terms}_{\varrho_\vartheta}(\Sigma)$;

(iii) if $t_1, t_2 \in \textit{Terms}_{\vartheta,\varrho_\vartheta}(\Sigma)$ then the *term application* $t_1\ t_2$ is an element of $\textit{Terms}_{\varrho_\vartheta}(\Sigma)$;

(iv) if $x \in \mathcal{X}_{term}$ and $t \in \mathit{Terms}_{\varrho_\vartheta}(\Sigma)$ then the *abstraction* $\lambda(x ::_{type} \varrho_\vartheta(x)).t$ is an element of $\mathit{Terms}_{\varrho_\vartheta}(\Sigma)$.

A *term over* Σ is then an element of $\mathit{Terms}_{\varrho_\vartheta}(\Sigma)$ for some assignments ϑ and ϱ_ϑ, i.e.

$$\mathit{Terms}(\Sigma) \stackrel{def}{=} \bigcup_{\vartheta:\mathcal{X}_{type}\to \mathit{Sorts}(\Sigma_{class}(\mathcal{T})),\varrho_\vartheta:\mathcal{X}_{term}\to \mathit{Types}_\vartheta(\Sigma_{type}(\mathcal{T}))} \mathit{Terms}_{\varrho_\vartheta}(\Sigma).$$

Occurrence of constants in a term. Any term $t \in \mathit{Terms}(\Sigma)$ with $\Sigma \subseteq \Sigma_{op}(\mathcal{T})$ defines the set $\mathit{consts\text{-}of}_\mathcal{T}(t) \subseteq \Sigma$ of constants occurring in t. In other words, $\mathit{consts\text{-}of}_\mathcal{T}(t)$ is the least Σ with $t \in \mathit{Terms}(\Sigma)$. In particular, we have $t \in \mathit{Terms}(\mathit{consts\text{-}of}_\mathcal{T}(t))$.

Occurrence of variables in a term. For any term $t \in \mathit{Terms}(\Sigma)$ we can define the finite set $\mathit{termvars}(t) \subset \mathcal{X}_{term}$ of term variables, which occur in t, and the finite set $\mathit{typevars}(t) \subset \mathcal{X}_{type}$ of type variables, which occur in t. The subsets of $\mathit{termvars}(t)$ of bound and free term variables are defined in the standard way, and t is called *closed* if all variables in $\mathit{termvars}(t)$ are bound.

Let $\mathcal{T}$ be a theory, $\Sigma \subseteq \Sigma_{op}(\mathcal{T})$ and $t \in \mathit{Terms}(\Sigma)$. Similarly as with sort assignments and types, we can choose some sort assignment ϑ over $\mathcal{T}$ and then some type assignment ϱ_ϑ with $t \in \mathit{Terms}_{\varrho_\vartheta}(\Sigma)$, and denote these by ϑ_t, ϱ_{ϑ_t}, respectively. Again, the values $\vartheta_t(\alpha)$ and $\varrho_{\vartheta_t}(x)$ do not depend on the particular choice as long as $\alpha \in \mathit{typevars}(t)$ and $x \in \mathit{termvars}(t)$ hold. As with types, terms here correspond to *certified terms* in Isabelle.

Substitution of type variables in a term. The substitution of type variables in a type is extended to substitution of type variables in a term $t \in \mathit{Terms}(\Sigma)$ with $\Sigma \subseteq \Sigma_{op}(\mathcal{T})$ as follows. A sequence $[\alpha_1 := \kappa_1, \ldots, \alpha_n := \kappa_n]$, where $\alpha_1, \ldots, \alpha_n \in \mathcal{X}_{type}$ and $\kappa_1, \ldots, \kappa_n \in \mathit{Types}_\vartheta(\Sigma_{type}(\mathcal{T}))$ is called a *type substitution in t* and denoted by $\overline{[\alpha_i := \kappa_i]_n}$, if the following conditions hold:

1. $\alpha_1, \ldots, \alpha_n$ are pairwise distinct type variables;
2. $\vartheta(\alpha) = \vartheta_t(\alpha)$, for any $\alpha \in (\mathit{typevars}(t) \setminus \{\alpha_1, \ldots, \alpha_n\})$;
3. κ_i *of-sort*$_\mathcal{T}$ $\vartheta_t(\alpha_i)$, for $1 \leq i \leq n$ and $\alpha_i \in \mathit{typevars}(t)$.

Then $t\overline{[\alpha_i := \kappa_i]_n} \in \mathit{Terms}_{\varrho_\vartheta}(\Sigma)$ denotes the term, which is obtained from t by simultaneous substitution of $\alpha_i ::_{sort} \vartheta_t(\alpha_i)$ by κ_i for $1 \leq i \leq n$, such that $\varrho_\vartheta(x) = \varrho_{\vartheta_t}(x)\overline{[\alpha_i := \kappa_i]_n}$ holds for all $x \in \mathit{termvars}(t)$.

α-convertible terms. Now, the α-conversion on types will be extended to terms. At this, only bijections between type variables will be considered. The reason is that in the following the use of the α-conversion on terms will be reduced to those cases where the argument terms are closed. So, the question of the existence of a bijection between term variables in two terms reduces to the question of equality on the terms, obtained from these replacing term variables by corresponding de Bruijn indices.

Let $\mathcal{T}$ be a theory and $\Sigma_1, \Sigma_2 \subseteq \Sigma_{op}(\mathcal{T})$. Terms $t_1 \in \mathit{Terms}(\Sigma_1)$ and $t_2 \in \mathit{Terms}(\Sigma_2)$ are *α-convertible* by a bijection $\theta : \mathit{typevars}(t_1) \to \mathit{typevars}(t_2)$, denoted by $t_1 \overset{\alpha}{=}_\theta t_2$, if the following conditions hold, where $\mathit{typevars}(t_1) = \{\alpha_1, \dots, \alpha_n\}$ is assumed.

1. $\vartheta_{t_1}(\alpha) = \vartheta_{t_2}(\theta(\alpha))$, for any $\alpha \in \{\alpha_1, \dots, \alpha_n\}$;
2. $t_2 = t_1\overline{[\alpha_i := \theta(\alpha_i) ::_{sort} \vartheta_{t_1}(\alpha_i)]_n}$.

As with the α-conversion on types, it remains to justify that the substitution above is well-defined. To this end we note, that for all $1 \leq i \leq n$, $\theta(\alpha_i) ::_{sort} \vartheta_{t_1}(\alpha_i) \in \mathit{Types}_{\vartheta_{t_2}}(\Sigma_{type}(\mathcal{T}))$, and $\theta(\alpha_i) ::_{sort} \vartheta_{t_1}(\alpha_i)$ *of-sort*$_\mathcal{T}$ $\vartheta_{t_1}(\alpha_i)$ hold.

Since the substitution of type variables in a term extends the substitution in types, also the statement of Lemma 2.4 can be extended to α-convertible terms as follows.

Lemma 2.5 *Let $t, t' \in \mathit{Terms}(\Sigma)$ with $\mathit{typevars}(t) \subseteq \{\alpha_1, \dots, \alpha_n\}$ and $t \overset{\alpha}{=}_\theta t'$. Then $t\overline{[\alpha_i := \kappa_i]_n} = t'\overline{[\theta'(\alpha_i) := \kappa_i]_n}$ holds as well, where $\theta' : \{\alpha_1, \dots, \alpha_n\} \to \mathcal{X}_{type}$ is a mapping, which extends θ as follows:*

$$\theta'(\alpha_i) = \begin{cases} \theta(\alpha_i) & \text{if } \alpha_i \in \mathit{typevars}(t), \\ \text{some } \beta \notin \mathit{typevars}(t') & \text{otherwise.} \end{cases}$$

of-type relation. As with types and sorts, the following definition relates terms and types.

Definition 6 (of-type relation)
Let $\mathcal{T}$ be a theory, ϑ a sort assignment and ϱ_ϑ a type assignment. The relation *of-type*$_\mathcal{T}(\varrho_\vartheta) \subseteq \mathit{Terms}_{\varrho_\vartheta}(\Sigma_{op}(\mathcal{T})) \times \mathit{Types}_\vartheta(\Sigma_{type}(\mathcal{T}))$ is inductively defined as follows:

(i) if $t = x ::_{type} \varrho_\vartheta(x)$ with $x \in \mathcal{X}_{term}$ then t *of-type*$_\mathcal{T}(\varrho_\vartheta)$ $\varrho_\vartheta(x)$;

(ii) if $t = f ::_{type} \kappa$ with $\kappa \in \mathit{Types}_\vartheta(\Sigma_{type}(\mathcal{T}))$ then t *of-type*$_\mathcal{T}(\varrho_\vartheta)$ κ;

(iii) if $t = t_1\ t_2$ with $t_1, t_2 \in \mathit{Terms}_{\varrho_\vartheta}(\Sigma_{op}(\mathcal{T}))$, t_i *of-type*$_\mathcal{T}(\varrho_\vartheta)$ κ_i for $i = 1, 2$ and $\kappa_1 = \kappa_2 \Rightarrow \kappa$ then t *of-type*$_\mathcal{T}(\varrho_\vartheta)$ κ;

(iv) if $t = \lambda(x ::_{type} \varrho_\vartheta(x)).t_1$ with $t_1 \in \mathit{Terms}_{\varrho_\vartheta}(\Sigma_{op}(\mathcal{T}))$, t_1 *of-type*$_\mathcal{T}(\varrho_\vartheta)$ κ_1 and $\kappa = \varrho_\vartheta(x) \Rightarrow \kappa_1$ then t *of-type*$_\mathcal{T}(\varrho_\vartheta)$ κ.

The relation *of-type*$_\mathcal{T} \subseteq \mathit{Terms}(\Sigma_{op}(\mathcal{T})) \times \mathit{Types}(\Sigma_{type}(\mathcal{T}))$ is then defined as the union of *of-type*$_\mathcal{T}(\varrho_\vartheta)$ for any sort assignment ϑ over $\mathcal{T}$ and type assignment over ϑ, i.e.

$$\textit{of-type}_\mathcal{T} \stackrel{def}{=} \bigcup_{\vartheta:\mathcal{X}_{type} \to \mathit{Sorts}(\Sigma_{class}(\mathcal{T})),\varrho_\vartheta:\mathcal{X}_{term} \to \mathit{Types}_\vartheta(\Sigma_{type}(\mathcal{T}))} \textit{of-type}_\mathcal{T}(\varrho_\vartheta).$$

The definition provides, that *of-type*$_\mathcal{T}$ assigns to any term at most one type, i.e. is a partial map $\mathit{Terms}(\Sigma_{op}(\mathcal{T})) \rightharpoonup \mathit{Types}(\Sigma_{type}(\mathcal{T}))$, as the following lemma shows.

Lemma 2.6 *Let $t \in \mathit{Terms}_{\varrho_\vartheta}(\Sigma_{op}(\mathcal{T}))$. Then*

$$\frac{t\ \textit{of-type}_\mathcal{T}\ \kappa \quad t\ \textit{of-type}_\mathcal{T}\ \kappa'}{\kappa = \kappa'}$$

holds for all $\kappa, \kappa' \in \mathit{Types}(\Sigma_{type}(\mathcal{T}))$.

Proof. (by structural induction on t)
(i) If $t = x ::_{type} \varrho_\vartheta(x)$ then $\kappa = \varrho_\vartheta(x)$. More precisely, t *of-type*$_\mathcal{T}$ κ primarily implies t *of-type*$_\mathcal{T}(\varrho'_{\vartheta'})$ κ. Then $t \in \mathit{Terms}_{\varrho'_{\vartheta'}}(\Sigma_{op}(\mathcal{T}))$ holds, such that $\varrho'_{\vartheta'}$ and ϱ_ϑ agree on variables occurring in t. Thus, t *of-type*$_\mathcal{T}$ κ' implies $\kappa' = \varrho_\vartheta(x)$ as well.

(ii) If $t = f ::_{type} \textit{type-of}_\mathcal{T}(f)\overline{[\alpha_i := \kappa_i]}_n$ then $\kappa = \textit{type-of}_\mathcal{T}(f)\overline{[\alpha_i := \kappa_i]}_n$. Further, t *of-type*$_\mathcal{T}$ κ' also implies $\kappa' = \textit{type-of}_\mathcal{T}(f)\overline{[\alpha_i := \kappa_i]}_n$, and hence $\kappa = \kappa'$.

(iii) Let $t = t_1\ t_2$. Then t_1 *of-type*$_\mathcal{T}(\varrho_\vartheta)$ $\kappa_1 \Rightarrow \kappa$, and t_1 *of-type*$_\mathcal{T}(\varrho'_{\vartheta'})$ $\kappa'_1 \Rightarrow \kappa'$. By the induction hypothesis we have $\kappa_1 \Rightarrow \kappa = \kappa'_1 \Rightarrow \kappa'$ and, hence $\kappa = \kappa'$.

(iv) Let $t = \lambda(x ::_{type} \varrho_\vartheta(x)).t_1$. Then $\kappa = \varrho_\vartheta(x) \Rightarrow \kappa_1$ and $\kappa' = \varrho'_{\vartheta'}(x) \Rightarrow \kappa'_1$ with t_1 *of-type*$_\mathcal{T}(\varrho_\vartheta)$ κ_1, t_1 *of-type*$_\mathcal{T}(\varrho'_{\vartheta'})$ κ'_1, and $\varrho'_{\vartheta'}(x) = \varrho_\vartheta(x)$. By the induction hypothesis we have $\kappa_1 = \kappa'_1$ and, hence, $\kappa = \kappa'$. □

On the other hand, for any type $\kappa \in \mathit{Types}(\Sigma_{type}(\mathcal{T}))$ there is always a term t such that t *of-type*$_\mathcal{T}$ κ, e.g. $t \stackrel{def}{=} x ::_{type} \kappa$ with some term variable x.

The next lemma states that substitutions of type variables preserve the of-type relation.

Lemma 2.7 *Let $\mathcal{T} \in \mathit{Th}$, $t \in \mathit{Terms}_{\varrho_\vartheta}(\Sigma_{op}(\mathcal{T}))$, and $\overline{[\alpha_i := \kappa_i]}_n$ be a substitution in t. Then*

$$\frac{t \ \textit{of-type}_\mathcal{T}(\varrho_\vartheta)\ \kappa}{t\overline{[\alpha_i := \kappa_i]}_n \ \textit{of-type}_\mathcal{T}(\varrho_{\vartheta'})\ \kappa\overline{[\alpha_i := \kappa_i]}_n} \tag{2.4}$$

holds for all $\kappa \in \mathit{Types}(\Sigma_{type}(\mathcal{T}))$.

Proof. (by structural induction on t)
(i) If $t = x ::_{type} \varrho_\vartheta(x)$ then $\kappa = \varrho_\vartheta(x)$. Clearly, $\overline{[\alpha_i := \kappa_i]}_n$ is then a substitution in κ. Further, we have $t\overline{[\alpha_i := \kappa_i]}_n = x ::_{type} \kappa\overline{[\alpha_i := \kappa_i]}_n$, such that (2.4) holds.
(ii) The case $t = f ::_{type} \kappa_f$ is treated analogously.
(iii) Let $t = t_1\ t_2$. From t *of-type*$_\mathcal{T}(\varrho_\vartheta)$ κ we obtain some $\kappa_1 \in \mathit{Types}_\vartheta(\Sigma_{type}(\mathcal{T}))$ such that t_1 *of-type*$_\mathcal{T}(\varrho_\vartheta)$ $\kappa_1 \Rightarrow \kappa$ and t_2 *of-type*$_\mathcal{T}(\varrho_\vartheta)$ κ_1 hold. Further, by the induction hypothesis we obtain

$$t_1\overline{[\alpha_i := \kappa_i]}_n \ \textit{of-type}_\mathcal{T}(\varrho_{\vartheta'})\ (\kappa_1 \Rightarrow \kappa)\overline{[\alpha_i := \kappa_i]}_n$$

and

$$t_2\overline{[\alpha_i := \kappa_i]}_n \ \textit{of-type}_\mathcal{T}(\varrho_{\vartheta'})\ \kappa_1\overline{[\alpha_i := \kappa_i]}_n.$$

Then, by the definition of the of-type relation, we can conclude

$$t\overline{[\alpha_i := \kappa_i]}_n \ \textit{of-type}_\mathcal{T}(\varrho_{\vartheta'})\ \kappa\overline{[\alpha_i := \kappa_i]}_n$$

using $t\overline{[\alpha_i := \kappa_i]}_n = t_1\overline{[\alpha_i := \kappa_i]}_n\ t_2\overline{[\alpha_i := \kappa_i]}_n$ and $(\kappa_1 \Rightarrow \kappa)\overline{[\alpha_i := \kappa_i]}_n = \kappa_1\overline{[\alpha_i := \kappa_i]}_n \Rightarrow \kappa\overline{[\alpha_i := \kappa_i]}_n$.
(iv) Let $t = \lambda(x ::_{type} \varrho_\vartheta(x)).t_1$. From t *of-type*$_\mathcal{T}(\varrho_\vartheta)$ κ we obtain some $\kappa_1 \in \mathit{Types}_\vartheta(\Sigma_{type}(\mathcal{T}))$ with t_1 *of-type*$_\mathcal{T}(\varrho_\vartheta)$ κ_1 such that $\kappa = \varrho_\vartheta(x) \Rightarrow \kappa_1$. By the induction hypothesis

$$t_1\overline{[\alpha_i := \kappa_i]}_n \ \textit{of-type}_\mathcal{T}(\varrho_{\vartheta'})\ \kappa_1\overline{[\alpha_i := \kappa_i]}_n$$

holds. Since $t\overline{[\alpha_i := \kappa_i]}_n = \lambda(x ::_{type} \varrho_\vartheta(x)\overline{[\alpha_i := \kappa_i]}_n)t_1\overline{[\alpha_i := \kappa_i]}_n$,

$$t\overline{[\alpha_i := \kappa_i]}_n \ \textit{of-type}_\mathcal{T}(\varrho_{\vartheta'})\ \kappa\overline{[\alpha_i := \kappa_i]}_n$$

holds as well, by the definition of the of-type relation. □

As a consequence for the α-conversion, we have

$$\frac{t_1 \stackrel{\alpha}{=}_\theta t_2 \quad t_1 \ \textit{of-type}_\mathcal{T}\ \kappa}{t_2 \ \textit{of-type}_\mathcal{T}\ \theta(\kappa)} \tag{2.5}$$

where $\theta(\kappa)$ denotes the term obtained by the application of the substitution of type variables, given by $t_1 \stackrel{\alpha}{=}_\theta t_2$, to κ.

Substitution of term variables in a term. Let $t \in \textit{Terms}(\Sigma_{op}(\mathcal{T}))$, $x \in \mathcal{X}_{term}$, and $t' \in \textit{Terms}_{\varrho\vartheta}(\Sigma_{op}(\mathcal{T}))$ such that

1. t' *of-type*$_\mathcal{T}$ $\varrho\vartheta_t(x)$, if $x \in \textit{termvars}(t)$;
2. $\varrho\vartheta(y) = \varrho\vartheta_t(y)$, for any $y \in (\textit{termvars}(t) \setminus \{x\})$.

Then $t[t'/x] \in \textit{Terms}_{\varrho\vartheta}(\Sigma_{op}(\mathcal{T}))$ denotes the term obtained by the substitution of all occurrences of $x ::_{type} \varrho\vartheta_t(x)$ in t by t'.

The requirements on the substitution ensure that the rule

$$\frac{t \ \textit{of-type}_\mathcal{T}\ \kappa}{t[t'/x] \ \textit{of-type}_\mathcal{T}\ \kappa} \tag{2.6}$$

holds for any substitution $[t'/x]$ in t.

Well-typed terms and type inference. A term $t \in \textit{Terms}(\Sigma_{op}(\mathcal{T}))$ is *well-typed* if there exists some $\kappa \in \textit{Types}(\Sigma_{type}(\mathcal{T}))$ with t *of-type*$_\mathcal{T}$ κ. Explicit sort and type assignment information will be omitted in term annotations as long it can be inferred. Actually, sort information can be inferred without additional assumptions only in the case when exactly one sort (i.e. the empty sort) exists in a theory (e.g. *Pure*).

Based on sort assignments, type information can be inferred using the *of-type* rules and unification. Moreover, *of-sort* rules are used to infer the type instances, assigned to the operation symbols in a term. For example, let S_1 be a super sort of S_2, and f an operation symbol having the declared type $(\alpha ::_{sort} S_1) \Rightarrow \alpha\ C$, where C is some type constructor. Then the term $f(x ::_{type} (\alpha ::_{sort} S_2))$ is well-formed having the type $(\alpha ::_{sort} S_2)\ C$. This holds, since the instance

$$f ::_{type} ((\alpha ::_{sort} S_1) \Rightarrow \alpha\ C)[\alpha := (\alpha ::_{sort} S_2)]$$

can be inferred, because $(\alpha ::_{sort} S_2)$ *of-sort*$_\mathcal{T}$ S_1 is derivable in this context.

2.8 Propositions and axioms

Let $\mathcal{T}$ be a theory, ϑ a sort assignment over $\mathcal{T}$ and ϱ_ϑ a type assignment. The set of *propositions over* ϱ_ϑ in $\mathcal{T}$ is defined by

$$\mathit{Props}_{\varrho_\vartheta}(\mathcal{T}) \stackrel{def}{=} \{p \mid p \ \mathit{of\text{-}type}_{\mathcal{T}}(\varrho_\vartheta) \ \mathtt{prop}\}$$

Consequently, the set $\mathit{Props}(\mathcal{T})$ of propositions in $\mathcal{T}$ is the union of $\mathit{Props}_{\varrho_\vartheta}(\mathcal{T})$ for any ϑ and ϱ_ϑ. Further, since operation signatures respect the ancestor relation, $\mathit{Props}(\mathcal{T}') \subseteq \mathit{Props}(\mathcal{T})$ holds if $\mathcal{T}' \in \mathit{Anc}(\mathcal{T})$.

Next, let $\mathcal{L}$ denote some infinite set of *proposition labels*. Any theory $\mathcal{T}$ has a finite set of *local axioms*, denoted by $\mathit{Ax}(\mathcal{T})$, with $\mathit{Ax}(\mathcal{T}) \subset \mathcal{L}$. The set $\mathit{AX}(\mathcal{T})$ of *axioms* of $\mathcal{T}$ is defined as the union of local axioms from all the ancestor theories of $\mathcal{T}$, i.e.

$$\mathit{AX}(\mathcal{T}) \stackrel{def}{=} \bigcup_{\mathcal{T}' \in \mathit{Anc}(\mathcal{T})} \mathit{Ax}(\mathcal{T}')$$

Furthermore, there is a function $\mathit{prop\text{-}of}_{\mathcal{T}} : \mathit{AX}(\mathcal{T}) \rightarrow \mathit{Props}(\mathcal{T})$, assigning to any axiom in $\mathit{AX}(\mathcal{T})$ a *closed* proposition w.r.t. the ancestor relation, i.e. if $\mathcal{T} \in \mathit{Anc}(\mathcal{T}')$ then $\mathit{prop\text{-}of}_{\mathcal{T}}(a) = \mathit{prop\text{-}of}_{\mathcal{T}'}(a)$ for all theories $\mathcal{T}, \mathcal{T}'$ and $a \in \mathit{AX}(\mathcal{T})$.

For any two theories $\mathcal{T}_1$ and $\mathcal{T}_2$, domain, codomain, and global axioms are defined as follows:

$$\begin{aligned}
\mathit{Dom}_{axiom}(\mathcal{T}_1, \mathcal{T}_2) &\stackrel{def}{=} \bigcup_{\mathcal{T} \in \mathit{Dom}_{thy}(\mathcal{T}_1, \mathcal{T}_2)} \mathit{AX}(\mathcal{T}) \\
\mathit{Cod}_{axiom}(\mathcal{T}_1, \mathcal{T}_2) &\stackrel{def}{=} \bigcup_{\mathcal{T} \in \mathit{Cod}_{thy}(\mathcal{T}_1, \mathcal{T}_2)} \mathit{AX}(\mathcal{T}) \\
\mathit{Global}_{axiom}(\mathcal{T}_1, \mathcal{T}_2) &\stackrel{def}{=} \bigcup_{\mathcal{T} \in \mathit{Global}_{thy}(\mathcal{T}_1, \mathcal{T}_2)} \mathit{AX}(\mathcal{T})
\end{aligned}$$

Meta logic. The theory *Pure* provides access to the meta rules, implemented by the logical kernel (see [Paulson 1989]), which establish

- introduction/elimination for $\bigwedge$,
- introduction/elimination for $\Longrightarrow$,
- $\equiv$ as an equivalence relation on terms,
- $\equiv$ for the type `prop` is the same as $\Longrightarrow$ and $\Longleftarrow$,

- β-conversion and extensionality (η-conversion) on terms w.r.t. $\equiv$, such that, e.g. the rule for beta conversion assigns to any well-typed term t, having the form $(\lambda(x ::_{type} \kappa).t_1)\ t_2$, the proposition $t \equiv t_1[t_2/x]$,
- abstraction and combination rules.

These are called *primitive* rules, in contrast to some further *derived* rules, implemented by the kernel using the primitive rules.

2.9 Definitions, logical and derived constants

In this section the concept of definitions as a special sort of axioms will be introduced. More precisely, these axioms have to satisfy some 'local' conditions, like the form *lhs* $\equiv$ *rhs* where *rhs* is a closed term, as well as some 'global' conditions, like that the considered set of axioms does not contain cycles. These restrictions will ensure that any definition in a theory always gives a conservative extension to the theory. So, definitions provide *the* tool for conservative developments in a framework, and their importance for Isabelle has been already emphasised in the introduction. On the other hand, definitions will play an important rôle for morphisms as well, since they define constants in an operation signature by means of other constants, i.e. uniquely determine their interpretations in dependency on a set of 'undefined' (these will be called *logical*) constants.

Final constants. Any constant $f \in \Sigma_{op}(\mathcal{T})$ can be labelled as a *final* constant in the theory $\mathcal{T}$. For instance, all meta-logical constants are final. The idea behind the finalisation is to protect constants specified via general axioms from additional specification via definitions. The reason is that without finalisation inconsistencies could be introduced by proper definitions, violating the desired property of these.

Pre-definitions. Let $\mathcal{T}$ be a theory, $f \in \Sigma_{op}(\mathcal{T})$ with $\kappa_f \stackrel{def}{=} \textit{type-of}_{\mathcal{T}}(f)$, such that there exists some sort $S \in \textit{Sorts}(\Sigma_{class}(\mathcal{T}))$ with κ_f $\textit{of-sort}_{\mathcal{T}}$ S, and $t \in \textit{Terms}(\Sigma_{op}(\mathcal{T}))$ a *closed* term with t $\textit{of-type}_{\mathcal{T}}$ κ_f. Then $(f \equiv t)$ $\textit{of-type}_{\mathcal{T}}$ $\texttt{prop}$ holds, where

- f is a shortening for the term $f ::_{type} \kappa_f\overline{[\alpha_i := (\alpha_i ::_{sort} \vartheta_{\kappa_f}(\alpha_i))]}_n$ where $\textit{typevars}(\kappa_f) = \{\alpha_1, \ldots, \alpha_n\}$, i.e. the identity instance, and
- $\equiv$ for the term $\equiv\ ::_{type} \textit{type-of}_{\mathsf{Pure}}(\equiv)[\alpha := \kappa_f]$, which is well-formed since κ_f $\textit{of-sort}_{\mathcal{T}}$ $\emptyset$ holds,

i.e. the term $f \equiv t$ is a proposition.

Further, an axiom $a \in AX(\mathcal{T})$ with $\mathit{prop\text{-}of}_{\mathcal{T}}(a) = f \equiv t$ is called the *pre-definition* of f in $\mathcal{T}$ if the following conditions are satisfied:

1. $\mathit{typevars}(t) \subseteq \mathit{typevars}(\kappa_f)$; in connection with t $\mathit{of\text{-}type}_{\mathcal{T}}$ κ_f we can conclude $\mathit{typevars}(t) = \mathit{typevars}(\kappa_f)$, because any type variable, which does not occur in t, cannot occur in the derived type κ_f;

2. f is *not* a final constant in $\mathcal{T}$;

3. for any axiom $a' \in AX(\mathcal{T})$ and $t' \in \mathit{Terms}(\Sigma_{op}(\mathcal{T}))$ with $\mathit{prop\text{-}of}_{\mathcal{T}}(a') = f \equiv t'$, $a = a'$ and $t = t'$ hold.

Order on pre-definitions in a theory. Let $P \subseteq AX(\mathcal{T})$ be a set of pre-definitions in $\mathcal{T}$. Then we define the relation $\prec^{P}_{AX(\mathcal{T})} \subseteq AX(\mathcal{T}) \times AX(\mathcal{T})$ as follows: $a_1 \prec^{P}_{AX(\mathcal{T})} a_2$ holds iff $a_1 \in P$ with $\mathit{prop\text{-}of}_{\mathcal{T}}(a_1) = f_1 \equiv t_1$ and $a_2 \in P$ with $\mathit{prop\text{-}of}_{\mathcal{T}}(a_2) = f_2 \equiv t_2$ such that $f_2 \in \mathit{consts\text{-}of}_{\mathcal{T}}(t_1)$ holds.

The set of definitions in a theory. Now, the set of *definitions* $\mathit{Defs}(\mathcal{T})$ is a set of pre-definitions in $\mathcal{T}$, for which the order $\prec^{\mathit{Defs}(\mathcal{T})}_{AX(\mathcal{T})}$ is acyclic.

Next, we define the (partial) function $\mathit{def\text{-}of}_{\mathcal{T}} : \Sigma_{op}(\mathcal{T}) \rightharpoonup \mathit{Terms}(\Sigma_{op}(\mathcal{T}))$, which maps any $f \in \Sigma_{op}(\mathcal{T})$ to $t \in \mathit{Terms}(\Sigma_{op}(\mathcal{T}))$ iff there is a definition $a \in \mathit{Defs}(\mathcal{T})$ with $\mathit{prop\text{-}of}_{\mathcal{T}}(a) = f \equiv t$. That is, $\mathit{prop\text{-}of}_{\mathcal{T}}(a) = f \equiv \mathit{def\text{-}of}_{\mathcal{T}}(f)$ holds for any $a \in \mathit{Defs}(\mathcal{T})$.

Logical and derived constants. Such constants in $\Sigma_{op}(\mathcal{T})$, which are in the domain of $\mathit{def\text{-}of}_{\mathcal{T}}$, will be called *derived* and their set will be denoted by $\Sigma^{der}_{op}(\mathcal{T})$. In contrast, the set $\Sigma^{log}_{op}(\mathcal{T}) \stackrel{\mathit{def}}{=} \Sigma_{op}(\mathcal{T}) \setminus \Sigma^{der}_{op}(\mathcal{T})$ contains *logical* constants in $\Sigma_{op}(\mathcal{T})$.

Order on constants in an operation signature. Similarly to the order on pre-definitions, we define the relation $\prec_{\Sigma_{op}(\mathcal{T})} \subseteq \Sigma_{op}(\mathcal{T}) \times \Sigma_{op}(\mathcal{T})$ on the constants in the operation signature $\Sigma_{op}(\mathcal{T})$ by: $f_1 \prec_{\Sigma_{op}(\mathcal{T})} f_2$ holds iff there is a term t with $\mathit{def\text{-}of}_{\mathcal{T}}(f_1) = t$ such that $f_2 \in \mathit{consts\text{-}of}_{\mathcal{T}}(t)$ holds.

Since $\mathit{def\text{-}of}_{\mathcal{T}}$ is defined only for definitions in $\mathcal{T}$, the order $\prec_{\Sigma_{op}(\mathcal{T})}$ inherits the acyclic property from $\prec^{\mathit{Defs}(\mathcal{T})}_{AX(\mathcal{T})}$, i.e. $\Sigma_{op}(\mathcal{T})$ together with $\prec_{\Sigma_{op}(\mathcal{T})}$ form a directed acyclic graph. As a consequence we obtain that $\mathit{def\text{-}of}_{\mathcal{T}}(f) \in \mathit{Terms}(\Sigma_{op}(\mathcal{T}) \setminus \{f\})$ holds for any $f \in \Sigma^{der}_{op}(\mathcal{T})$, i.e. any derived f.

Finally, since operation signatures respect the ancestor relation we also have $\{f' \mid f \prec^*_{\Sigma_{op}(\mathcal{T}_2)} f'\} \subseteq \Sigma_{op}(\mathcal{T}_1)$, where $\mathcal{T}_1 \in Anc(\mathcal{T}_2)$ and $f \in \Sigma_{op}(\mathcal{T}_1)$.

Rewriting by definitions. The set $\Sigma^{der}_{op}(\mathcal{T}) = \{f_1, \ldots, f_n\}$ constitutes the set $\mathcal{R}_\mathcal{T} \stackrel{def}{=} \{f_1 \rightarrow_\mathcal{T} \textit{def-of}_\mathcal{T}(f_1), \ldots, f_n \rightarrow_\mathcal{T} \textit{def-of}_\mathcal{T}(f_n)\}$, which can be seen as a terminating and confluent term rewriting system. So, contracting a redex f_i $(1 \leq i \leq n)$ in a term $t \in \textit{Terms}(\Sigma_{op}(\mathcal{T}))$ corresponds to a single unfold of the definition of f_i in t.

2.10 Proof terms and derived propositions

Next, the proof terms will be introduced. As already mentioned in the introduction, proof terms correspond to the meta rules under the Curry-Howard isomorphism, and the abstract syntax of proof terms reflects this as follows:

- term variable abstraction – $\bigwedge$ -introduction,
- proof variable abstraction – $\Longrightarrow$ -introduction,
- application of a proof term to a term – $\bigwedge$ -elimination (instantiation),
- application of a proof term to a proof term – $\Longrightarrow$ -elimination (modus ponens).

Global context in a theory. A tuple $G(\mathcal{T}) = \langle D, \textit{thm-of}_\mathcal{T}\rangle$ comprises a finite set $D \subset \mathcal{L}$ and a mapping $\textit{thm-of}_\mathcal{T} : D \rightarrow \textit{Props}(\mathcal{T})$, and is called a *global context* in $\mathcal{T}$. The notions of proof terms and derivability, defined below, will be parameterised over a global context. So, for instance, $G_{AX}(\mathcal{T}) \stackrel{def}{=} \langle AX(\mathcal{T}), \textit{prop-of}_\mathcal{T}\rangle$ forms a global context in any theory $\mathcal{T}$.

Proof terms. Using the notion of global contexts, proof terms are defined as follows.

Definition 7 (proof terms)
Let $\mathcal{X}_{proof}$ denote a fixed infinite countable set of so-called *proof* variables, $\mathcal{T}$ a theory, ϑ a sort assignment over $\mathcal{T}$ and ϱ_ϑ a type assignment. Further, let $G = \langle D, \textit{thm-of}_\mathcal{T}\rangle$ be a global context in $\mathcal{T}$. The set $\textit{Proofs}_{\varrho_\vartheta}(G, \mathcal{T})$ of *proof terms* in $\mathcal{T}$ over ϱ_ϑ and G is inductively defined as follows:

(i) if $X \in \mathcal{X}_{proof}$ then $X \in \textit{Proofs}_{\varrho_\vartheta}(G, \mathcal{T})$;

(ii) if $d \in D$, $\textit{thm-of}_{\mathcal{T}}(d) = p$, $\textit{typevars}(p) \subseteq \{\alpha_1, \dots, \alpha_n\}$, $\kappa_1, \dots, \kappa_n \in \textit{Types}_{\vartheta}(\Sigma_{type}(\mathcal{T}))$, and $[\alpha_i := \kappa_i]_n$ is a substitution in p then $d_{\overline{[\alpha_i := \kappa_i]_n}} \in \textit{Proofs}_{\varrho_\vartheta}(G, \mathcal{T})$;

(iii) if $\pi \in \textit{Proofs}_{\varrho_\vartheta}(G, \mathcal{T})$ and $x \in \mathcal{X}_{term}$, then the *term variable abstraction* $\overline{\lambda}(x ::_{type} \varrho_\vartheta(x)).\pi$ is an element of $\textit{Proofs}_{\varrho_\vartheta}(G, \mathcal{T})$;

(iv) if $X \in \mathcal{X}_{proof}$, $p \in \textit{Terms}_{\varrho_\vartheta}(\Sigma_{op}(\mathcal{T}))$ such that $p \in \textit{Props}(\mathcal{T})$, and $\pi \in \textit{Proofs}_{\varrho_\vartheta}(G, \mathcal{T})$ then the *proof variable abstraction* $\overline{\overline{\lambda}}(X ::_{prop} p).\pi$ is an element of $\textit{Proofs}_{\varrho_\vartheta}(G, \mathcal{T})$;

(v) if $\pi \in \textit{Proofs}_{\varrho_\vartheta}(G, \mathcal{T})$ and $t \in \textit{Terms}_{\varrho_\vartheta}(\Sigma_{op}(\mathcal{T}))$ then the *application of a proof term to term* πt is an element of $\textit{Proofs}_{\varrho_\vartheta}(G, \mathcal{T})$;

(vi) if $\pi_1, \pi_2 \in \textit{Proofs}_{\varrho_\vartheta}(G, \mathcal{T})$ then the *application of a proof term to a proof term* $\pi_1 \pi_2$ is an element of $\textit{Proofs}_{\varrho_\vartheta}(G, \mathcal{T})$.

A *proof term* over G in $\mathcal{T}$ is then an element of $\textit{Proofs}_{\varrho_\vartheta}(G, \mathcal{T})$ for some ϱ_ϑ, i.e.

$$\textit{Proofs}(G, \mathcal{T}) \stackrel{def}{=} \bigcup_{\vartheta : \mathcal{X}_{type} \to \textit{Sorts}(\Sigma_{class}(\mathcal{T})), \varrho_\vartheta : \mathcal{X}_{term} \to \textit{Types}_{\vartheta}(\Sigma_{type}(\mathcal{T}))} \textit{Proofs}_{\varrho_\vartheta}(G, \mathcal{T})$$

Proof names. Any proof term $\pi \in \textit{Proofs}(G, \mathcal{T})$, where $G = \langle D, \textit{thm-of}_{\mathcal{T}} \rangle$, determines the finite set $\textit{proof-names}(\pi) \subseteq D$ of *global proof labels* occurring in π.

Derivability relation. Proof terms are used to characterise derivable propositions in a theory and some global context. This is explicitly formalised by means of the derivability relation between proof terms and propositions in a theory $\mathcal{T}$, which basically connects the syntactic operations on proof terms with their meta-logical counterparts.

The following definition uses a *local context* H_{ϱ_ϑ}, which can be seen as a partial map $\mathcal{X}_{proof} \rightharpoonup \textit{Props}_{\varrho_\vartheta}(\mathcal{T})$ with a finite domain. Further, let $H[X \mapsto p]$ denote the update of H on X, such that $H[X \mapsto p](X) = p$, and $H[X \mapsto \bot]$ denote deletion of X from the domain of H, i.e. such that $H[X \mapsto \bot](X)$ is undefined. Moreover, a variable $x \in \mathcal{X}_{term}$ *occurs* in H if there is some X such that x occurs free in $H(X)$.

Definition 8 (derivability relation)

Let $\mathcal{T}$ be a theory, ϑ a sort assignment over $\mathcal{T}$, ϱ_ϑ a type assignment, and $G = \langle D,\ \textit{thm-of}_\mathcal{T}\rangle$ a global context in $\mathcal{T}$. The relation $\textit{proof-of}_\mathcal{T}(G, \varrho_\vartheta) \subseteq (\mathcal{X}_{proof} \rightharpoonup \textit{Props}_{\varrho_\vartheta}(\mathcal{T})) \times \textit{Proofs}_{\varrho_\vartheta}(G, \mathcal{T}) \times \textit{Props}_{\varrho_\vartheta}(\mathcal{T})$ is denoted by $H \vdash \pi\ \textit{proof-of}_\mathcal{T}(G, \varrho_\vartheta)\ p$ and inductively defined as follows:

(i) if $\pi = X$ with $X \in \mathcal{X}_{proof}$ and $H(X) = p$ then $H \vdash \pi\ \textit{proof-of}_\mathcal{T}(G, \varrho_\vartheta)\ p$;

(ii) if $\pi = d_{\overline{[\alpha_i := \kappa_i]_n}}$ with $d \in D$, $\textit{thm-of}_\mathcal{T}(d) = p$ then $H \vdash \pi\ \textit{proof-of}_\mathcal{T}(G, \varrho_\vartheta)$ $p\overline{[\alpha_i := \kappa_i]_n}$;

(iii) if $\pi = \overline{\lambda}(x ::_{type} \varrho_\vartheta(x)).\pi'$, $H \vdash \pi'\ \textit{proof-of}_\mathcal{T}(G, \varrho_\vartheta)\ p'$, and x does not occur in H then $H \vdash \pi\ \textit{proof-of}_\mathcal{T}(G, \varrho_\vartheta)\ \bigwedge(x ::_{type} \varrho_\vartheta(x)).p'$;

(iv) if $\pi = \overline{\overline{\lambda}}(X ::_{prop} p).\pi'$, $H[X \mapsto p] \vdash \pi'\ \textit{proof-of}_\mathcal{T}(G, \varrho_\vartheta)\ p'$, and $H' = H[X \mapsto \bot]$ then $H' \vdash \pi\ \textit{proof-of}_\mathcal{T}(G, \varrho_\vartheta)\ p \Longrightarrow p'$;

(v) if $\pi = \pi'\ t$, $H \vdash \pi'\ \textit{proof-of}_\mathcal{T}(G, \varrho_\vartheta)\ \bigwedge(x ::_{type} \varrho_\vartheta(x)).p$, and moreover $t\ \textit{of-type}_\mathcal{T}(\varrho_\vartheta)\ \varrho_\vartheta(x)$ holds, then $H \vdash \pi\ \textit{proof-of}_\mathcal{T}(G, \varrho_\vartheta)\ p[t/x]$;

(vi) if $\pi = \pi_1\pi_2$, $H \vdash \pi_1\ \textit{proof-of}_\mathcal{T}(G, \varrho_\vartheta)\ p' \Longrightarrow p$, $H \vdash \pi_2\ \textit{proof-of}_\mathcal{T}(G, \varrho_\vartheta)\ p'$ then $H \vdash \pi\ \textit{proof-of}_\mathcal{T}(G, \varrho_\vartheta)\ p$;

Further, $H \vdash \pi\ \textit{proof-of}_\mathcal{T}(G)\ p$ holds iff there exist ϑ and ϱ_ϑ such that $H \vdash \pi\ \textit{proof-of}_\mathcal{T}(G, \varrho_\vartheta)\ p$ holds.

Any proof term π determines at most one proposition p such that $H \vdash \pi\ \textit{proof-of}_\mathcal{T}(G)\ p$ holds, as the following lemma shows.

Lemma 2.8 *If $H \vdash \pi\ \textit{proof-of}_\mathcal{T}(G, \varrho_\vartheta)\ p_1$ and $H \vdash \pi\ \textit{proof-of}_\mathcal{T}(G)\ p_2$ then $p_1 = p_2$ for any local context H and $p_1, p_2 \in \textit{Props}(\mathcal{T})$.*

Proof. (by structural induction on π)
(i) If $\pi = X$ with $X \in \mathcal{X}_{proof}$ then $p_1 = H(X) = p_2$.

(ii) If $\pi = d_{\overline{[\alpha_i := \kappa_i]_n}}$ then $p_1 = \textit{thm-of}_\mathcal{T}(d)\overline{[\alpha_i := \kappa_i]_n} = p_2$.

(iii) Let $\pi = \overline{\lambda}(x ::_{type} \varrho_\vartheta(x)).\pi'$. Then $p_1 = \bigwedge(x ::_{type} \varrho_\vartheta(x)).p_1'$ and $p_2 = \bigwedge(x ::_{type} \varrho_\vartheta(x)).p_2'$ with $H \vdash \pi'\ \textit{proof-of}_\mathcal{T}(G, \varrho_\vartheta)\ p_1'$ and $H \vdash \pi'\ \textit{proof-of}_\mathcal{T}(G)\ p_2'$. By the induction hypothesis $p_1' = p_2'$ holds, and, hence $p_1 = p_2$.

(iv) Let $\pi = \overline{\overline{\lambda}}(X ::_{prop} p_X).\pi'$. Then $p_1 = p_X \Longrightarrow p_1'$, $p_2 = p_X \Longrightarrow p_2'$, and $H' = H[X \mapsto \bot]$ with $H'[X \mapsto p_X] \vdash \pi'\ \textit{proof-of}_\mathcal{T}(G, \varrho_\vartheta)\ p_1'$ and $H'[X \mapsto p_X] \vdash \pi'\ \textit{proof-of}_\mathcal{T}(G)\ p_2'$. By the induction hypothesis $p_1' = p_2'$ holds, and, hence

$p_1 = p_2$ as well.

(v) Let $\pi = \pi'\ t$. Then $H \vdash \pi'$ *proof-of*$_\mathcal{T}(G, \varrho_\vartheta)$ $\bigwedge(x ::_{type} \varrho_\vartheta(x)).p_1'$ $H \vdash \pi'$ *proof-of*$_\mathcal{T}(G)$ $\bigwedge(x ::_{type} \varrho'_{\vartheta'}(x)).p_2'$ (we can use the same variable x in both cases, because it is bound), t *of-type*$_\mathcal{T}(G, \varrho_\vartheta)$ $\varrho_\vartheta(x)$, t *of-type*$_\mathcal{T}(G, \varrho'_{\vartheta'})$ $\varrho'_{\vartheta'}(x)$, and $p_1 = p_1'[t/x]$, $p_2 = p_2'[t/x]$. Since $\bigwedge(x ::_{type} \varrho_\vartheta(x)).p_1' = \bigwedge(x ::_{type} \varrho'_{\vartheta'}(x)).p_2'$, by the induction hypothesis, we have $p_1' = p_2'$ and so $p_1 = p_2$.

(vi) If $\pi = \pi_1\pi_2$ then $H \vdash \pi_1$ *proof-of*$_\mathcal{T}(G)$ $p' \Longrightarrow p_2$, $H \vdash \pi_1$ *proof-of*$_\mathcal{T}(G, \varrho_\vartheta)$ $p \Longrightarrow p_1$, such that by the induction hypothesis $p \Longrightarrow p_1 = p' \Longrightarrow p_2$ holds, and so $p_1 = p_2$. □

Admissible global contexts in a theory. Let $\mathcal{T}$ be a theory. The set of *admissible global contexts* in $\mathcal{T}$, denoted by $Adm(\mathcal{T})$, is inductively defined by the following two rules:

(i) $A \in Adm(\mathcal{T})$ for any $A \subseteq G_{AX}(\mathcal{T})$;

(ii) if $G \in Adm(\mathcal{T})$ with $G = \langle D, \textit{thm-of}_\mathcal{T}\rangle$, $\emptyset \vdash \pi$ *proof-of*$_\mathcal{T}(G)$ p, and $d \in \mathcal{L}$ with $d \notin D$ then $\langle \overline{D}, \overline{\textit{thm-of}_\mathcal{T}}\rangle \in Adm(\mathcal{T})$, where $\overline{D} \stackrel{def}{=} D \cup \{d\}$, $\overline{\textit{thm-of}_\mathcal{T}}(d) \stackrel{def}{=} p$ and $\overline{\textit{thm-of}_\mathcal{T}}(x) \stackrel{def}{=} \textit{thm-of}_\mathcal{T}(x)$ for all $x \in D$.

So, admissible global contexts are essentially nothing else as the particular steps of the process, closing the set of axioms of a theory under the derivability, and the closure $\mathcal{T}^*$ itself is thus given by $\bigcup\{\textit{thm-of}(D) \mid \langle D, \textit{thm-of}\rangle \in Adm(\mathcal{T})\}$ for any theory $\mathcal{T}$.

The next lemma is a basic observation about global contexts, which is especially used in the extension of theories, presented below.

Lemma 2.9 *For any $\mathcal{T}, \mathcal{T}' \in \mathcal{T}h$ with $\mathcal{T} \in Anc(\mathcal{T}')$, $Adm(\mathcal{T}) \subseteq Adm(\mathcal{T}')$ holds.*

Proof. (by induction on the structure of $Adm(\mathcal{T})$)
(i) If $A \subseteq G_{AX}(\mathcal{T})$ then $A \subseteq G_{AX}(\mathcal{T}')$, and thus $A \in Adm(\mathcal{T}')$;

(ii) Let $G \in Adm(\mathcal{T})$ with $G = \langle D, \textit{thm-of}_\mathcal{T}\rangle$. By the induction hypothesis $G \subset Adm(\mathcal{T}')$ holds as well. Further, let $\emptyset \vdash \pi$ *proof-of*$_\mathcal{T}(G)$ p and $d \in \mathcal{L}$ with $d \notin D$. Using $\mathcal{T} \in Anc(\mathcal{T}')$ we can firstly conclude $\emptyset \vdash \pi$ *proof-of*$_{\mathcal{T}'}(G)$ p, and, secondly, by the definition of admissible contexts, $\langle \overline{D}, \overline{\textit{thm-of}_\mathcal{T}}\rangle \in Adm(\mathcal{T}')$. □

Further, any theory $\mathcal{T} \in \mathcal{T}h$ has its fixed admissible global context $\mathcal{G}(\mathcal{T}) \in Adm(\mathcal{T})$, such that for any two theories $\mathcal{T}, \mathcal{T}'$ with $\mathcal{T}' \in Anc(\mathcal{T})$ and $\mathcal{G}(\mathcal{T}) = \langle D, \mathit{thm\text{-}of}_{\mathcal{T}} \rangle$, $\mathcal{G}(\mathcal{T}') = \langle D', \mathit{thm\text{-}of}_{\mathcal{T}'} \rangle$, the following conditions hold:

1. $D \subseteq D'$;

2. $\mathit{thm\text{-}of}_{\mathcal{T}'}(d) = \mathit{thm\text{-}of}_{\mathcal{T}}(d)$, for any $d \in D$.

In particular, for the meta-logical theory, $\mathcal{G}(\mathsf{Pure}) \stackrel{def}{=} G_{AX}(\mathsf{Pure})$ is defined. Furthermore, the requirements on $\mathcal{G}(\mathcal{T})$ are conservative in the sense, that they can be satisfied in the primitive way by setting $\mathcal{G}(\mathcal{T})$ to $G_{AX}(\mathcal{T})$ for any theory $\mathcal{T}$.

Using the admissible global context of a theory $\mathcal{T}$, the set of *admissible proofs* in $\mathcal{T}$ is defined by $\mathit{Proofs}^{Adm}(\mathcal{T}) \stackrel{def}{=} \mathit{Proofs}(\mathcal{G}(\mathcal{T}), \mathcal{T})$. Moreover, the notation $H \vdash \pi \ \mathit{proof\text{-}of}_{\mathcal{T}} \ p$ is used for $H \vdash \pi \ \mathit{proof\text{-}of}_{\mathcal{T}}(\mathcal{G}(\mathcal{T})) \ p$. So, the set of *derivable* propositions in $\mathcal{T}$, denoted by $Der(\mathcal{T})$, contains all propositions $p \in Props(\mathcal{T})$ for which a proof $\pi \in \mathit{Proofs}^{Adm}(\mathcal{T})$ with $\emptyset \vdash \pi \ \mathit{proof\text{-}of}_{\mathcal{T}} \ p$ exists.

Theorems. Let $\mathcal{T}$ be a theory and $\mathcal{G}(\mathcal{T}) = \langle D, \mathit{thm\text{-}of}_{\mathcal{T}} \rangle$. The (finite) set $D \setminus AX(\mathcal{T})$ is called the set of *theorems* of $\mathcal{T}$ and is denoted by $Thms(\mathcal{T})$. By the definition of the admissible global contexts, for each $d \in Thms(\mathcal{T})$ we have the derivable proposition $\mathit{thm\text{-}of}_{\mathcal{T}}(d) \in Der(\mathcal{T})$.

Moreover, let $\mathit{prf}_{\mathcal{T}} : Thms(\mathcal{T}) \to \mathit{Proofs}^{Adm}(\mathcal{T})$ be a mapping assigning to any $d \in Thms(\mathcal{T})$ a proof term π with $\emptyset \vdash \pi \ \mathit{proof\text{-}of}_{\mathcal{T}} \ \mathit{thm\text{-}of}_{\mathcal{T}}(d)$. By the construction of admissible contexts, $d \notin \mathit{proof\text{-}names}(\mathit{prf}_{\mathcal{T}}(d))$ holds, independently of the choice of $\mathit{prf}_{\mathcal{T}}(d)$. Using this, the acyclic relation $\prec_{Thms(\mathcal{T})} \subseteq D \times D$ is defined as follows: $d_1 \prec_{Thms(\mathcal{T})} d_2$ holds iff $d_2 \in \mathit{proof\text{-}names}(\mathit{prf}_{\mathcal{T}}(d_1))$ hold.

Introduction and elimination of free variables. Let $\mathcal{T}$ be a theory and $p \in Der(\mathcal{T})$ such that $\emptyset \vdash \pi \ \mathit{proof\text{-}of}_{\mathcal{T}}(\mathcal{G}(\mathcal{T}), \varrho_\vartheta) \ p$ holds. Additionally, suppose $x \in \mathcal{X}_{term}$ occurs free in p. Then

$$\emptyset \vdash \overline{\lambda}(x ::_{type} \varrho_\vartheta(x)).\pi \ \mathit{proof\text{-}of}_{\mathcal{T}}(\mathcal{G}(\mathcal{T}), \varrho_\vartheta) \ \bigwedge(x ::_{type} \varrho_\vartheta(x)).p$$

holds as well, i.e. $\bigwedge(x ::_{type} \varrho_\vartheta(x)).p \in Der(\mathcal{T})$, such that x becomes bound by the leading meta-quantification.

On the other hand, if $\emptyset \vdash \pi \ \mathit{proof\text{-}of}_{\mathcal{T}}(\mathcal{G}(\mathcal{T})) \ \bigwedge(x ::_{type} \kappa_x).p$ holds then

$$\emptyset \vdash \pi \ (y ::_{type} \kappa_x) \ \mathit{proof\text{-}of}_{\mathcal{T}}(\mathcal{G}(\mathcal{T})) \ p[(y ::_{type} \kappa_x)/x]$$

holds as well, where $y \in \mathcal{X}_{term}$ denotes some fresh variable which does not occur free in p.

Altogether, this justifies that any proposition can be restated without free variables preserving the derivability relation, which will be used below in the specification of the extension of a theory by an axiom.

Consistency and conservative extensions. Using the notion of derivability, we can define consistency as follows: a theory $\mathcal{T}$ is called *consistent* if there exists a proposition $p \in \mathit{Props}(\mathcal{T})$ which is not derivable, i.e. $p \notin \mathit{Der}(\mathcal{T})$, or in other words $\mathit{Der}(\mathcal{T}) \subset \mathit{Props}(\mathcal{T})$. Further, a theory $\mathcal{T}$ is a *conservative* extension of a theory $\mathcal{T}'$ if $\mathcal{T}' \in \mathit{Anc}(\mathcal{T})$ and $\mathcal{T}$ is consistent iff $\mathcal{T}'$ is.

Now, we can reason that definitional axioms are conservative extensions. To this end, let $\mathcal{T}, \mathcal{T}'$ be theories, such that $\mathcal{T}'$ only extends $\mathcal{T}$ by a definition a with $\mathit{prop\text{-}of}_{\mathcal{T}'}(a) = f \equiv t$, where $f \in \Sigma_{op}^{log}(\mathcal{T})$ and $t \in \mathit{Terms}(\Sigma_{op}(\mathcal{T}))$, such that $\mathit{def\text{-}of}_{\mathcal{T}'}(f) = t$ holds. Let us also assume, that the global context of $\mathcal{T}'$ differs from the global context of $\mathcal{T}$ only by a, i.e. $\mathcal{G}(\mathcal{T}) = \langle D, \mathit{thm\text{-}of}_{\mathcal{T}}\rangle$ and $\mathcal{G}(\mathcal{T}') = \langle D \cup \{a\}, \mathit{thm\text{-}of}_{\mathcal{T}'}\rangle$.

Further, let $p \in \mathit{Der}(\mathcal{T})$. Then $p \in \mathit{Der}(\mathcal{T}')$ holds as well, since any admissible proof of p in $\mathcal{T}$ is also an admissible proof of p in $\mathcal{T}'$. In the other direction, let $p \in \mathit{Der}(\mathcal{T}')$ provided by a proof $\pi \in \mathit{Proofs}^{Adm}(\mathcal{T}')$. If a does not occur in π then $\pi \in \mathit{Proofs}^{Adm}(\mathcal{T})$ holds, and hence $p \in \mathit{Der}(\mathcal{T})$. On the other hand, if a occurs in π then there is a sub-proof of π of the form

$$(\bar{\bar{\lambda}}(X ::_{prop} (f \equiv t)\overline{[\alpha_i := \kappa_i]_n}).\pi')\ a_{\overline{[\alpha_i := \kappa_i]_n}}$$

The sub-proof π' can then use the proof variable X in order to extend a derivation of a proposition $P\ t$ to a derivation of $P\ f$ using, e.g. applications of the combination rule to X. Such extensions can be eliminated from π', which yields a proof π'' where X does not occur and only propositions involving t are derived. The restrictions on definitions ensure that $\pi'' \in \mathit{Proofs}^{Adm}(\mathcal{T})$ holds. Altogether, from the original proof term $\pi \in \mathit{Proofs}^{Adm}(\mathcal{T}')$ we obtain a proof term $\overline{\pi} \in \mathit{Proofs}^{Adm}(\mathcal{T})$ from which the proposition $p' \in \mathit{Der}(\mathcal{T})$, obtained from p replacing any sub-term of the form $f ::_{type} \mathit{type\text{-}of}_{\mathcal{T}}(f)\overline{[\alpha_i := \kappa_i]_n}$ by the corresponding term $t\overline{[\alpha_i := \kappa_i]_n}$, is derivable.

2.11 Theory extensions

This section presents the operations for extension of theories by particular elements described in the previous sections, as well as how these operations are represented at the top-level of Isabelle. Using these extension operations we will be able to formulate procedures, which construct new theories from existing, satisfying some properties specified by means of morphisms.

Extension by type classes. The operation $\mathcal{T} +_{class} (c, \langle c_1, \ldots, c_n \rangle)$, where $n \in \mathbb{N}$, $c \notin \Sigma_{class}(\mathcal{T})$ and $c_i \in \Sigma_{class}(\mathcal{T})$ for $1 \leq i \leq n$, assigns to any theory $\mathcal{T}$ the theory $\mathcal{T}'$ defined by:

1. $Par(\mathcal{T}') \stackrel{def}{=} \{\mathcal{T}\}$;
2. the set of type classes of $\mathcal{T}'$ is $\{c\}$;
3. $\Sigma_{class}(\mathcal{T}')$ extends $\Sigma_{class}(\mathcal{T})$ by the relations $c \prec_{\Sigma_{class}(\mathcal{T}')} c_i$ for all $1 \leq i \leq n$;
4. the sets of type constructors, constants and axioms of $\mathcal{T}'$ are empty;
5. $\mathcal{G}(\mathcal{T}') \stackrel{def}{=} \mathcal{G}(\mathcal{T})$.

This definition of $\mathcal{T}'$ allows us to conclude:

1. $c \in \Sigma_{class}(\mathcal{T} +_{class} (c, \langle c_1, \ldots, c_n \rangle))$;
2. $c \prec_{\Sigma_{class}(\mathcal{T} +_{class} (c, \langle c_1, \ldots, c_n \rangle))} c_i$ for $1 \leq i \leq n$;
3. respective type and operation signatures stay unchanged;
4. $AX(\mathcal{T} +_{class} (c, \langle c_1, \ldots, c_n \rangle)) = AX(\mathcal{T})$, i.e. the set of axioms stays unchanged as well.

At the top-level of Isabelle this operation has the form:
classes $c \subseteq c_1, \ldots, c_n$
or just
classes c
if $n = 0$.

Extension by type constructors. The operation $\mathcal{T} +_{tcons} (C, n)$, where $C \notin \Sigma_{type}(\mathcal{T})$ and $n \in \mathbb{N}$ assigns to any theory $\mathcal{T}$ the theory $\mathcal{T}'$ defined by:

1. $Par(\mathcal{T}') \stackrel{def}{=} \{\mathcal{T}\}$;
2. the set of type constructors of $\mathcal{T}'$ is $\{C\}$ with $rank_{\mathcal{T}'}(C) \stackrel{def}{=} n$;

3. the sets of type classes, constants and axioms of $\mathcal{T}'$ are empty;
4. $\mathcal{G}(\mathcal{T}') \stackrel{def}{=} \mathcal{G}(\mathcal{T})$.

This definition of $\mathcal{T}'$ allows us to conclude:

1. $\Sigma_{type}(\mathcal{T} +_{tcons} (C, n)) = \Sigma_{type}(\mathcal{T}) \cup C$;
2. $rank_{\mathcal{T} +_{tcons} (C,n)}(C) = n$;
3. respective class and operation signatures, as well as the sets of axioms stay unchanged;

At the top-level of Isabelle, this operation is represented by
typedecl $(\alpha_1, \ldots, \alpha_n)$ C
where α_i are distinct type variables.

Extension by arities. The operation $\mathcal{T} +_{arity} (C, (S_1, \ldots, S_n, S_{n+1}))$, where $C \in \Sigma_{type}(\mathcal{T})$ with the rank n, and $S_i \in Sorts(\Sigma_{class}(\mathcal{T}))$ for $i = 1, \ldots, n+1$, assigns to any theory $\mathcal{T}$ the theory $\mathcal{T}'$ defined by:

1. $Par(\mathcal{T}') \stackrel{def}{=} \{\mathcal{T}\}$;
2. the set of arities of C in $\Sigma_{type}(\mathcal{T})$ is extended by $(S_1, \ldots, S_n, S_{n+1})$;
3. the sets of type classes, type constructors, constants and axioms of $\mathcal{T}'$ are empty;
4. $\mathcal{G}(\mathcal{T}') \stackrel{def}{=} \mathcal{G}(\mathcal{T})$.

So, we have

$$arities\text{-}of_{\mathcal{T} +_{arity} (C,(S_1,\ldots,S_n,S_{n+1}))}(C) = arities\text{-}of_{\mathcal{T}}(C) \cup \{(S_1, \ldots, S_n, S_{n+1})\},$$

while everything else stays unchanged. Furthermore, let

$$\begin{aligned} \mathcal{T} +_{arities} (C, \langle \ & (S_{1,1}, \ldots, S_{1,n+1}), \\ & \ldots, \\ & (S_{a,1}, \ldots, S_{a,n+1})\rangle) \end{aligned}$$

be the operation applying $+_{arity}$ successively a-times. At the top-level of Isabelle, this is represented by
arities
$C :: (S_{1,1}, \ldots, S_{1,n})\ S_{1,n+1}$
...
$C :: (S_{a,1}, \ldots, S_{a,n})\ S_{a,n+1}$

Extension by constants. The operation $\mathcal{T} +_{op} (f, \kappa)$, where $f \notin \Sigma_{op}(\mathcal{T})$ and $\kappa \in \mathit{Types}(\Sigma_{type}(\mathcal{T}))$, assigns to any theory $\mathcal{T}$ the theory $\mathcal{T}'$ defined by:

1. $\mathit{Par}(\mathcal{T}') \stackrel{def}{=} \{\mathcal{T}\}$;
2. the set of constants of $\mathcal{T}'$ is $\{f\}$ with $\mathit{type\text{-}of}_{\mathcal{T}'}(f) \stackrel{def}{=} \kappa$;
3. the sets of classes, type constructors and axioms of $\mathcal{T}'$ are empty;
4. $\mathcal{G}(\mathcal{T}') \stackrel{def}{=} \mathcal{G}(\mathcal{T})$.

This definition of $\mathcal{T}'$ allows us to conclude:

1. $\Sigma_{op}(\mathcal{T} +_{op} (f, \kappa)) = \Sigma_{op}(\mathcal{T}) \cup f$;
2. $\mathit{type\text{-}of}_{\mathcal{T} +_{op} (f,\kappa)}(f) = \kappa$;
3. respective class and type signatures, as well as the sets of axioms stay unchanged;

At the top-level of Isabelle, this operation is represented by
consts $f :: \kappa$

Extension by axioms. Let $\mathcal{T}$ be a theory with $\mathcal{G}(\mathcal{T}) = \langle D, \mathit{thm\text{-}of}_{\mathcal{T}} \rangle$, $a \in \mathcal{L}$, $a \notin D$. The operation $\mathcal{T} +_{axiom} (a, p)$, where $p \in \mathit{Props}(\mathcal{T})$, assigns to any theory $\mathcal{T}$ the theory $\mathcal{T}'$ defined by:

1. $\mathit{Par}(\mathcal{T}') \stackrel{def}{=} \{\mathcal{T}\}$;
2. the set of axioms of $\mathcal{T}'$ is $\{a\}$;
3. $\mathit{prop\text{-}of}_{\mathcal{T}'}(a) \stackrel{def}{=} p'$, where p' is the same proposition as p except, that free term variables of p are closed by leading meta-quantifications, as already mentioned;
4. $\mathit{prop\text{-}of}_{\mathcal{T}'}(x) \stackrel{def}{=} \mathit{prop\text{-}of}_{\mathcal{T}}(x)$ for any $x \in AX(\mathcal{T})$;
5. the sets of type classes, type constructors, and constants of $\mathcal{T}'$ are empty;
6. $\mathcal{G}(\mathcal{T}') \stackrel{def}{=} \langle D \cup \{a\}, \mathit{thm\text{-}of}_{\mathcal{T}'} \rangle$, where

$$\mathit{thm\text{-}of}_{\mathcal{T}'}(d) \stackrel{def}{=} \begin{cases} \mathit{prop\text{-}of}_{\mathcal{T}'}(d) & \text{if } d = a \\ \mathit{thm\text{-}of}_{\mathcal{T}}(d) & \text{otherwise.} \end{cases}$$

This definition of $\mathcal{T}'$ allows us to conclude:

1. $AX(\mathcal{T} +_{axiom} (a,p)) = AX(\mathcal{T}) \cup a$;
2. $\mathit{prop\text{-}of}_{\mathcal{T} +_{axiom} (a,p)}(a) = p'$, as mentioned above;
3. respective class, type and operation signatures stay unchanged;

At the top-level of Isabelle, this operation has the form
axioms $a \ : \ p$

Extension by definitions. The operation $\mathcal{T} +_{def} (a,p)$ is a special case of the extension of a theory by an axiom, additionally checking the definitional restrictions for p, especially that $\prec^{Defs(\mathcal{T}) \cup \{a\}}_{AX(\mathcal{T} +_{axiom} (a,p))}$ is acyclic.

So, if $p = f \ \equiv \ t$ is a proper definition in $\mathcal{T}$, additionally to the properties of $\mathcal{T} +_{axiom} (a,p)$, we can conclude

1. $\mathit{def\text{-}of}_{\mathcal{T} +_{def} (a,p)}(f) = t$;
2. $f \in \Sigma^{der}_{op}(\mathcal{T} +_{def} (a,p))$

i.e. the logical constant f becomes derived in $\mathcal{T} +_{def} (a,p)$. So, adding a definition to a logical constant is a kind of finalisation, since a constant cannot be defined more than once. At the top-level, this operation is represented by
defs $a \ : \ f \ x_1 \ \ldots \ x_n \ \equiv \ t$
where $n \in \mathbb{N}$, $x_1, \ldots, x_n \in \mathcal{X}_{term}$ are distinct, and $\mathit{termvars}(t) \subseteq \{x_1, \ldots, x_n\}$. This form is then internally transformed into the form $a \ : \ f \ \equiv \ \lambda x_1 \ \ldots \ \lambda x_n . \, t$, provided by the extensionality and abstraction rules, such that the original form can be derived via combination and beta-conversion rules.

Extension by theorems. Let $\mathcal{T}$ be a theory with $\mathcal{G}(\mathcal{T}) = \langle D, \mathit{thm\text{-}of}_{\mathcal{T}} \rangle$, $d \in \mathcal{L}$, $d \notin D$ and $\pi \in \mathit{Proofs}^{Adm}(\mathcal{T})$ such that there exists a closed proposition p with $\emptyset \vdash \pi \ \mathit{proof\text{-}of}_{\mathcal{T}}(\mathcal{G}(\mathcal{T})) \ p$. Then $G \stackrel{def}{=} \langle \overline{D}, \overline{\mathit{thm\text{-}of}_{\mathcal{T}}} \rangle$ is an admissible context in $\mathcal{T}$ as well, where $\overline{D} \stackrel{def}{=} D \cup \{d\}$, $\overline{\mathit{thm\text{-}of}_{\mathcal{T}}}(d) \stackrel{def}{=} p$ and $\overline{\mathit{thm\text{-}of}_{\mathcal{T}}}(x) \stackrel{def}{=} \mathit{thm\text{-}of}_{\mathcal{T}}(x)$ for all $x \in D$.

So, the operation $\mathcal{T} +_{thm} (d, \pi)$ assigns to $\mathcal{T}$ the theory $\mathcal{T}'$ defined by:

1. $\mathit{Par}(\mathcal{T}') \stackrel{def}{=} \{\mathcal{T}\}$;
2. the sets of type classes, type constructors, constants and axioms of $\mathcal{T}'$ are empty;

3. $\mathcal{G}(\mathcal{T}') \stackrel{def}{=} G$, which is well-defined since $\mathcal{T} \in Anc(\mathcal{T}')$ and $G \in Adm(\mathcal{T})$;

4. $prf_{\mathcal{T}'}(d) \stackrel{def}{=} \pi$.

This definition of $\mathcal{T}'$ allows us to conclude:

1. $Thms(\mathcal{T} \ +_{thm} \ (d, \pi)) = Thms(\mathcal{T}) \cup \{d\}$;

2. $thm\text{-}of_{\mathcal{T} \ +_{thm} \ (d,\pi)}(d) = p$;

3. respective class, type and operation signatures stay unchanged;

4. moreover, $AX(\mathcal{T} \ +_{thm} \ (d, \pi)) = AX(\mathcal{T})$ holds as well.

At the top-level of Isabelle this operation is represented by
theorem $d : p$
`<proof script>`
where *<proof script>* has to be filled by a proof script in the Isar language, constructed interactively, constituting a proof term π, from which the proposition p can be derived. This usually proceeds by the backward resolution, refining the conclusion C of the proposition p to subgoals $C_1, \ldots, C_n$, which have to be then proved under the premises of p. This step is, of course, provided by the derived proposition $C_1 \Longrightarrow \ldots \Longrightarrow C_n \Longrightarrow C$, which is, in turn, obtained by (in general higher order) unification with some existing theorem or axiom. This is a very rough sketch of a quite sophisticated subject (e.g. [Berghofer and Nipkow 2000]). Actually, the support of the construction of proofs by various tactics, tools, procedures, etc., forms one of the core topics for research and development in the context of Isabelle.

Finally a technical note. If the proposition p above is not closed, the constructed proof π is extended to a proof π', which replaces free term variables in p by leading meta-quantifications. Moreover, the Isar commands **lemma** and **corollary** can be used as synonyms for **theorem**.

2.12 Object logics

Formalisations of object logics start in the theories, having *Pure* as its parent theory, as sketched in Figure 2.1. Object logics are not necessarily formalised by a single theory. So, for instance, the first order logic is formalised by two theories, where the first formalises intuitionistic fragment of FOL and the second extends it by classical axioms.

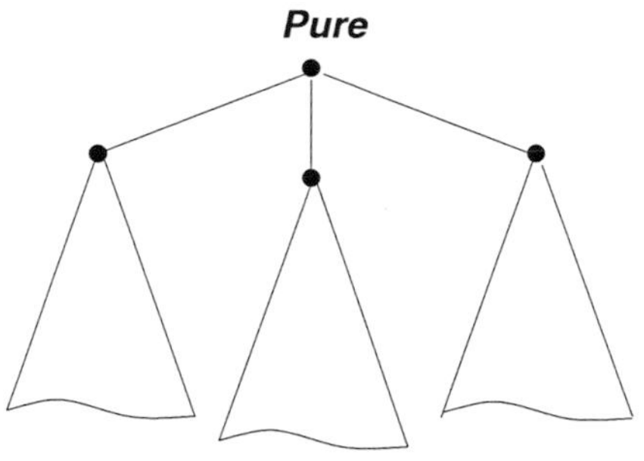

Figure 2.1: Meta logic and object logics

Referring again to Figure 1.1 in the introduction, a *conservative* theory development in an object logic is a theory development $\mathcal{D}$, where for any theory $\mathcal{T} \in \mathcal{D}$ the set of axioms $AX(\mathcal{T})$ comprises

- either the meta-logical axioms,
- or the object-logical axioms, e.g. rules for introduction/elimination of the object-logical connectives,
- or axioms, which form conservative extensions, e.g. definitions.

As already mentioned in the introduction, the most developed object logic in Isabelle is the higher order logic (HOL). Since all the developments, presented in this document, are in HOL, the next section emphasises its structure.

2.12.1 Higher order logic

The formalisation of HOL is started in the theory *HOL*, which then has *Pure* as the parent theory.

Type classes. The theory *HOL* introduces the type class *type* without super classes, together with the declaration that the sort $\{type\}$ is the *default* sort in the logic. This declaration provides the default sort assignment in HOL: it maps any type variable to the default sort. So, the sort assignment annotations for type variables in a type declaration are explicitly used only if the sort assignment for the particular type variable differs from the default.

Type constructors. Any declaration of a type constructor C with the rank n via **typedecl** in HOL, additionally introduces the standard HOL arity for it, having the form

$$\underbrace{(\{type\},\ldots,\{type\})}_{n+1-\text{times}}.$$

Since *HOL* extends *Pure*, its type signature contains the type constructors `prop` and $\Rightarrow$, both having no arities in *Pure*. The type constructor $\Rightarrow$ is 'adapted' for the higher order logic by adding the standard HOL arity $(\{type\},\{type\},\{type\})$ to $\Rightarrow$ via the command **arities** in *HOL*. Further, a sort $S \in Sorts(\Sigma_{class}(\mathcal{T}))$ with $HOL \in Anc(\mathcal{T})$ will be called a *proper* HOL-sort if $S \preceq^{\mathcal{T}}_{sort} \{type\}$ holds.

Altogether, the default sort and standard arities provide the following 'of-sort derivability' for any theory $\mathcal{T}$ with $HOL \in Anc(\mathcal{T})$: if $\kappa \in Types(\Sigma_{type}(\mathcal{T}))$, $\vartheta_\kappa(\alpha)$ is a proper HOL-sort for any $\alpha \in typevars(\kappa)$, and `prop` does not occur in κ then κ *of-sort*$_{\mathcal{T}}$ $\{type\}$ holds.

So, using the HOL version of **typedecl**, the type constructor *bool* with the rank 0 is introduced by
typedecl *bool*
in the theory *HOL*, i.e. the arity $\{type\}$ is provided for *bool* as well. The type *bool* is used to represent HOL-propositions. The connection to the type of meta-logical proposition `prop` is achieved by the special constant (also called judgement) *Trueprop* having the type $bool \Rightarrow$ `prop`. So, *Trueprop* P is used to express 'P is true', i.e. that some HOL-proposition P holds. *Trueprop* is basically not explicitly annotated, but is always implicitly present in HOL-propositions, i.e. terms of type *bool*.

Constants and axioms. The basic logical constants introduced in *HOL* are

- the polymorphic HOL-equality $= \;::\; \alpha \Rightarrow \alpha \Rightarrow bool$,
- the HOL-implication $\longrightarrow \;::\; bool \Rightarrow bool \Rightarrow bool$,
- the polymorphic description operator $The :: (\alpha \Rightarrow bool) \Rightarrow \alpha$.

These are specified via axioms and finalised. So, e.g. for the HOL-equality there are the following axioms:
axioms
`refl :` $t = t$
`subst :` $s = t \Longrightarrow P\ s \Longrightarrow P\ t$
Note that, e.g. the reflexivity axiom basically denotes the closed proposition $\bigwedge(t ::_{type} \alpha).\ t = t$.

For instance the symmetry property of = is then a derived proposition in *HOL*:

theorem *sym* : $a = b \Longrightarrow b = a$

and is shown via the instantiations of t by a in *refl* and s by a, t by b, P by $\lambda x.(x = a)$ in *subst*. Moreover, in a similar manner one can also show the proposition $a \equiv b \Longrightarrow a = b$, connecting the meta-logical and object-logical equalities.

Further logical connectives are represented by derived constants, e.g. *True*, *False*, the negation $\neg$, the conjunction $\wedge$, the disjunction $\vee$, and the quantifiers $\forall$, $\exists$ are specified via definitions in *HOL*.

However, to formalise the classical higher order logic, the additional axiom

axioms

True_or_False : $(P = True) \vee (P = False)$

is introduced in *HOL*.

HOL development. The structure of the HOL development is sketched in Figure 2.2 where

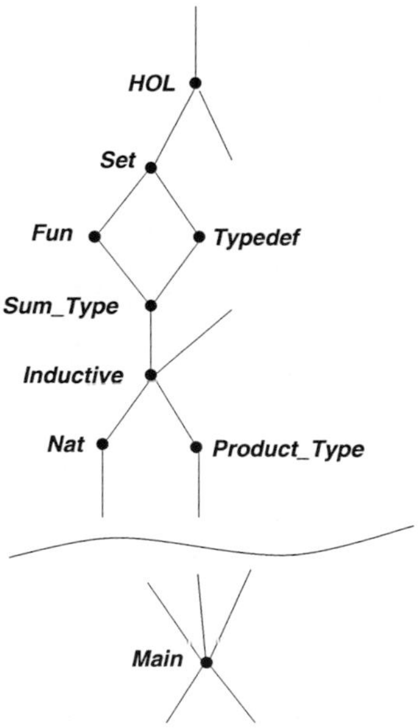

Figure 2.2: A fragment of the HOL development

- *HOL* forms the basis of the formalisation as described above.
- *Set* – sets and standard operations on these. In HOL, a set of elements of a type α is basically a function with the type $\alpha \Rightarrow bool$, i.e. a polymorphic HOL-predicate. So, the type $\alpha\ set$ is an abbreviation for the type $\alpha \Rightarrow bool$, as well as $x \in P$ an abbreviation for $P\ x$. The operations on sets (union, intersection, etc.) are then defined in terms of the object-logical connectives introduced in *HOL*.
- *Fun* – functions. It introduces the polymorphic identity ($id :: \alpha \Rightarrow \alpha$) and the composition ($\circ :: (\beta \Rightarrow \gamma) \Rightarrow (\alpha \Rightarrow \beta) \Rightarrow \alpha \Rightarrow \gamma$) operators as well as convenient concrete syntax for functions, e.g. that the composition is used as an infix operator.
- *Typedef* – provides the construction of new types from nonempty sets.
- *Sum_Type* – introduces the type constructor $\alpha + \beta$, which corresponds to the disjoint union on sets.
- *Inductive* – provides inductive definitions of sets.
- *Product_Type* – introduces the type constructor $\alpha \times \beta$, which corresponds to the Cartesian product on sets.
- *Nat* – introduces the inductive set of natural numbers and abstracts this by the freely generated type nat with the constructors $0 :: nat$, $Suc :: nat \Rightarrow nat$; this construction requires the *axiom of infinity*, which is introduced in this theory as well.
- *Main* – has any theory in the HOL-development as an ancestor, summarising this development of the higher order logic.

2.13 Conclusion

In this chapter a description of an LCF-style logical framework, based on the setup of the theorem prover Isabelle, has been given. So, the central notion of a theory as well as the theory hierarchy have been introduced. Further class, type and operation signatures have been assigned to any theory w.r.t. the theory hierarchy. These signatures together form the *signature* of a theory. Based on these concepts and notions, signature morphisms will relate two theories relating their particular signatures. This will be the subject of the next chapter.

Furthermore, theory morphisms will extend signature morphisms, additionally relating the admissible global contexts of two theories. Finally, the presented operations for extension of theories will allow us to formulate procedures, which systematically construct new theories from already existing, e.g. computing the instantiation of a parameterised theory by a theory morphism.

Chapter 3

Signature Morphisms

In this setting the signature of a theory comprises its class, type, and operation signatures. Consequently, signature morphisms will relate two theories 'point-wise', relating their respective signatures by means of class, type and operation mappings. Moreover, the presented notion of signature morphisms will also reflect the distinction between logical and derived constants in an operation signature, based on a relation between definitions in the involved theories. Since definitions form a subset of axioms of a theory and relating axioms from two theories will be further a task for theory morphisms, signature morphisms will thus anticipate and reduce this task for theory morphisms to 'proper' axioms, i.e. those which are not associated with a derived constant as its definition. This treatment of definitions by signature morphisms is also justified by the fact that derived constants have a fixed interpretation w.r.t. the logical constants in an operation signature, such that an appropriate mapping of the logical constants to constants in another signature will also determine a mapping between derived constants in these signatures. The presented normalisation procedure for signature morphisms is also based on this fact.

The mappings between class, type, and operation signatures are hierarchic in the sense that a mapping between type signatures is parameterised by a mapping between respective class signatures, as well as a mapping between operation signatures is parameterised by both a class and a type map. The most sophisticated of these three kinds of mappings are type maps. In contrast to standard formalisations of signatures and morphisms, e.g. in [Burstall and Goguen 1980], where a signature consists of a set of sorts (types) S and a set of operations indexed by $S^* \times S$, here the set of constants in a signature is indexed by the types over the particular type signature, which can also contain type variables. In the former case, a signature morphism relates two operations f and f' by means of a sort mapping $\phi : S \to S'$ as follows: if f is indexed by $S^n \times S$ then f' has to be indexed by $\overline{\phi}(S^n) \times \phi(S)$, where $\overline{\phi}$ extends ϕ to strings over sorts. In the current situation we could similarly take a rank-preserving mapping of type constructors $\phi : \Sigma_{type}(\mathcal{T}) \to \Sigma_{type}(\mathcal{T}')$ and relate constants by means of its extension $\overline{\phi} : Types(\Sigma_{type}(\mathcal{T})) \to Types(\Sigma_{type}(\mathcal{T}'))$. So, constants with the types $\alpha\, C_1 \Rightarrow \alpha\, C_1$ and $\alpha\, C_2 \Rightarrow \alpha\, C_2$ can be related if $\phi(C_1) = C_2$. On the other hand, we could not relate $\alpha\, C_1 \Rightarrow \alpha\, C_1$ with e.g. $(\alpha, \alpha)\, C_3 \Rightarrow (\alpha, \alpha)\, C_3$ or $C_4 \Rightarrow C_4$ this way.

Hence, it turns out to be a good approach to specify a class of functions which map types in $Types(\Sigma_{type}(\mathcal{T}))$ to those in $Types(\Sigma_{type}(\mathcal{T}'))$ satisfying certain structure-preserving properties (these will be called *type transformers*) and, thus to abstract over their construction. Constants in two operation signatures are then related by means of a type transformer between the types over respective type signatures.

In particular, we obtain that the extension of type constructor maps, as ϕ above, generates only a relatively small subset of possible type transformers. So, this construction has been extended to a more involved, based on so-called *type schemes*, which will be presented in the sequel.

3.1 Relating class signatures

Since any class signature forms a directed acyclic graph with type classes as nodes and class relations as edges, it is natural to relate these in terms of mappings preserving the respective graph structures.

Definition 9 (related class signatures and class map)
Let $\mathcal{T}$ and $\mathcal{T}'$ be theories, and $\Sigma_{class} \subseteq \Sigma_{class}(\mathcal{T})$ such that $Super_{\Sigma_{class}(\mathcal{T})}(c) \subseteq \Sigma_{class}$ holds for any $c \in \Sigma_{class}$. Then $\Sigma_{class}(\mathcal{T})$ and $\Sigma_{class}(\mathcal{T}')$ are *related* by a mapping $\varphi : \Sigma_{class} \to \Sigma_{class}(\mathcal{T}')$ w.r.t. Σ_{class} if $c \prec^*_{\Sigma_{class}(\mathcal{T})} c'$ implies $\varphi(c) \prec^*_{\Sigma'_{class}} \varphi(c')$ for any $c, c' \in \Sigma_{class}$. This circumstance is denoted by $\Sigma_{class}(\mathcal{T}) \overset{\varphi}{\longmapsto}_{\Sigma_{class}} \Sigma_{class}(\mathcal{T}')$, and φ is called a *class map* w.r.t. Σ_{class}. In the case, when $\Sigma_{class} = \Sigma_{class}(\mathcal{T})$, the notation $\Sigma_{class}(\mathcal{T}) \overset{\varphi}{\longmapsto} \Sigma_{class}(\mathcal{T}')$ is used, and φ is called a *class map*.

For example, let $\mathcal{T}$ be a theory with $\Sigma_{class}(\mathcal{T}) = \{c_1, c_2, c_3\}$, and $\mathcal{T}'$ a theory with $\Sigma_{class}(\mathcal{T}') = \{c'_1, c'_2, c'_3\}$. Further, let $c_2 \prec_{\Sigma_{class}(\mathcal{T})} c_1$, $c_3 \prec_{\Sigma_{class}(\mathcal{T})} c_1$

in $\Sigma_{class}(\mathcal{T})$, and $c_2' \prec_{\Sigma_{class}(\mathcal{T}')} c_1'$ in $\Sigma_{class}(\mathcal{T}')$. Then for $\Sigma_{class} \stackrel{def}{=} \{c_1, c_2\}$, $Super_{\Sigma_{class}(\mathcal{T})}(c_i) \subseteq \Sigma_{class}$ holds for $i = 1, 2$. Moreover, $\varphi : \Sigma_{class} \to \Sigma_{class}(\mathcal{T}')$, where $\varphi(c_1) \stackrel{def}{=} c_1'$ and $\varphi(c_2) \stackrel{def}{=} c_2'$, is clearly a class map w.r.t. Σ_{class}, such that $\Sigma_{class}(\mathcal{T}) \stackrel{\varphi}{\longmapsto}_{\Sigma_{class}} \Sigma_{class}(\mathcal{T}')$ holds. Now, φ can be extended to $\Sigma_{class}(\mathcal{T})$ in two different ways, defining either $\varphi(c_3) \stackrel{def}{=} c_1'$ or $\varphi(c_3) \stackrel{def}{=} c_2'$ such that $\Sigma_{class}(\mathcal{T}) \stackrel{\varphi}{\longmapsto} \Sigma_{class}(\mathcal{T}')$ holds. In contrast, the extension by $\varphi(c_3) \stackrel{def}{=} c_3'$ does not give a class map, because $c_3 \prec^*_{\Sigma_{class}(\mathcal{T})} c_1$ holds, whereas $c_3' \prec^*_{\Sigma_{class}(\mathcal{T}')} c_1'$ does not hold.

3.1.1 Class map extension

Generally, if Σ_{class} and Σ'_{class} are class signatures and $\varphi : \Sigma_{class} \to \Sigma'_{class}$ a mapping, then φ can be extended to the mapping $\overline{\varphi} : Sorts(\Sigma_{class}) \to Sorts(\Sigma'_{class})$ simply by $\overline{\varphi}(S) \stackrel{def}{=} \varphi\langle S\rangle$, where $\varphi\langle S\rangle$ denotes the image of S under φ. Let us call $\overline{\varphi}$ the *extension* of φ. The next lemma shows the connection between super classes and extensions of class maps.

Lemma 3.1 *Let $\Sigma_{class} \subseteq \Sigma_{class}(\mathcal{T})$ such that $Super_{\Sigma_{class}(\mathcal{T})}(c) \subseteq \Sigma_{class}$ holds for any $c \in \Sigma_{class}$. Then the condition $\Sigma_{class}(\mathcal{T}) \stackrel{\varphi}{\longmapsto}_{\Sigma_{class}} \Sigma_{class}(\mathcal{T}')$ is equivalent to the condition*

$$\overline{\varphi}(Super_{\Sigma_{class}(\mathcal{T})}(c)) \subseteq Super_{\Sigma_{class}(\mathcal{T}')}(\varphi(c)) \text{ for all } c \in \Sigma_{class} \tag{3.1}$$

Proof. (by cases)
(i) Suppose $\Sigma_{class}(\mathcal{T}) \stackrel{\varphi}{\longmapsto}_{\Sigma_{class}} \Sigma_{class}(\mathcal{T}')$, and $c_1 \in \overline{\varphi}(Super_{\Sigma_{class}(\mathcal{T})}(c))$ with arbitrary $c \in \Sigma_{class}$. Since $c \in \Sigma_{class}$ we have $Super_{\Sigma_{class}(\mathcal{T})}(c) \subseteq \Sigma_{class}$ such that $\overline{\varphi}(Super_{\Sigma_{class}(\mathcal{T})}(c))$ is well-defined. Hence, there is some $c_2 \in Super_{\Sigma_{class}(\mathcal{T})}(c)$ with $c_1 = \varphi(c_2)$. Further, since φ is a class map from $c \prec^*_{\Sigma_{class}(\mathcal{T})} c_2$ we have $\varphi(c) \prec^*_{\Sigma_{class}(\mathcal{T}')} \varphi(c_2)$, i.e. $c_1 \in Super_{\Sigma_{class}(\mathcal{T}')}(\varphi(c))$.
(ii) If we assume $c \prec^*_{\Sigma_{class}(\mathcal{T})} c'$ for some $c, c' \in \Sigma_{class}$ then we have $\varphi(c') \in \overline{\varphi}(Super_{\Sigma_{class}(\mathcal{T})}(c))$. Hence, $\varphi(c') \in Super_{\Sigma_{class}(\mathcal{T}')}(\varphi(c))$ holds by (3.1), i.e. $\varphi(c) \prec^*_{\Sigma_{class}(\mathcal{T}')} \varphi(c')$. □

If φ is a class map then $\overline{\varphi}$ is also called a *sort* map. The following lemma states that any sort map preserves the particular subsort relations.

Lemma 3.2 *Let $\mathcal{T}$ and $\mathcal{T}'$ be theories with $\Sigma_{class}(\mathcal{T}) \stackrel{\varphi}{\longmapsto} \Sigma_{class}(\mathcal{T}')$. Then the property*

$$\frac{S \preceq^{\mathcal{T}}_{sort} S'}{\overline{\varphi}(S) \preceq^{\mathcal{T}'}_{sort} \overline{\varphi}(S')}$$

holds for all sorts S and S' in $Sorts(\Sigma_{class}(\mathcal{T}))$.

Proof. Let S and S' be sorts over $\Sigma_{class}(\mathcal{T})$ with $S \preceq^{\mathcal{T}}_{sort} S'$. We have to show, that for any $c' \in \overline{\varphi}(S')$ there exists $c \in \overline{\varphi}(S)$ with $c \prec^{*}_{\Sigma_{class}(\mathcal{T}')} c'$. So, let $c' \in \overline{\varphi}(S')$. Then $c' = \varphi(c'_1)$ for some $c'_1 \in S'$. By $S \preceq^{\mathcal{T}}_{sort} S'$ we have $c_1 \in S$ with $c_1 \prec^{*}_{\Sigma_{class}(\mathcal{T})} c'_1$. Since φ is a class map, $\varphi(c_1) \prec^{*}_{\Sigma_{class}(\mathcal{T}')} \varphi(c'_1)$ holds, which shows the goal, because $\varphi(c_1) \in \overline{\varphi}(S)$. □

3.2 Relating type signatures

A central rôle in this section is played by functions which transform types over one type signature into types over another type signature, satisfying the properties stated by the following definition.

Definition 10 (type transformer)
Let $\mathcal{T}$ and $\mathcal{T}'$ be theories and φ a class map with $\Sigma_{class}(\mathcal{T}) \overset{\varphi}{\longmapsto} \Sigma_{class}(\mathcal{T}')$. A function $f^{\varphi} : Types(\Sigma_{type}(\mathcal{T})) \rightarrow Types(\Sigma_{type}(\mathcal{T}'))$ is called a *type transformer* w.r.t. φ if for any $\kappa \in Types(\Sigma_{type}(\mathcal{T}))$ and $S \in Sorts(\Sigma_{class}(\mathcal{T}))$ the following conditions hold:

1. $typevars(f^{\varphi}(\kappa)) \subseteq typevars(\kappa)$,
2. $\vartheta_{f^{\varphi}(\kappa)}(\alpha) = \overline{\varphi}(\vartheta_{\kappa}(\alpha))$ is satisfied for all $\alpha \in typevars(f^{\varphi}(\kappa))$,
3. f^{φ} preserves the particular of-sort relations, i.e.

$$\frac{\kappa \ \textit{of-sort}_{\mathcal{T}} \ S}{f^{\varphi}(\kappa) \ \textit{of-sort}_{\mathcal{T}'} \ \overline{\varphi}(S)}$$

So, for instance, any mapping of type constructors from $\Sigma_{type}(\mathcal{T}_1)$ to those with the same rank in $\Sigma_{type}(\mathcal{T}_2)$ (satisfying certain arity properties) can be extended to a type transformer, which replaces type constructors for type constructors in any type, according to the mapping. This possibility has been already sketched at the beginning of this chapter. However, type transformers obtained from such simple type constructor mappings form only a small subset of the possible type transformers. But we can refine this idea of type transformer generation introducing the notion of type schemes, which are actually closely related to types, except that here we abstract over the set of type variables and get rid of sort assignments.

Definition 11 (type schemes over a type signature and a set)
Let $\mathcal{T}$ be a theory. The set of *type schemes* over the type signature $\Sigma_{type}(\mathcal{T})$ and an infinite countable set X of *indices*, denoted by $\widehat{Types}_X(\Sigma_{type}(\mathcal{T}))$, is defined inductively as follows:

(i) if $x \in X$ then $`x \in \widehat{Types}_X(\Sigma_{type}(\mathcal{T}))$;

(ii) if $C \in \Sigma_{type}(\mathcal{T})$, $\varsigma_1, \ldots, \varsigma_n \in \widehat{Types}_X(\Sigma_{type}(\mathcal{T}))$ where $n = rank_{\mathcal{T}}(C)$, then the ***application*** $(\varsigma_1, \ldots, \varsigma_n) \cdot C$ is an element of $\widehat{Types}_X(\Sigma_{type}(\mathcal{T}))$.

As with types we write the application of type constructors with the rank 0 by C instead of $() \cdot C$ and $\varsigma \cdot C$ instead of $(\varsigma) \cdot C$. Moreover, the application operator $\cdot$ is primarily used to avoid confusion between types and type schemes occurring in the same term, and can be omitted if possible. So, a type scheme $`\alpha \cdot C \in \widehat{Types}_{\mathcal{X}_{type}}(\Sigma_{type}(\mathcal{T}))$ could be also denoted just by $\alpha\ C$.

Let the function $Inds_X : \widehat{Types}_X(\Sigma_{type}(\mathcal{T})) \to \mathcal{P}_{fin}(X)$ be the unique homomorphism, which interprets the application $\cdot$ as the union of sets, and extends the function $\iota : X \to \mathcal{P}_{fin}(X)$ with $\iota(x) \stackrel{def}{=} \{x\}$, such that $Inds_X(`x) = \iota(x)$ holds for all $x \in X$. So, $Inds_X(\varsigma)$ is the finite subset of elements of X occurring in ς.

Let $\mathbb{N}_+ \stackrel{def}{=} \mathbb{N} \setminus \{0\}$. In the following, a central rôle will play the set

$$\widehat{Types}(\Sigma_{type}(\mathcal{T})) \stackrel{def}{=} \widehat{Types}_{\mathbb{N}_+}(\Sigma_{type}(\mathcal{T}))$$

i.e. the set of type schemes with natural numbers greater 0 as indices. So, for any $\varsigma \in \widehat{Types}(\Sigma_{type}(\mathcal{T}))$ we can now define

$$Bound(\varsigma) \stackrel{def}{=} \begin{cases} \text{Max}(Inds_{\mathbb{N}_+}(\varsigma)), & \text{if } Inds_{\mathbb{N}_+}(\varsigma) \neq \emptyset \\ 0, & \text{otherwise.} \end{cases}$$

In other words, if some indices occur in ς then $Bound(\varsigma)$ is the maximum of these, and, hence an element of $\mathbb{N}_+$ as well. On the other hand, $Bound(\varsigma) = 0$ holds iff no indices occur in ς.

Using *Bound*, we define how type schemes over a type signature act on types over this type signature.

Definition 12 (application of a type scheme to a type sequence)
Let $\mathcal{T}$ be a theory and $\varsigma \in \widehat{Types}(\Sigma_{type}(\mathcal{T}))$. Further, let $\langle \kappa_1, \ldots, \kappa_b \rangle \in Types_{\vartheta}(\Sigma_{type}(\mathcal{T}))^b$ be a type sequence with $b \in \mathbb{N}$ such that $Bound(\varsigma) \leq$

b (if $b = 0$ then $\langle \kappa_1, \ldots, \kappa_b \rangle$ is the empty sequence $\langle\rangle$). The *application* $\varsigma \bullet \langle \kappa_1, \ldots, \kappa_b \rangle \in \mathit{Types}_\vartheta(\Sigma_{type}(\mathcal{T}))$ is recursively defined as follows:

(i) $`i \bullet \langle \kappa_1, \ldots, \kappa_b \rangle = \kappa_i$;

(ii) $(\varsigma_1, \ldots, \varsigma_n) \cdot C \bullet \langle \kappa_1, \ldots, \kappa_b \rangle = (\varsigma_1 \bullet \langle \kappa_1, \ldots, \kappa_b \rangle, \ldots, \varsigma_n \bullet \langle \kappa_1, \ldots, \kappa_b \rangle)C$

The basic observation about this function is that for any $\varsigma \in \widehat{\mathit{Types}}(\Sigma_{type}(\mathcal{T}))$ with $\mathit{Bound}(\varsigma) = 0$, $\varsigma \bullet s$ is a well-defined type without type variables for any sequence of types s. Moreover, the value $\varsigma \bullet s$ does not depend on the choice of s, because in this case ς consists only of applications of type constructors and, thus, e.g. $\varsigma \bullet \langle\rangle$ translates ς to the corresponding monomorphic type.

Type signatures will be now related using a mapping of type constructors in one type signature to type schemes over another type signature.

Definition 13 (related type signatures and type map)
Let $\mathcal{T}$ and $\mathcal{T}'$ be theories and $\Sigma_{class}(\mathcal{T}) \overset{\varphi}{\longmapsto} \Sigma_{class}(\mathcal{T}')$. The type signatures $\Sigma_{type}(\mathcal{T})$ and $\Sigma_{type}(\mathcal{T}')$ are *related* by $\phi : \Sigma_{type}(\mathcal{T}) \to \widehat{\mathit{Types}}(\Sigma_{type}(\mathcal{T}'))$ w.r.t. φ, if the following *type map conditions* hold for any $C \in \Sigma_{type}(\mathcal{T})$ with $\mathit{rank}_{\mathcal{T}}(C) = n$:

1. $\mathit{Bound}(\phi(C)) \leq n$; moreover, if $\mathit{Bound}(\phi(C)) < n$ holds then the assignment $C \mapsto \phi(C)$ will be called *contracting*;

2. for any arity $(S_1, \ldots, S_n, S_{n+1}) \in \mathit{arities\text{-}of}_{\mathcal{T}}(C)$ and any sort assignment ϑ, the rule

$$\frac{\kappa_i \ \mathit{of\text{-}sort}_{\mathcal{T}'} \ \overline{\varphi}(S_i) \ \kappa_i \in \mathit{Types}_\vartheta(\Sigma_{type}(\mathcal{T}')) \text{ for all } 1 \leq i \leq n}{\phi(C) \bullet \langle \kappa_1, \ldots, \kappa_n \rangle \ \mathit{of\text{-}sort}_{\mathcal{T}'} \ \overline{\varphi}(S_{n+1})} \tag{3.2}$$

   holds.

Additionally, there are two *meta-logical requirements* on ϕ:

1. $\phi(\texttt{prop}) = \texttt{prop}$

2. $\phi(\Rightarrow) = (`1, `2) \cdot \Rightarrow$

The fact that the type signatures are related by ϕ w.r.t. φ is denoted by $\Sigma_{type}(\mathcal{T}) \overset{\phi}{\longmapsto}_{\varphi} \Sigma_{type}(\mathcal{T}')$ and ϕ is called a *type map* w.r.t. φ.

In particular, if $C \in \Sigma_{type}(\mathcal{T})$ is a type constructor with $\mathit{rank}_{\mathcal{T}}(C) = 0$ then $\mathit{Bound}(\phi(C)) = 0$ holds for any type map $\phi : \Sigma_{type}(\mathcal{T}) \to \widehat{\mathit{Types}}(\Sigma_{type}(\mathcal{T}'))$.

So, in this case the condition (3.2) becomes equivalent to the condition: $\phi(C) \bullet \langle\rangle$ *of-sort*$_{\mathcal{T}'}$ $\overline{\varphi}(S_j)$ for all $1 \leq j \leq a$, where *arities-of*$_{\mathcal{T}}(C) = \{S_1, \ldots, S_a\}$. This condition can be verified directly for each arity $S \in$ *arities-of*$_{\mathcal{T}}(C)$.

On the other hand, if $n > 0$, we cannot in general verify the condition (3.2) for C directly, since this would require to consider infinitely many possible types in *Types*$(\Sigma_{type}(\mathcal{T}'))$. The condition given by the following lemma allows to consider only a fixed finite subset of type variables:

Lemma 3.3 *The condition* (3.2) *is equivalent to the condition*

$$\frac{\alpha_1, \ldots, \alpha_n \in \mathcal{X}_{type} \textit{ distinct} \quad (S_1, \ldots, S_n, S_{n+1}) \in \textit{arities-of}_{\mathcal{T}}(C)}{\phi(C) \bullet \langle \alpha_1 ::_{sort} \overline{\varphi}(S_1), \ldots, \alpha_n ::_{sort} \overline{\varphi}(S_n)\rangle \textit{ of-sort}_{\mathcal{T}'} \overline{\varphi}(S_{n+1})} \tag{3.3}$$

Proof. We first show that (3.2) implies (3.3). To this end, let $\alpha_1, \ldots, \alpha_n \in \mathcal{X}_{type}$ be distinct type variables and $(S_1, \ldots, S_n, S_{n+1})$ an arity of C. There exists a sort assignment ϑ such that $\vartheta(\alpha_i) = \overline{\varphi}(S_i)$, since $\alpha_1, \ldots, \alpha_n$ are distinct. Hence, $(\alpha_i ::_{sort} \overline{\varphi}(S_i)) \in$ *Types*$_{\vartheta}(\Sigma_{type}(\mathcal{T}'))$ holds for all $1 \leq i \leq n$. We have to show $\phi(C) \bullet \langle \alpha_1 ::_{sort} \overline{\varphi}(S_1), \ldots, \alpha_n ::_{sort} \overline{\varphi}(S_n)\rangle$ *of-sort*$_{\mathcal{T}'}$ $\overline{\varphi}(S_{n+1})$. So, by (3.2), it is then sufficient to show $\alpha_i ::_{sort} \overline{\varphi}(S_i)$ *of-sort*$_{\mathcal{T}'}$ $\overline{\varphi}(S_i)$, for all $1 \leq i \leq n$. This holds by Definition 4, since $\alpha_i ::_{sort} \overline{\varphi}(S_i)$ *of-sort*$_{\mathcal{T}'}(\vartheta)$ $\overline{\varphi}(S_i)$ holds.

Next, we show that (3.3) implies (3.2). So, let ϑ be a sort assignment over $\mathcal{T}$ and $(S_1, \ldots, S_n, S_{n+1}) \in$ *arities-of*$_{\mathcal{T}}(C)$. Further, we can assume $\kappa_i \in$ *Types*$_{\vartheta}(\Sigma_{type}(\mathcal{T}'))$ and κ_i *of-sort*$_{\mathcal{T}'}$ $\overline{\varphi}(S_i)$ for all $1 \leq i \leq n$ and have to show

$$\phi(C) \bullet \langle \kappa_1, \ldots, \kappa_n\rangle \textit{ of-sort}_{\mathcal{T}'} \overline{\varphi}(S_{n+1}).$$

To this end we note that $\phi(C) \bullet \langle \kappa_1, \ldots, \kappa_n\rangle$ can be written as

$$(\phi(C) \bullet \langle \alpha_1 ::_{sort} \overline{\varphi}(S_1), \ldots, \alpha_n ::_{sort} \overline{\varphi}(S_n)\rangle)\overline{[\alpha_i := \kappa_i]_n}$$

provided by some fixed distinct type variables $\alpha_1, \ldots, \alpha_n$, which is well-defined due to the assumptions on κ_i and using the fact that

$$\textit{typevars}(\phi(C) \bullet \langle \alpha_1 ::_{sort} \overline{\varphi}(S_1), \ldots, \alpha_n ::_{sort} \overline{\varphi}(S_n)\rangle) \subseteq \{\alpha_1, \ldots, \alpha_n\}.$$

Then, by (3.3) we obtain

$$\phi(C) \bullet \langle \alpha_1 ::_{sort} \overline{\varphi}(S_1), \ldots, \alpha_n ::_{sort} \overline{\varphi}(S_n)\rangle \textit{ of-sort}_{\mathcal{T}'} \overline{\varphi}(S_{n+1}).$$

Hence, by (2.2),

$$(\phi(C) \bullet \langle \alpha_1 ::_{sort} \overline{\varphi}(S_1), \ldots, \alpha_n ::_{sort} \overline{\varphi}(S_n)\rangle)\overline{[\alpha_i := \kappa_i]_n} \textit{ of-sort}_{\mathcal{T}'} \overline{\varphi}(S_{n+1})$$

which proves the claim. □

This lemma justifies the following procedure to verify the type map conditions for a given type constructor $C \in \Sigma_{type}(\mathcal{T})$ and a type scheme $\varsigma \in \widehat{Types}(\Sigma_{type}(\mathcal{T}'))$.

Procedure *type constructor assignment check*

Input:

1. theories $\mathcal{T}$ and $\mathcal{T}'$;
2. class map $\Sigma_{class}(\mathcal{T}) \overset{\varphi}{\longmapsto} \Sigma_{class}(\mathcal{T}')$;
3. $C \in \Sigma_{type}(\mathcal{T})$ and $\varsigma \in \widehat{Types}(\Sigma_{type}(\mathcal{T}'))$.

Output: The assignment $C \mapsto \varsigma$ is either accepted or rejected.

Description: Let $n = rank_{\mathcal{T}}(C)$ and $a = |arities\text{-}of_{\mathcal{T}}(C)|$. If $n < Bound(\varsigma)$ holds then $C \mapsto \varsigma$ is rejected. Otherwise, if $a = 0$ then $C \mapsto \varsigma$ is accepted, since in this case the condition (3.3) holds obviously. Otherwise, let

$$(S_{1,1}, \ldots, S_{1,n}, S_{1,n+1}), \ldots, (S_{a,1}, \ldots, S_{a,n}, S_{a,n+1})$$

be some sequence of arities from $arities\text{-}of_{\mathcal{T}}(C)$. Furthermore, let $\alpha_1, \ldots, \alpha_n$ be some fixed distinct type variables. Then, for all $1 \leq j \leq a$ the type scheme applications $\varsigma \bullet \langle \alpha_1 ::_{sort} \overline{\varphi}(S_{j,1}), \ldots, \alpha_n ::_{sort} \overline{\varphi}(S_{j,n}) \rangle$ is evaluated yielding a sequence of types $\kappa_1, \ldots, \kappa_a$ from $Types(\Sigma_{type}(\mathcal{T}'))$. If κ_j *of-sort*$_{\mathcal{T}'}$ $\overline{\varphi}(S_{j,n+1})$ holds for all $1 \leq j \leq a$ then $C \mapsto \varsigma$ is accepted, otherwise rejected.

This procedure will be used later in the context of signature morphism construction for the decision whether a given mapping of type constructors in a type signature constitutes a correct type map.

3.2.1 Type map extension

The goal of this section is to show that from any type map $\Sigma_{type}(\mathcal{T}) \overset{\phi}{\longmapsto}_{\varphi} \Sigma_{type}(\mathcal{T}')$ a type transformer $\overline{\phi}^{\varphi} : Types(\Sigma_{type}(\mathcal{T})) \to Types(\Sigma_{type}(\mathcal{T}'))$, called *type map extension*, can be derived. The main step towards this result is the proof of the following

Lemma 3.4 *Let $\Sigma_{type}(\mathcal{T}) \overset{\phi}{\longmapsto}_{\varphi} \Sigma_{type}(\mathcal{T}')$ and ϑ a sort assignment over $\mathcal{T}$. Then for all $\kappa \in Types_{\vartheta}(\Sigma_{type}(\mathcal{T}))$ there exists $\kappa' \in Types_{\vartheta'}(\Sigma_{type}(\mathcal{T}'))$, where $\vartheta'(\alpha) \overset{def}{=} \overline{\varphi}(\vartheta(\alpha))$, such that*

1. $typevars(\kappa') \subseteq typevars(\kappa)$, *and*

2. for all $S \in Sorts(\Sigma_{class}(\mathcal{T}))$ *the rule*

$$\frac{\kappa \textit{ of-sort}_{\mathcal{T}}\ S}{\kappa' \textit{ of-sort}_{\mathcal{T}'}\ \overline{\varphi}(S)} \tag{3.4}$$

hold.

Proof. (by structural induction on κ)

(i) Let $\kappa = \alpha ::_{sort} \vartheta(\alpha)$ with $\alpha \in \mathcal{X}_{type}$. We define $\kappa' \stackrel{def}{=} \alpha ::_{sort} \vartheta'(\alpha)$ such that even $typevars(\kappa') = typevars(\kappa)$ holds. For (3.4), let $S \in Sorts(\Sigma_{class}(\mathcal{T}))$ with κ *of-sort*$_{\mathcal{T}}$ S. Hence, $\vartheta(\alpha) \preceq^{\mathcal{T}}_{sort} S$, and so $\overline{\varphi}(\vartheta(\alpha)) \preceq^{\mathcal{T}'}_{sort} \overline{\varphi}(S)$, by Lemma 3.2. Thus, κ' *of-sort*$_{\mathcal{T}'}$ $\overline{\varphi}(S)$.

(ii) Let $\kappa = (\kappa_1, \ldots, \kappa_n)C$ where $C \in \Sigma_{type}(\mathcal{T})$, $n = rank_{\mathcal{T}}(C)$, $\kappa_1, \ldots, \kappa_n \in Types_{\vartheta}(\Sigma_{type}(\mathcal{T}))$. By the induction hypothesis, for κ_i with $1 \leq i \leq n$ we obtain $\kappa'_i \in Types_{\vartheta'}(\Sigma_{type}(\mathcal{T}'))$ satisfying the required conditions. Since ϕ satisfies the type map conditions, $\kappa' \stackrel{def}{=} \phi(C) \bullet \langle \kappa'_1, \ldots, \kappa'_n \rangle$, is well-defined such that $\kappa' \in Types_{\vartheta'}(\Sigma_{type}(\mathcal{T}'))$. Further, using the fact that a type variable occurs in κ' iff it occurs in some κ'_i, and the induction hypothesis, we obtain $typevars(\kappa') \subseteq typevars(\kappa)$. It remains to show (3.4). To this end, let κ *of-sort*$_{\mathcal{T}}$ S for some fixed sort $S \in Sorts(\Sigma_{class}(\mathcal{T}))$. Then κ *of-sort*$_{\mathcal{T}}(\vartheta)$ S holds as well, since $\kappa \in Types_{\vartheta}(\Sigma_{type}(\mathcal{T}))$. Hence, by Definition 4, there is an arity $(S_1, \ldots, S_n, S_{n+1}) \in$ *arities-of*$_{\mathcal{T}}(C)$ such that κ_i *of-sort*$_{\mathcal{T}}(\vartheta)$ S_i for $1 \leq i \leq n$ and $S_{n+1} \preceq^{\mathcal{T}}_{sort} S$. We have to show κ' *of-sort*$_{\mathcal{T}'}$ $\overline{\varphi}(S)$. This is implied by κ' *of-sort*$_{\mathcal{T}'}$ $\overline{\varphi}(S_{n+1})$, by Lemma 2.3, and since $\overline{\varphi}(S_{n+1}) \preceq^{\mathcal{T}'}_{sort} \overline{\varphi}(S)$ holds by Lemma 3.2. By definition of κ', κ' *of-sort*$_{\mathcal{T}'}$ $\overline{\varphi}(S_{n+1})$ is equivalent to

$$\phi(C) \bullet \langle \kappa'_1, \ldots, \kappa'_n \rangle \textit{ of-sort}_{\mathcal{T}'}\ \overline{\varphi}(S_{n+1})$$

Provided by the type map condition (3.2), it suffices to show κ'_i *of-sort*$_{\mathcal{T}'}$ $\overline{\varphi}(S_i)$ for all $1 \leq i \leq n$. For $n > 0$ this follows from the induction hypothesis applied respectively to the facts κ_i *of-sort*$_{\mathcal{T}}$ S_i for all $1 \leq i \leq n$. □

The proof establishes that the recursive function

(i) $\overline{\phi}^{\varphi}(\vartheta)(\alpha ::_{sort} \vartheta(\alpha)) = \alpha ::_{sort} \overline{\varphi}(\vartheta(\alpha))$;

(ii) $\overline{\phi}^{\varphi}(\vartheta)((\kappa_1, \ldots, \kappa_n)C) = \phi(C) \bullet \langle \overline{\phi}^{\varphi}(\vartheta)(\kappa_1), \ldots, \overline{\phi}^{\varphi}(\vartheta)(\kappa_n) \rangle$

maps types in $Types_\vartheta(\Sigma_{type}(\mathcal{T}))$ to types in $Types_{\vartheta'}(\Sigma_{type}(\mathcal{T}))$, such that the conditions stated in the lemma hold. Since the value $\overline{\phi}^\varphi(\vartheta)(\kappa)$ does not depend on the type variables which does not occur in κ, the extension

$$\overline{\phi}^\varphi(\kappa) \stackrel{def}{=} \overline{\phi}^\varphi(\vartheta_\kappa)(\kappa)$$

is well-defined. Regarding sort assignments, the equation $\vartheta_{\overline{\phi}^\varphi(\kappa)}(\alpha) = \overline{\varphi}(\vartheta_\kappa(\alpha))$ holds for all $\alpha \in \mathit{typevars}(\overline{\phi}^\varphi(\kappa))$, as required for type transformers, but not for all type variables, since $\vartheta_{\overline{\phi}^\varphi(\kappa)}$ is not unique.

Moreover, if $\kappa_1, \ldots, \kappa_n \in Types_\vartheta(\Sigma_{type}(\mathcal{T}))$ then there exists ϑ' such that $\overline{\phi}^\varphi(\kappa_1), \ldots, \overline{\phi}^\varphi(\kappa_n) \in Types_{\vartheta'}(\Sigma_{type}(\mathcal{T}'))$, because any $\alpha \in \mathit{typevars}(\overline{\phi}^\varphi(\kappa_i))$ occurs also in κ_i and hence $\vartheta_{\kappa_i}(\alpha) = \vartheta(\alpha)$ holds, such that $\vartheta'(\alpha) \stackrel{def}{=} \overline{\varphi}(\vartheta(\alpha))$ is well-defined. The following equation, expressing the connection between the extension $\overline{\phi}^\varphi$ and the substitution of type variables, is based on this fact:

$$\overline{\phi}^\varphi(\kappa\overline{[\alpha_i := \kappa_i]}_n) = \overline{\phi}^\varphi(\kappa)\overline{[\alpha_i := \overline{\phi}^\varphi(\kappa_i)]}_n \tag{3.5}$$

Finally, the following lemma is important as it emphasises that α-equality is compatible with any type map extension.

Lemma 3.5 *For any $\kappa \in Types_\vartheta(\Sigma_{type}(\mathcal{T}))$*

$$\frac{\kappa \stackrel{\alpha}{=}_\theta \kappa'}{\overline{\phi}^\varphi(\kappa) \stackrel{\alpha}{=}_{\theta'} \overline{\phi}^\varphi(\kappa')}$$

holds for all $\kappa' \in Types(\Sigma_{type}(\mathcal{T}))$, where θ' is the bijection restricting θ to $\mathit{typevars}(\overline{\phi}^\varphi(\kappa))$.

Proof. (by structural induction on κ)
(i) Let $\kappa = \alpha ::_{sort} \vartheta(\alpha)$. From $\kappa \stackrel{\alpha}{=}_\theta \kappa'$ we have $\kappa' = \theta(\alpha) ::_{sort} \vartheta(\alpha)$, such that $\overline{\phi}^\varphi(\kappa) = \alpha ::_{sort} \overline{\varphi}(\vartheta(\alpha)) \stackrel{\alpha}{=}_\theta \theta(\alpha) ::_{sort} \overline{\varphi}(\vartheta(\alpha)) = \overline{\phi}^\varphi(\kappa')$.
(ii) Let $\kappa = (\kappa_1, \ldots, \kappa_n)C$. Then $\kappa \stackrel{\alpha}{=}_\theta \kappa'$ implies $\kappa' = (\kappa'_1, \ldots, \kappa'_n)C$ with $\kappa_i \stackrel{\alpha}{=}_{\theta_i} \kappa'_i$ for $1 \leq i \leq n$, where θ_i is the restriction of θ to $\mathit{typevars}(\kappa_i)$ and a bijection as well. Further, we can also observe that $\theta_i(\beta) = \theta_j(\beta)$ for all $1 \leq i, j \leq n$ and any $\beta \in \mathit{typevars}(\kappa_i) \cap \mathit{typevars}(\kappa_j)$. By the induction hypothesis we obtain $\overline{\phi}^\varphi(\kappa_i) \stackrel{\alpha}{=}_{\theta'_i} \overline{\phi}^\varphi(\kappa'_i)$ for $1 \leq i \leq n$, where θ'_i is the restriction of θ_i to $\mathit{typevars}(\overline{\phi}^\varphi(\kappa_i))$. Moreover, we also obtain $\theta'_i(\beta) = \theta'_j(\beta)$ for any $\beta \in \mathit{typevars}(\overline{\phi}^\varphi(\kappa_i)) \cap \mathit{typevars}(\overline{\phi}^\varphi(\kappa_j))$ and $1 \leq i, j \leq n$, such that θ' can be defined

as the union of θ'_i for $1 \leq i \leq n$, which gives a bijection from *typevars*($\overline{\phi}^{\varphi}(\kappa)$) to *typevars*($\overline{\phi}^{\varphi}(\kappa')$). Finally, we can reason:

$$
\begin{array}{ll}
\overline{\phi}^{\varphi}(\kappa') & = \\
\phi(C) \bullet \langle \overline{\phi}^{\varphi}(\kappa'_1), \ldots, \overline{\phi}^{\varphi}(\kappa'_n) \rangle & = \\
\phi(C) \bullet \langle \overline{\phi}^{\varphi}(\kappa_1)\overline{[\alpha_{1,j} := \theta'_1(\alpha_{1,j}) ::_{sort} \vartheta(\alpha_{1,j})]_{k_1}}, \ldots, & \\
\quad \overline{\phi}^{\varphi}(\kappa_n)\overline{[\alpha_{n,j} := \theta'_1(\alpha_{n,j}) ::_{sort} \vartheta(\alpha_{n,j})]_{k_n}} \rangle & = \\
\qquad \text{where } \{\alpha_{i,1}, \ldots, \alpha_{i,k_i}\} = typevars(\overline{\phi}^{\varphi}(\kappa_i)) & \\
(\phi(C) \bullet \langle \overline{\phi}^{\varphi}(\kappa_1), \ldots, \overline{\phi}^{\varphi}(\kappa_n) \rangle)\overline{[\alpha_i := \theta'(\alpha_i) ::_{sort} \vartheta(\alpha_i)]_k} & \overset{\alpha}{=}_{\theta'} \\
\qquad \text{where } \{\alpha_1, \ldots, \alpha_k\} = \bigcup_{1 \leq i \leq n} typevars(\overline{\phi}^{\varphi}(\kappa_i)) & \\
\overline{\phi}^{\varphi}(\kappa) &
\end{array}
$$

□

3.2.2 Example: sets as monads

Now we will start a stepwise development of a concrete example in order to illustrate how the introduced concepts practically work. To this end we take a well-known category-theoretical structure, namely monads [Mac Lane 1998]. First of all, the general definition is the following.

Definition 14 (monad)
Let **C** be a category, $T : \mathbf{C} \longrightarrow \mathbf{C}$ an endo-functor and

1. $\eta : 1_{\mathbf{C}} \dot{\to} T$ (where $1_{\mathbf{C}}$ is the identity functor)
2. $\mu : T^2 \dot{\to} T$

two natural transformations. Then $\langle T, \eta, \mu \rangle$ is called a *monad* with the *unit* η and the *multiplication* μ, if the following rules hold:

$$\mu \circ T\mu = \mu \circ \mu T \qquad (3.6)$$

$$\mu \circ T\eta = 1_T \quad \text{(where } 1_T \text{ is the identity transformation)} \qquad (3.7)$$

$$\mu \circ \eta T = 1_T \qquad (3.8)$$

These rules require some explanations. So, if X is an object of **C** then $\mu_X : T(TX) \to TX$ is an arrow in **C** such that $T\mu$ is the arrow $T(\mu_X) : T(T(TX)) \to T(TX)$. On the other hand, μT denotes the arrow $\mu_{TX} : T(T(TX)) \to T(TX)$. In other words, the rule (3.6) essentially expresses the identity $\mu_X \circ T(\mu_X) = \mu_X \circ \mu_{TX}$. Similarly, $T\eta$ is the arrow $T(\eta_X) : TX \to T(TX)$, whereas ηT – the arrow $\eta_{TX} : TX \to T(TX)$.

In this example we will consider monads over the category **Set**. More precisely, in the framework given in the previous chapter, the corresponding category with types as objects and functions as arrows has been already introduced. That is, the set of terms having a type $\kappa_1 \Rightarrow \kappa_2$, where κ_1, κ_2 are some types, forms this category: a κ-identity is the term $\lambda(x ::_{type} \kappa).x$, and the composition is $\lambda f\ g\ x.f(g\ x)$. Using the polymorphism, the already mentioned theory *Fun* (Section 2.12.1) formalises this introducing the derived constants

- $id :: \alpha \Rightarrow \alpha$, and
- $\_ \circ \_ :: (\beta \Rightarrow \gamma) \Rightarrow (\alpha \Rightarrow \beta) \Rightarrow (\alpha \Rightarrow \gamma)$

Thus, a theory axiomatising a monad over this category takes *Fun* as its single parent theory and might look as follows.

typedecl $\alpha\ T$
consts
$T\ ::\ (\alpha \Rightarrow \beta) \Rightarrow (\alpha\ T \Rightarrow \beta\ T)$
$\eta\ ::\ \alpha \Rightarrow \alpha\ T$
$\mu\ ::\ (\alpha\ T)T \Rightarrow \alpha\ T$

axioms (* functor *)
F1 : $T\ id = id$
F2 : $T(f \circ g) = T\ f \circ T\ g$

axioms (* natural transformations *)
N1 : $T\ f \circ \eta = \eta \circ f$
N2 : $T\ f \circ \mu = \mu \circ T(T\ f)$

axioms (* monad *)
M1 : $\mu \circ T\ \mu = \mu \circ \mu$
M2 : $\mu \circ T\ \eta = id$
M3 : $\mu \circ \eta = id$

Let us refer to the theory as *Monad*. Notice, how the axioms M1 – M3 correspond to the rules (3.6) – (3.8). So, for instance the arrow μT in (3.6) is just expressed by μ in M1, and we let the type system figure out that $\mu \circ \mu$ basically denotes the term $(\mu ::_{type} (\alpha\ T)T \Rightarrow \alpha\ T) \circ (\mu ::_{type} ((\alpha\ T)T)T \Rightarrow (\alpha\ T)T)$.

The ultimate goal of this example is to establish in this framework the well-known fact that the power-set functor $\mathcal{P}$: **Set** $\longrightarrow$ **Set** with the union as multiplication and where unit embeds elements of a set as singletons, forms a monad. This will be done by a stepwise construction of a theory *PSet_Monad* together with a theory morphism from *Monad* to *PSet_Monad*.

First of all, we take the theory *Fun* as the single parent theory of *PSet_Monad*, i.e. we have the hierarchy

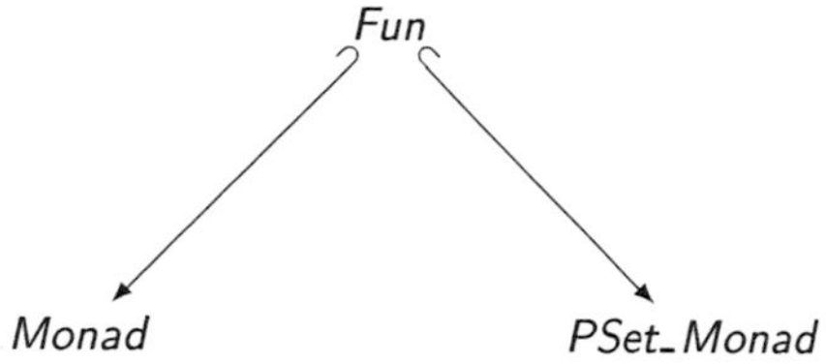

However, at this point we can only relate the type signatures by means of a type map $\phi_{PSet\_Monad} : \Sigma_{type}(Monad) \to \widehat{Types}(\Sigma_{type}(PSet\_Monad))$. In order to focus on the essential, we firstly define the class map φ_{PSet_Monad} : $\Sigma_{class}(Monad) \to \Sigma_{class}(PSet\_Monad)$ to be an identity, and secondly assume that the global type constructors from $\Sigma_{type}(Fun)$ are also mapped by 'identity' by ϕ_{PSet_Monad} (this treatment will be further explained and justified in Section 3.7).

Thus, the only type constructor which has to be treated is $T \in \Sigma_{type}(Monad)$, having the rank 1. This is done by the assignment $T \mapsto (`1, bool) \cdot \Rightarrow$, i.e. any (sub-)type $\kappa\ T$ will be transformed to $\kappa \Rightarrow bool$ which is nothing else as a predicate type, usually abbreviated by $\kappa\ set$. In other words, the type transformer $\overline{\phi}_{PSet\_Monad}$, obtained this way, replaces any occurrence of T by *set* for any type in $Types(\Sigma_{type}(Monad))$. In particular, for the types of the logical constants in *Monad* this means:

1. the type of the functor T is transformed to $\kappa_T \stackrel{def}{=} (\alpha \Rightarrow \beta) \Rightarrow (\alpha\ set \Rightarrow \beta\ set)$;

2. the type of the unit η is transformed to $\kappa_\eta \stackrel{def}{=} \alpha \Rightarrow \alpha\ set$;

3. the type of the multiplication μ is transformed to $\kappa_\mu \stackrel{def}{=} (\alpha\, set) set \Rightarrow \alpha\, set$.

The obtained types already hint to a possibility to treat the logical constants T, η and μ, and this will be formally defined in the next section.

3.3 Relating operation signatures

Based on the concepts of related class and type signatures, introduced in the previous sections, we can define a relation on operation signatures in a similar manner.

Definition 15 (related operation signatures and operation map)
Let $\mathcal{T}$ and $\mathcal{T}'$ be theories, $\Sigma_{class}(\mathcal{T}) \stackrel{\varphi}{\longmapsto} \Sigma_{class}(\mathcal{T}')$, $\Sigma_{type}(\mathcal{T}) \stackrel{\phi}{\longmapsto}_{\varphi} \Sigma_{type}(\mathcal{T}')$, and $\Sigma \subseteq \Sigma_{op}(\mathcal{T})$. The operation signatures $\Sigma_{op}(\mathcal{T})$ and $\Sigma_{op}(\mathcal{T}')$ are *related* by a mapping $\psi : \Sigma \to \Sigma_{op}(\mathcal{T}')$ w.r.t. φ and ϕ, denoted by $\Sigma \stackrel{\psi}{\longmapsto}_{\varphi,\phi} \Sigma_{op}(\mathcal{T}')$, if the following *operation map condition* holds for any $f \in \Sigma$:

$$\textit{type-of}_{\mathcal{T}'}(\psi(f)) \stackrel{\alpha}{=} \overline{\phi}^{\varphi}(\textit{type-of}_{\mathcal{T}}(f)). \tag{3.9}$$

As with type maps, there are also *meta-logical requirements* on ψ, which are simply that the meta-logical constants are mapped to themselves by ψ, i.e. $\psi(\equiv) = \equiv$, $\psi(\Longrightarrow) = \Longrightarrow$, and $\psi(\bigwedge) = \bigwedge$. Due to the meta-logical requirements on ϕ, this requirement is compatible with the operation map condition. So, e.g. $\overline{\phi}^{\varphi}(\textit{type-of}_{\mathcal{T}}(\equiv)) = \textit{type-of}_{\mathcal{T}}(\equiv)$ holds in any theory $\mathcal{T}$. Finally, if ψ satisfies $\Sigma \stackrel{\psi}{\longmapsto}_{\varphi,\phi} \Sigma_{op}(\mathcal{T}')$, it is called an *operation map* for Σ w.r.t. φ and ϕ.

Similarly to type maps, for any ψ with $\Sigma \stackrel{\psi}{\longmapsto}_{\varphi,\phi} \Sigma_{op}(\mathcal{T}')$ we can derive a mapping from terms in $\textit{Terms}(\Sigma)$ to terms in $\textit{Terms}(\Sigma_{op}(\mathcal{T}'))$ preserving the of-type relations.

Lemma 3.6 *Suppose $\Sigma \subseteq \Sigma_{op}(\mathcal{T})$ and $\Sigma \stackrel{\psi}{\longmapsto}_{\varphi,\phi} \Sigma_{op}(\mathcal{T}')$. Then for all $t \in \textit{Terms}_{\varrho_\vartheta}(\Sigma)$ there exists $t' \in \textit{Terms}_{\varrho'_{\vartheta'}}(\Sigma_{op}(\mathcal{T}'))$, where $\vartheta'(\alpha) \stackrel{def}{=} \overline{\varphi}(\vartheta(\alpha))$ and $\varrho'_{\vartheta'}(x) \stackrel{def}{=} \overline{\phi}^{\varphi}(\varrho_\vartheta(x))$, such that*

1. *$\textit{termvars}(t') = \textit{termvars}(t)$ and $\textit{typevars}(t') \subseteq \textit{typevars}(t)$, and*

2. *for all $\kappa \in \textit{Types}(\Sigma_{type}(\mathcal{T}))$ the property*

$$\frac{t \ \textit{of-type}_{\mathcal{T}}\ \kappa}{t' \ \textit{of-type}_{\mathcal{T}'}\ \overline{\phi}^{\varphi}(\kappa)} \tag{3.10}$$

   holds.

Proof. (by structural induction on t)
(i) Let $t = x ::_{type} \varrho_\vartheta(x)$ with $x \in \mathcal{X}_{term}$. We define $t' \stackrel{def}{=} x ::_{type} \varrho'_{\vartheta'}(x)$. For the proof of (3.10), let $\kappa \in \textit{Types}(\Sigma_{type}(\mathcal{T}))$ with t *of-type*$_{\mathcal{T}}$ κ. So, $\kappa = \varrho_\vartheta(x)$ and, thus, t' *of-type*$_{\mathcal{T}'}$ $\overline{\phi}^{\varphi}(\kappa)$.
(ii) Let $t = f ::_{type} \kappa$ with $f \in \Sigma$, $\kappa = \textit{type-of}_{\mathcal{T}}(f)\overline{[\alpha_i := \kappa_i]}_n$ with $\{\alpha_1, \ldots, \alpha_n\} \supseteq \textit{typevars}(\textit{type-of}_{\mathcal{T}}(f))$ and $\kappa_i \in \textit{Types}_\vartheta(\Sigma_{type}(\mathcal{T}))$. Let $t' \stackrel{def}{=} \psi(f) ::_{type} \overline{\phi}^{\varphi}(\kappa)$.

Since ψ is an operation map, we have $\overline{\phi}^{\varphi}(\textit{type-of}_{\mathcal{T}}(f)) \stackrel{\alpha}{=}_{\theta} \textit{type-of}_{\mathcal{T}'}(\psi(f))$, and, hence, by Lemma 2.4,

$$\overline{\phi}^{\varphi}(\textit{type-of}_{\mathcal{T}}(f))\overline{[\alpha_i := \overline{\phi}^{\varphi}(\kappa_i)]_n} = \textit{type-of}_{\mathcal{T}'}(\psi(f))\overline{[\theta'(\alpha_i) := \overline{\phi}^{\varphi}(\kappa_i)]_n}$$

where θ' extends the domain of θ to $\{\alpha_1, \ldots, \alpha_n\}$. Further, using (3.5), we can conclude

$$\overline{\phi}^{\varphi}(\textit{type-of}_{\mathcal{T}}(f))\overline{[\alpha_i := \overline{\phi}^{\varphi}(\kappa_i)]_n} = \overline{\phi}^{\varphi}(\textit{type-of}_{\mathcal{T}}(f)\overline{[\alpha_i := \kappa_i]_n})$$

which means, that $\overline{\phi}^{\varphi}(\kappa)$ is obtained by a substitution in $\textit{type-of}_{\mathcal{T}'}(\psi(f))$, and justifies the well formedness of t' as a term. For the proof of (3.10), let $\kappa' \in \textit{Types}(\Sigma_{type}(\mathcal{T}))$ be some type with t $\textit{of-type}_{\mathcal{T}}$ κ', which means $\kappa' = \kappa$ by the definition of $\textit{of-type}_{\mathcal{T}}$. Thus, t' $\textit{of-type}_{\mathcal{T}'}$ $\overline{\phi}^{\varphi}(\kappa')$ holds.

(iii) Let $t = t_1\ t_2$ with $t_1, t_2 \in \textit{Terms}_{\varrho_\vartheta}(\Sigma)$. So, by the induction hypothesis, we obtain $t_1', t_2' \in \textit{Terms}_{\varrho'_{\vartheta'}}(\Sigma_{op}(\mathcal{T}'))$ satisfying the required conditions w.r.t. t_1 and t_2. Thus, defining $t' \stackrel{def}{=} t_1'\ t_2'$, it remains to show (3.10). To this end, let t $\textit{of-type}_{\mathcal{T}}$ κ_B with an arbitrary $\kappa_B \in \textit{Types}(\Sigma_{type}(\mathcal{T}))$. Then, by the definition of $\textit{of-type}_{\mathcal{T}}$, we have t_1 $\textit{of-type}_{\mathcal{T}}$ $(\kappa_A \Rightarrow \kappa_B)$ and t_2 $\textit{of-type}_{\mathcal{T}}$ κ_A. Then for t_1' we obtain t_1' $\textit{of-type}_{\mathcal{T}'}$ $\overline{\phi}^{\varphi}(\kappa_A \Rightarrow \kappa_B)$, or equivalently t_1' $\textit{of-type}_{\mathcal{T}'}$ $\phi(\Rightarrow) \bullet \langle \overline{\phi}^{\varphi}(\kappa_A), \overline{\phi}^{\varphi}(\kappa_B)\rangle$, which is in turn equivalent to t_1' $\textit{of-type}_{\mathcal{T}'}$ $\overline{\phi}^{\varphi}(\kappa_A) \Rightarrow \overline{\phi}^{\varphi}(\kappa_B)$ by the meta-logical requirements on ϕ. For t_2' we obtain t_2' $\textit{of-type}_{\mathcal{T}'}$ $\overline{\phi}^{\varphi}(\kappa_A)$, and can conclude t' $\textit{of-type}_{\mathcal{T}'}$ $\overline{\phi}^{\varphi}(\kappa_B)$.

(iv) Let $t = \lambda(x ::_{type} \varrho_\vartheta(x)).t_1$ with $t_1 \in \textit{Terms}_{\varrho_\vartheta}(\Sigma)$. By the induction hypothesis we obtain $t_1' \in \textit{Terms}_{\varrho'_{\vartheta'}}(\Sigma_{op}(\mathcal{T}'))$ satisfying the required conditions w.r.t. t_1, and define $t' \stackrel{def}{=} \lambda(x ::_{type} \varrho'_{\vartheta'}(x)).t_1' = \lambda(x ::_{type} \overline{\phi}^{\varphi}(\varrho_\vartheta(x))).t_1'$. For the proof of (3.10), let t $\textit{of-type}_{\mathcal{T}}$ κ with an arbitrary $\kappa \in \textit{Types}(\Sigma_{type}(\mathcal{T}))$. By the definition of $\textit{of-type}_{\mathcal{T}}$ we obtain $\kappa = \varrho_\vartheta(x) \Rightarrow \kappa_1$ with $\kappa_1 \in \textit{Types}_\vartheta(\Sigma_{type}(\mathcal{T}))$ and t_1 $\textit{of-type}_{\mathcal{T}}$ κ_1. Then for t_1' we have, that t_1' $\textit{of-type}_{\mathcal{T}'}$ $\overline{\phi}^{\varphi}(\kappa_1)$ holds, and have to show t' $\textit{of-type}_{\mathcal{T}'}$ $\overline{\phi}^{\varphi}(\varrho_\vartheta(x) \Rightarrow \kappa_1)$. By the meta-logical requirements on ϕ, this is equivalent to t' $\textit{of-type}_{\mathcal{T}'}$ $\varrho'_{\vartheta'}(x) \Rightarrow \overline{\phi}^{\varphi}(\kappa_1)$, which is true, provided by the properties of $\overline{\phi}^{\varphi}$, of t' and of the $\textit{of-type}_{\mathcal{T}'}$ relation. □

The proof establishes that the recursive function $\overline{\psi}^{\varphi,\phi}(\varrho_\vartheta)$ satisfying the equations

(i) $\overline{\psi}^{\varphi,\phi}(\varrho_\vartheta)(x ::_{type} \varrho_\vartheta(x)) = x ::_{type} \overline{\phi}^{\varphi}(\vartheta)(\varrho_\vartheta(x))$

(ii) $\overline{\psi}^{\varphi,\phi}(\varrho_\vartheta)(f ::_{type} \kappa) = \psi(f) ::_{type} \overline{\phi}^{\varphi}(\vartheta)(\kappa)$

(iii) $\overline{\psi}^{\varphi,\phi}(\varrho_\vartheta)(t_1\, t_2) = \overline{\psi}^{\varphi,\phi}(\varrho_\vartheta)(t_1)\, \overline{\psi}^{\varphi,\phi}(\varrho_\vartheta)(t_2)$

(iv) $\overline{\psi}^{\varphi,\phi}(\varrho_\vartheta)(\lambda(x ::_{type} \varrho_\vartheta(x)).t_1) = \lambda(x ::_{type} \overline{\phi}^{\varphi}(\vartheta)(\varrho_\vartheta(x))).\overline{\psi}^{\varphi,\phi}(\varrho_\vartheta)(t_1)$

maps terms such that the conditions stated in the lemma hold. As with the type map extension, we can use the fact that $\overline{\psi}^{\varphi,\phi}(\varrho_\vartheta)(t)$ does not depend on the type and term variables, which does not occur in t, and define the mapping

$$\overline{\psi}^{\varphi,\phi}(t) \stackrel{def}{=} \overline{\psi}^{\varphi,\phi}(\varrho_{\vartheta_t})(t)$$

called the *extension* of ψ, which maps any term t in $Terms(\Sigma)$ to the term $\overline{\psi}^{\varphi,\phi}(t)$ in $Terms(\Sigma_{op}(\mathcal{T}'))$ satisfying the conditions of Lemma 3.6 for the particular sort and type assignments ϑ_t and ϱ_{ϑ_t}.

The property (3.5) of type map extensions can be reformulated for operation maps as follows:

$$\overline{\psi}^{\varphi,\phi}(t\overline{[\alpha_i := \kappa_i]_n}) = \overline{\psi}^{\varphi,\phi}(t)\overline{[\alpha_i := \overline{\phi}^{\varphi}(\kappa_i)]_n} \tag{3.11}$$

which is also based on the property $typevars(\overline{\psi}^{\varphi,\phi}(t)) \subseteq typevars(t)$.

Using this, the property that α-equality on types is compatible with any type map extension can be also extended to α-equality on terms and operation map extensions as follows.

Lemma 3.7 *Let $\Sigma, \Sigma' \subseteq \Sigma_{op}(\mathcal{T})$. Then for any $t \in Terms_{\varrho_\vartheta}(\Sigma)$ the rule*

$$\frac{t \stackrel{\alpha}{=}_\theta t'}{\overline{\psi}^{\varphi,\phi}(\varrho_\vartheta)(t) \stackrel{\alpha}{=}_{\theta'} \overline{\psi}^{\varphi,\phi}(t')}$$

holds for all $t' \in Terms(\Sigma')$, where θ' is the restriction of θ to $typevars(\overline{\psi}^{\varphi,\phi}(t))$.

The proof proceeds by the structural induction on t, and is similar to the proof of Lemma 3.5. The following rule is derived from Lemma 3.7:

$$\frac{t \stackrel{\alpha}{=} t'}{\overline{\psi}^{\varphi,\phi}(t) \stackrel{\alpha}{=} \overline{\psi}^{\varphi,\phi}(t')} \tag{3.12}$$

for any $t, t' \in Terms(\Sigma_{op}(\mathcal{T}))$.

Finally, since term variables can be substituted in terms as well, we need the property that the substitution factors through the operation map extension. To this end the following lemma proves this for the function $\overline{\psi}^{\varphi,\phi}(\varrho_\vartheta)$ with a fixed arbitrary type assignment ϱ_ϑ.

Lemma 3.8 *Let $\Sigma, \Sigma' \subseteq \Sigma_{op}(\mathcal{T})$ and $t \in Terms_{\varrho_\vartheta}(\Sigma)$. Then*

$$\overline{\psi}^{\varphi,\phi}(\varrho_\vartheta)(t[t'/x]) = \overline{\psi}^{\varphi,\phi}(\varrho_\vartheta)(t)[\overline{\psi}^{\varphi,\phi}(t')/x]$$

holds for any $t' \in Terms(\Sigma')$ and $x \in \mathcal{X}_{term}$ with t' of-type$_{\mathcal{T}}$ $\varrho_\vartheta(x)$.

Proof. (by structural induction on t)
(i) Let $t = (y ::_{type} \varrho_\vartheta(y))$. If $y = x$ then $t[t'/x] = t'$ and, so

$$\begin{array}{ll}
\overline{\psi}^{\varphi,\phi}(\varrho_\vartheta)(t)[\overline{\psi}^{\varphi,\phi}(\varrho_\vartheta)(t')/x] & = \\
(x ::_{type} \overline{\phi}^{\varphi}(\vartheta)(\varrho_\vartheta(x)))[\overline{\psi}^{\varphi,\phi}(\varrho_\vartheta)(t')/x] & = \\
\overline{\psi}^{\varphi,\phi}(\varrho_\vartheta)(t') & = \\
\overline{\psi}^{\varphi,\phi}(\varrho_\vartheta)(t[t'/x]). &
\end{array}$$

If $y \neq x$ then $t[t'/x] = t$, and, so

$$\begin{array}{ll}
\overline{\psi}^{\varphi,\phi}(\varrho_\vartheta)(t[t'/x]) & = \\
\overline{\psi}^{\varphi,\phi}(\varrho_\vartheta)(t) & = \\
\overline{\psi}^{\varphi,\phi}(\varrho_\vartheta)(t)[\overline{\psi}^{\varphi,\phi}(\varrho_\vartheta)(t')/x] &
\end{array}$$

holds.
(ii) If $t = f ::_{type} \kappa$ with $f \in \Sigma$ then $t[t'/x] = t$, and, so

$$\begin{array}{ll}
\overline{\psi}^{\varphi,\phi}(\varrho_\vartheta)(t[t'/x]) & = \\
\overline{\psi}^{\varphi,\phi}(\varrho_\vartheta)(t) & = \\
\psi(f) ::_{type} \overline{\phi}^{\varphi}(\vartheta)(\kappa) & = \\
(\psi(f) ::_{type} \overline{\phi}^{\varphi}(\vartheta)(\kappa))[\overline{\psi}^{\varphi,\phi}(\varrho_\vartheta)(t')/x]. &
\end{array}$$

(iii) Let $t = t_1\ t_2$ with $t_1, t_2 \in Terms_{\varrho_\vartheta}(\Sigma)$. Then

$$\begin{array}{ll}
\overline{\psi}^{\varphi,\phi}(\varrho_\vartheta)(t[t'/x]) & = \\
\overline{\psi}^{\varphi,\phi}(\varrho_\vartheta)(t_1[t'/x]\ t_2[t'/x]) & = \\
\overline{\psi}^{\varphi,\phi}(\varrho_\vartheta)(t_1[t'/x])\ \overline{\psi}^{\varphi,\phi}(\varrho_\vartheta)(t_2[t'/x]) & = \\
& \text{by ind. hyp.} \\
\overline{\psi}^{\varphi,\phi}(\varrho_\vartheta)(t_1)[\overline{\psi}^{\varphi,\phi}(\varrho_\vartheta)(t')/x]\ \overline{\psi}^{\varphi,\phi}(\varrho_\vartheta)(t_2)[\overline{\psi}^{\varphi,\phi}(\varrho_\vartheta)(t')/x] & = \\
\overline{\psi}^{\varphi,\phi}(\varrho_\vartheta)(t)[\overline{\psi}^{\varphi,\phi}(\varrho_\vartheta)(t')/x]. &
\end{array}$$

(iv) Let $t = \lambda(y ::_{type} \varrho_\vartheta(y)).t_1$ with $t_1 \in \mathit{Terms}_{\varrho_\vartheta}(\Sigma)$. If $y = x$ then $t[t'/x] = t$, so

$$\begin{array}{ll}
\overline{\psi}^{\varphi,\phi}(\varrho_\vartheta)(t[t'/x]) & = \\
\overline{\psi}^{\varphi,\phi}(\varrho_\vartheta)(t) & = \\
\lambda(y ::_{type} \overline{\phi}^{\varphi}(\vartheta)(\varrho_\vartheta(x))).\overline{\psi}^{\varphi,\phi}(\varrho_\vartheta)(t_1) & = \\
(\lambda(y ::_{type} \overline{\phi}^{\varphi}(\vartheta)(\varrho_\vartheta(x))).\overline{\psi}^{\varphi,\phi}(\varrho_\vartheta)(t_1))[\overline{\psi}^{\varphi,\phi}(\varrho_\vartheta)(t')/x]. &
\end{array}$$

If $y \neq x$ then

$$\begin{array}{ll}
\overline{\psi}^{\varphi,\phi}(\varrho_\vartheta)(t[t'/x]) & = \\
\overline{\psi}^{\varphi,\phi}(\varrho_\vartheta)(\lambda(y ::_{type} \varrho_\vartheta(y)).t_1[t'/x]) & = \\
\lambda(y ::_{type} \overline{\phi}^{\varphi}(\vartheta)(\varrho_\vartheta(y))).\overline{\psi}^{\varphi,\phi}(\varrho_\vartheta)(t_1[t'/x]) & = \\
 & \text{by ind. hyp.} \\
\lambda(y ::_{type} \overline{\phi}^{\varphi}(\vartheta)(\varrho_\vartheta(y))).\overline{\psi}^{\varphi,\phi}(\varrho_\vartheta)(t_1)[\overline{\psi}^{\varphi,\phi}(\varrho_\vartheta)(t')/x] & = \\
\overline{\psi}^{\varphi,\phi}(\varrho_\vartheta)(t)[\overline{\psi}^{\varphi,\phi}(\varrho_\vartheta)(t')/x]. &
\end{array}$$

□

From this, we can now deduce the following property of the operation map extension $\overline{\psi}^{\varphi,\phi}$ in connection with the substitution of variables:

$$\overline{\psi}^{\varphi,\phi}(t[t'/x]) = \overline{\psi}^{\varphi,\phi}(t)[\overline{\psi}^{\varphi,\phi}(t')/x] \tag{3.13}$$

holds for any $\Sigma, \Sigma' \subseteq \Sigma_{op}(\mathcal{T})$, $t \in \mathit{Terms}(\Sigma)$, $t' \in \mathit{Terms}(\Sigma')$, $x \in \mathit{termvars}(t)$ with t' *of-type*$_{\mathcal{T}}$ $\varrho_{\vartheta_t}(x)$. In the case when $x \notin \mathit{termvars}(t)$, (3.13) holds obviously for any $t' \in \mathit{Terms}(\Sigma')$.

3.3.1 Example: sets as monads

Now we can continue our monad example, started in Section 3.2.2. There, the class map $\varphi_{\mathsf{PSet\_Monad}} : \Sigma_{class}(\mathsf{Monad}) \to \Sigma_{class}(\mathsf{PSet\_Monad})$ and the type map $\phi_{\mathsf{PSet\_Monad}} : \Sigma_{type}(\mathsf{Monad}) \to \widehat{\mathit{Types}}(\Sigma_{type}(\mathsf{PSet\_Monad}))$ have been already constructed. Here we extend the theory *PSet_Monad* such that we can define the operation map $\psi_{\mathsf{PSet\_Monad}} : \Sigma_{op}(\mathsf{Monad}) \to \Sigma_{op}(\mathsf{PSet\_Monad})$.

Similarly to the class map, we define that any global constant from $\Sigma_{op}(\mathsf{Fun})$ is mapped by $\psi_{\mathsf{PSet\_Monad}}$ to itself. Thus, it remains to give assignments for the constants $T, \eta, \mu \in \Sigma_{op}(\mathsf{Monad})$. To this end, we extend the theory *PSet_Monad* as follows:

consts

$$\begin{array}{lll} T^{\mathsf{PSet}} & :: & (\alpha \Rightarrow \beta) \Rightarrow (\alpha\ set \Rightarrow \beta\ set) \\ \eta^{\mathsf{PSet}} & :: & \alpha \Rightarrow \alpha\ set \\ \mu^{\mathsf{PSet}} & :: & (\alpha\ set)set \Rightarrow \alpha\ set \end{array}$$

Notice, that the assigned types are nothing else as the transformed types κ_T, κ_η, and κ_μ from Section 3.2.2, respectively.

Now it is clear, that with

$$\begin{array}{lll} \psi_{\mathsf{PSet\_Monad}}(T) & \stackrel{def}{=} & T^{\mathsf{PSet}} \\ \psi_{\mathsf{PSet\_Monad}}(\eta) & \stackrel{def}{=} & \eta^{\mathsf{PSet}} \\ \psi_{\mathsf{PSet\_Monad}}(\mu) & \stackrel{def}{=} & \mu^{\mathsf{PSet}} \end{array}$$

$\psi_{\mathsf{PSet\_Monad}}$ is an operation map, which relates $\Sigma_{op}(\mathit{Monad})$ and $\Sigma_{op}(\mathit{PSet\_Monad})$ w.r.t. $\varphi_{\mathsf{PSet\_Monad}}$ and $\phi_{\mathsf{PSet\_Monad}}$. Introducing signature morphisms in the next section, we will return to this example and show how these mappings give us a signature morphism.

3.4 Signature morphisms

Since the signature of a theory comprises the class, type and operations signatures of the theory, for any two theories their class, type and operation signatures can be related as defined in the previous sections. Consequently, the notion of signature morphism combines the particular relations.

Definition 16 (signature morphism)
A *signature morphism* $\sigma(\Sigma) = \langle \mathcal{T}_1, \mathcal{T}_2, \sigma_{class}, \sigma_{type}, \Sigma, \sigma_{op}\rangle$ is given by

1. Two theories $\mathcal{T}_1, \mathcal{T}_2$, called the *source* and the *target* of $\sigma(\Sigma)$, respectively;

2. A mapping σ_{class} satisfying $\Sigma_{class}(\mathcal{T}_1) \stackrel{\sigma_{class}}{\longmapsto} \Sigma_{class}(\mathcal{T}_2)$, called the *class map* of $\sigma(\Sigma)$;

3. A mapping σ_{type} satisfying $\Sigma_{type}(\mathcal{T}_1) \stackrel{\sigma_{type}}{\longmapsto}_{\sigma_{class}} \Sigma_{type}(\mathcal{T}_2)$, called the *type map* of $\sigma(\Sigma)$;

4. $\Sigma^{log}_{op}(\mathcal{T}_1) \subseteq \Sigma \subsetneq \Sigma_{op}(\mathcal{T}_1)$ is a part of the operation signature of $\mathcal{T}_1$ which is closed under the order on constants in $\Sigma_{op}(\mathcal{T}_1)$, i.e. if $f \in \Sigma$ and $f \prec_{\Sigma_{op}(\mathcal{T}_1)} f'$ then $f' \in \Sigma$;

5. A (partial) map $\sigma_{op} : \Sigma_{op}(\mathcal{T}_1) \rightharpoonup \Sigma_{op}(\mathcal{T}_2)$, called the *operation map* of $\sigma(\Sigma)$, satisfying $\Sigma \overset{\sigma_{op}}{\longmapsto}_{\sigma_{class},\sigma_{type}} \Sigma_{op}(\mathcal{T}_2)$ such that for any derived constant $f \in \Sigma \cap \Sigma_{op}^{der}(\mathcal{T}_1)$ additionally the both *definitional conditions*

   (a) $\sigma_{op}(f) \in \Sigma_{op}^{der}(\mathcal{T}_2)$,

   (b) and

$$\textit{def-of}_{\mathcal{T}_2}(\sigma_{op}(f)) \overset{\alpha}{=} \overline{\sigma}(\Sigma)(\textit{def-of}_{\mathcal{T}_1}(f)) \tag{3.14}$$

   hold. In (3.14), $\overline{\sigma}(\Sigma) : \mathit{Terms}(\Sigma) \to \mathit{Terms}(\Sigma_{op}(\mathcal{T}_2))$ denotes the extension $\overline{\sigma}_{op}^{\sigma_{class},\sigma_{type}}$ of σ_{op}. $\overline{\sigma}(\Sigma)$ is also called the *homomorphic extension* of the signature morphism $\sigma(\Sigma)$. The requirement, that Σ has to be closed under $\prec_{\Sigma_{op}(\mathcal{T}_1)}$, ensures $\textit{def-of}_{\mathcal{T}_1}(f) \in \mathit{Terms}(\Sigma)$, such that the application $\overline{\sigma}(\Sigma)(\textit{def-of}_{\mathcal{T}_1}(f))$ is well-defined.

Signature morphisms will be also denoted by $\sigma(\Sigma) : \mathcal{T}_1 \longrightarrow \mathcal{T}_2$, while $\sigma : \mathcal{T}_1 \longrightarrow \mathcal{T}_2$ will be used as a shortening for $\sigma(\Sigma_{op}(\mathcal{T}_1)) : \mathcal{T}_1 \longrightarrow \mathcal{T}_2$. $\sigma(\Sigma_{op}(\mathcal{T}_1))$ will be also called a *complete* signature morphism.

3.4.1 Example: sets as monads

Returning to our sets as monads example, we can now define the complete signature morphism

$$\langle \mathsf{Monad}, \mathsf{PSet\_Monad}, \varphi_{\mathsf{PSet\_Monad}}, \phi_{\mathsf{PSet\_Monad}}, \Sigma_{op}(\mathsf{Monad}), \psi_{\mathsf{PSet\_Monad}} \rangle$$

which will be called $\sigma_{\mathsf{PSet\_Monad}}$. That $\sigma_{\mathsf{PSet\_Monad}}$ is complete can be seen as follows. Since *Monad* does not introduce definitions, $\Sigma_{op}^{der}(\mathsf{Monad}) = \Sigma_{op}^{der}(\mathsf{Fun})$ holds. This means $\Sigma_{op}(\mathsf{Monad}) \cap \Sigma_{op}^{der}(\mathsf{Monad}) = \Sigma_{op}^{der}(\mathsf{Fun})$, i.e. the definitional conditions have to be ensured only for global derived constants. By our definition of class, type and operation maps on global constants (i.e. identity mappings), these conditions hold.

Thus, the homomorphic extension $\overline{\sigma}_{\mathsf{PSet\_Monad}}$ of the signature morphism translates terms in $\mathit{Terms}(\Sigma_{op}(\mathsf{Monad}))$ to terms in $\mathit{Terms}(\Sigma_{op}(\mathsf{PSet\_Monad}))$ according to the defined class, type and operation maps. Roughly, this means that T is replaced by T^{PSet}, η by η^{PSet}, and μ by μ^{PSet} in any term $t \in \mathit{Terms}(\Sigma_{op}(\mathsf{Monad}))$.

In particular, we can consider the propositions in $\mathit{Props}(\mathsf{PSet\_Monad})$, obtained via this translation from the axioms of the theory *Monad* on page 70. So, for instance the functor axioms F1 and F2 give the following propositions in $\mathit{Props}(\mathsf{PSet\_Monad})$:

1. $T^{\mathsf{PSet}}\ id = id$, and

2. $\bigwedge f\ g.\ T^{\mathsf{PSet}}(f \circ g) = T^{\mathsf{PSet}}\ f \circ T^{\mathsf{PSet}}\ g$

Generally, such images of axioms under the extension of a signature morphism will play a central rôle for the construction of theory morphisms, and we will return to this example in the context of theory morphisms again.

3.4.2 Base and normalisable signature morphisms

By the definition, the mappings $\Sigma_{class}(\mathcal{T}_1) \stackrel{\varphi}{\longmapsto} \Sigma_{class}(\mathcal{T}_2)$, $\Sigma_{type}(\mathcal{T}_1) \stackrel{\phi}{\longmapsto}_{\varphi} \Sigma_{type}(\mathcal{T}_2)$ and $\Sigma_{op}^{log}(\mathcal{T}_1) \stackrel{\psi}{\longmapsto}_{\varphi,\phi} \Sigma_{op}(\mathcal{T}_2)$ already determine the *base* signature morphism $\sigma(\Sigma_{op}^{log}(\mathcal{T}_1)) : \mathcal{T}_1 \longrightarrow \mathcal{T}_2$. Its homomorphic extension $\overline{\sigma}(\Sigma_{op}^{log}(\mathcal{T}_1))$ maps the terms in $Terms(\Sigma_{op}^{log}(\mathcal{T}_1))$ to terms in $Terms(\Sigma_{op}(\mathcal{T}_2))$ preserving the particular of-type relations, according to (3.10). On the other hand, we can take into account the fact that any $t \in Terms(\Sigma_{op}(\mathcal{T}_1))$ can be reduced to the normal form $t{\downarrow} \in Terms(\Sigma_{op}^{log}(\mathcal{T}_1))$, replacing any derived constant f by its definition $def\text{-}of_{\mathcal{T}_1}(f)$ in t. Since $f \equiv def\text{-}of_{\mathcal{T}_1}(f)$ is an axiom, the proposition $t \equiv t{\downarrow}$ is well-formed and derivable in $\mathcal{T}_1$, provided by the basic meta-rules for $\equiv$ like transitivity and combination. In particular, this means that the mapping $t \mapsto t{\downarrow}$ preserves the $type\text{-}of_{\mathcal{T}_1}$ relation. Altogether, this allows us to extend $\overline{\sigma}(\Sigma_{op}^{log}(\mathcal{T}_1))$ to a mapping of terms in $Terms(\Sigma_{op}(\mathcal{T}_1))$ to terms in $Terms(\Sigma_{op}(\mathcal{T}_2))$ (because logical constants in the source could be mapped to derived constants in the target) satisfying (3.10). However, this mapping in general does not maintain the structure of terms over the source signature, since it unfolds the source definitions. In contrast, the homomorphic extension $\overline{\sigma}$ of a complete signature morphism $\sigma : \mathcal{T}_1 \longrightarrow \mathcal{T}_2$ gives a mapping of terms in $Terms(\Sigma_{op}(\mathcal{T}_1))$ to terms in $Terms(\Sigma_{op}(\mathcal{T}_2))$, which not only satisfies (3.10) but also preserves the structure of terms, i.e. maintains all constants derived in the source theory. Therefore we are basically interested in complete signature morphisms.

The basic observation regarding complete signature morphisms is, that if there is some signature morphism $\sigma(\Sigma) : \mathcal{T}_1 \longrightarrow \mathcal{T}_2$ then a complete signature from $\mathcal{T}_1$ to $\mathcal{T}_2$ not necessarily exists. This is exactly because the set $\Sigma_{op}^{der}(\mathcal{T}_2)$ not necessarily provides appropriate images for derived constants, for which the operation map is not defined, i.e. in $\Sigma_{op}(\mathcal{T}_1) \setminus \Sigma$.

The central question now is, whether a theory extending $\mathcal{T}_2$ exists, such that $\sigma(\Sigma)$ can be extended to a complete signature morphism with a new target theory. As will be shown in the next section, this is not always the case, which justifies the following definition.

Definition 17 (normalisable signature morphisms)
Let $\sigma(\Sigma) : \mathcal{T}_1 \longrightarrow \mathcal{T}_2$ be a signature morphism. Then $\sigma(\Sigma)$ is called *normalisable* if there exists a theory $\mathcal{T}_2'$ such that the following conditions hold:

1. $\mathcal{T}_2 \in Anc(\mathcal{T}_2')$;
2. if there is an axiom $a \in AX(\mathcal{T}_2')$ with $a \notin AX(\mathcal{T}_2)$ then $a \in Defs(\mathcal{T}_2')$ holds, i.e. $\mathcal{T}_2'$ is a definitional extension of $\mathcal{T}_2$;
3. $\sigma(\Sigma)$ can be extended to a complete signature morphism $\sigma' : \mathcal{T}_1 \longrightarrow \mathcal{T}_2'$, i.e. the class and type maps of $\sigma(\Sigma)$ and σ' are equal, whereas the operation map of σ' contains the operation map of $\sigma(\Sigma)$.

The next section gives a procedure, called *normalisation*, which not only decides whether a given signature morphism is normalisable, but also computes the 'least' new target theory and the extended signature morphism in positive cases.

3.5 Normalisation

The core of the normalisation forms the following
Procedure *extension of operation map and target theory by a derived constant*
Input:

1. two theories $\mathcal{T}_1$, $\mathcal{T}_2$;
2. a signature morphism $\sigma(\Sigma) : \mathcal{T}_1 \longrightarrow \mathcal{T}_2$ with the class map σ_{class} and the type map σ_{type};
3. a derived constant $f \in \Sigma_{op}^{der}(\mathcal{T}_1) \setminus \Sigma$ with $\textit{def-of}_{\mathcal{T}_1}(f) \in \textit{Terms}(\Sigma)$;

Output: either by a new derived constant extended target theory $\mathcal{T}_2''$, and the signature morphism $\sigma(\Sigma \cup \{f\}) : \mathcal{T}_1 \longrightarrow \mathcal{T}_2''$, or the output: `not normalisable`;
Description: let $f' \notin \Sigma_{op}(\mathcal{T}_2)$ be a fresh operation symbol. Firstly, we extend $\mathcal{T}_2$ by the constant f' as follows:

$$\mathcal{T}_2' \overset{def}{=} \mathcal{T}_2 +_{op} (f', \overline{\sigma}_{type}^{\sigma_{class}}(\textit{type-of}_{\mathcal{T}_1}(f)))$$

Next, we consider the image of $\textit{def-of}_{\mathcal{T}_1}(f)$ under $\overline{\sigma}(\Sigma)$. Since $\textit{def-of}_{\mathcal{T}_1}(f) \in \textit{Terms}(\Sigma)$ holds by assumption, $\overline{\sigma}(\Sigma)(\textit{def-of}_{\mathcal{T}_1}(f))$ is a well-defined term in $\textit{Terms}(\Sigma_{op}(\mathcal{T}_2))$. Furthermore,

$$\overline{\sigma}(\Sigma)(\textit{def-of}_{\mathcal{T}_1}(f)) \ \textit{of-type}_{\mathcal{T}_2'} \ \overline{\sigma}_{type}^{\sigma_{class}}(\textit{type-of}_{\mathcal{T}_1}(f)) \qquad (3.15)$$

holds by Lemma 3.6, since $\textit{def-of}_{\mathcal{T}_1}(f)$ $\textit{of-type}_{\mathcal{T}_1}$ $\textit{type-of}_{\mathcal{T}_1}(f)$ holds.

Now, the crucial point is the condition on type variables on *lhs* and *rhs* in definitions, namely:

$$\textit{typevars}(\overline{\sigma}(\Sigma)(\textit{def-of}_{\mathcal{T}_1}(f))) \subseteq \textit{typevars}(\overline{\sigma}^{\sigma_{class}}_{type}(\textit{type-of}_{\mathcal{T}_1}(f))) \tag{3.16}$$

So, we distinguish the cases:

1. If (3.16) holds then the extension

   $$\mathcal{T}_2'' \overset{def}{=} \mathcal{T}_2' +_{def} (d', f' \equiv \overline{\sigma}(\Sigma)(\textit{def-of}_{\mathcal{T}_1}(f)))$$

   is well-defined, where d' is a fresh proposition label in $\mathcal{T}_2$, because in this case the term $f' \equiv \overline{\sigma}(\Sigma)(\textit{def-of}_{\mathcal{T}_1}(f))$ forms a pre-definition, and moreover does not introduce cycles, since $f' \notin \Sigma_{op}(\mathcal{T}_2)$.

   Finally, $\sigma(\Sigma \cup \{f\})$ is obtained by the extension of the operation map of $\sigma(\Sigma)$ by $f \mapsto f'$. So, the operation map condition (3.9) and the definitional condition (3.14) for this assignment are satisfied by construction.

2. The case, when the condition (3.16) does not hold, produces the output `not normalisable`.

Firstly, we justify that if the procedure returns `not normalisable` then $\sigma(\Sigma)$ is not normalisable in the sense of Definition 17.

Assuming that $\sigma(\Sigma)$ is normalisable, we obtain a theory $\mathcal{T}_2'$ extending $\mathcal{T}_2$ and the complete signature morphism $\sigma' : \mathcal{T}_1 \longrightarrow \mathcal{T}_2'$, extending $\sigma(\Sigma)$. This would imply that the operation map of σ' assigns to f a derived constant f' in $\Sigma^{der}_{op}(\mathcal{T}_2')$ satisfying (3.14), i.e. $\textit{def-of}_{\mathcal{T}_2'}(f') \overset{\alpha}{=} \overline{\sigma}'(\textit{def-of}_{\mathcal{T}_1}(f))$. Now,

$$\begin{array}{ll} \overline{\sigma}'(\textit{def-of}_{\mathcal{T}_1}(f)) & = \\ & \text{since } \textit{def-of}_{\mathcal{T}_1}(f) \in \textit{Terms}(\Sigma) \\ \overline{\sigma}'(\Sigma)(\textit{def-of}_{\mathcal{T}_1}(f)) & = \\ \overline{\sigma}(\Sigma)(\textit{def-of}_{\mathcal{T}_1}(f)) & \end{array}$$

holds, and we obtain

$$\textit{def-of}_{\mathcal{T}_2'}(f') \overset{\alpha}{=}_{\theta} \overline{\sigma}(\Sigma)(\textit{def-of}_{\mathcal{T}_1}(f))$$

or, equivalently, $\overline{\sigma}(\Sigma)(\textit{def-of}_{\mathcal{T}_1}(f))) = \theta(\textit{def-of}_{\mathcal{T}_2'}(f'))$. Moreover, since f' is a derived constant

$$\textit{def-of}_{\mathcal{T}_2'}(f') \ \textit{of-type}_{\mathcal{T}_2'} \ \textit{type-of}_{\mathcal{T}_2'}(f')$$

holds, and by (2.5) we have

$$\overline{\sigma}(\Sigma)(\textit{def-of}_{\mathcal{T}_1}(f)) \ \textit{of-type}_{\mathcal{T}_2'} \ \theta(\textit{type-of}_{\mathcal{T}_2'}(f'))$$

On the other hand, similarly to (3.15) above, we also have

$$\overline{\sigma}(\Sigma)(\textit{def-of}_{\mathcal{T}_1}(f)) \ \textit{of-type}_{\mathcal{T}_2'} \ \overline{\sigma}_{\textit{type}}^{\sigma_{\textit{class}}}(\textit{type-of}_{\mathcal{T}}(f))$$

i.e. $\overline{\sigma}_{\textit{type}}^{\sigma_{\textit{class}}}(\textit{type-of}_{\mathcal{T}_1}(f)) = \theta(\textit{type-of}_{\mathcal{T}_2'}(f'))$ holds. Also because f' is a derived constant, $\textit{typevars}(\textit{def-of}_{\mathcal{T}_2'}(f')) \subseteq \textit{typevars}(\textit{type-of}_{\mathcal{T}_2'}(f'))$ holds, and hence

$$\textit{typevars}(\theta(\textit{def-of}_{\mathcal{T}_2'}(f'))) \subseteq \textit{typevars}(\theta(\textit{type-of}_{\mathcal{T}_2'}(f')))$$

i.e.

$$\textit{typevars}(\overline{\sigma}(\Sigma)(\textit{def-of}_{\mathcal{T}_1}(f))) \subseteq \textit{typevars}(\overline{\sigma}_{\textit{type}}^{\sigma_{\textit{class}}}(\textit{type-of}_{\mathcal{T}_1}(f)))$$

which is a contradiction to the assumption that (3.16) does not hold.

The condition (3.16) allows to describe the situations quite exactly, when a signature morphism cannot be normalised. The case

$$\textit{typevars}(\overline{\sigma}(\Sigma)(\textit{def-of}_{\mathcal{T}_1}(f))) \not\subseteq \textit{typevars}(\overline{\sigma}_{\textit{type}}^{\sigma_{\textit{class}}}(\textit{type-of}_{\mathcal{T}_1}(f))) \tag{3.17}$$

for a derived constant f can occur if

1. there exists a type constructor $C \in \Sigma_{\textit{type}}(\mathcal{T}_1)$ with $\textit{rank}_{\mathcal{T}_1}(C) > 0$, which occurs in $\textit{type-of}_{\mathcal{T}_1}(f)$;
2. there exists a constant $f_0 \in \Sigma_{\textit{op}}(\mathcal{T}_1)$ with $f \prec_{\Sigma_{\textit{op}}(\mathcal{T}_1)} f_0$, i.e. which occurs in $\textit{def-of}_{\mathcal{T}_1}(f)$;
3. there exists a type constructor $C_0 \in \Sigma_{\textit{type}}(\mathcal{T}_1)$ with $\textit{rank}_{\mathcal{T}_1}(C_0) > 0$, which occurs in $\textit{type-of}_{\mathcal{T}_1}(f_0)$ but not in $\textit{type-of}_{\mathcal{T}_1}(f)$, and will be called a *loose* type constructor in the definition of f;
4. the type map $\sigma_{\textit{type}}$ assigns to C a type scheme ς with $\textit{Bound}(\varsigma) < \textit{rank}_{\mathcal{T}_1}(C)$, such that by the application $\overline{\sigma}_{\textit{type}}^{\sigma_{\textit{class}}}(\textit{type-of}_{\mathcal{T}}(f))$ some type variable arguments, occurring in $\textit{type-of}_{\mathcal{T}}(f)$, disappear;
5. the type map $\sigma_{\textit{type}}$ assigns to C_0 a type scheme ς_0 with $\textit{Bound}(\varsigma_0) = \textit{rank}_{\mathcal{T}_1}(C)$, such that by the application $\overline{\sigma}(\Sigma)(\textit{def-of}_{\mathcal{T}_1}(f))$ the type variable arguments, which have disappeared in the previous point, stay here.

Let, for example, $\mathit{type\text{-}of}_{\mathcal{T}_1}(f) = (\alpha ::_{sort} \emptyset)\, C \Rightarrow (\alpha ::_{sort} \emptyset)\, C$ and $\mathit{type\text{-}of}_{\mathcal{T}_1}(f_0) = (\alpha ::_{sort} \emptyset)\, C \Rightarrow (\alpha ::_{sort} \emptyset)\, C_0$. Additionally, let f_1 be some constant with $\mathit{type\text{-}of}_{\mathcal{T}_1}(f_1) = (\alpha ::_{sort} \emptyset)\, C_0 \Rightarrow (\alpha ::_{sort} \emptyset)\, C$ such that f is defined by $f \equiv f_0 \odot f_1$, where a global $\odot$-operator composes f_0 and f_1 in some way. So, by this setting, C_0 is a loose type constructor. Further, let $\sigma_{type}(C) = C'$ and $\sigma_{type}(C_0) = \text{‘}1 \cdot C_0'$, such that the assignment for C is contracting, in contrast to the assignment for C_0. Evaluating the condition (3.16) now yields

- $\mathit{typevars}(\overline{\sigma}(\Sigma)(\mathit{def\text{-}of}_{\mathcal{T}_1}(f))) = \{\, \alpha \,\}$, and
- $\mathit{typevars}(\overline{\sigma}^{\sigma_{class}}_{type}(\mathit{type\text{-}of}_{\mathcal{T}_1}(f))) = \mathit{typevars}(C' \Rightarrow C') = \emptyset$

i.e. the case (3.17), since the type variable α stays on *rhs* and disappears on *lhs* of the definition $f \equiv f_0 \odot f_1$. Another interesting point is the type of the constant $\sigma_{op}(f_0)$, for which

$$\begin{array}{ll} \mathit{type\text{-}of}_{\mathcal{T}_2}(\sigma_{op}(f_0)) & \stackrel{\alpha}{=} \\ \overline{\sigma}^{\sigma_{class}}_{type}(\mathit{type\text{-}of}_{\mathcal{T}_1}(f_0)) & = \\ C' \Rightarrow (\alpha ::_{sort} \emptyset)\, C_0' & \end{array}$$

holds, due to (3.9). This means that the constant $\sigma_{op}(f_0)$ cannot have a meaningful definition in $\mathcal{T}_2$.

Altogether, the essential observation is that whether a signature morphism is normalisable depends on how the particular type map contracts loose type constructors in domain definitions.

The normalisation procedure. Finally, the actual normalisation procedure is obtained by iteration of the operation map and target theory extension procedure above as follows. Let $\sigma(\Sigma^{log}_{op}(\mathcal{T}_1)) : \mathcal{T}_1 \longrightarrow \mathcal{T}_2$ be some base signature morphism and $N \stackrel{def}{=} |\Sigma^{der}_{op}(\mathcal{T}_1)|$. If $N = 0$ then $\sigma(\Sigma^{log}_{op}(\mathcal{T}_1))$ is already complete. Otherwise, let $\theta : \Sigma^{der}_{op}(\mathcal{T}_1) \rightarrowtail \{1, \ldots, N\}$ be a bijection such that $f \prec_{\Sigma_{op}(\mathcal{T}_1)} f'$ implies $\theta(f') < \theta(f)$ for all $f, f' \in \Sigma^{der}_{op}(\mathcal{T}_1)$. Such a bijection exists, since the order $\prec_{\Sigma_{op}(\mathcal{T}_1)}$ is acyclic. So, we can define a sequence $s = \langle f_1, \ldots, f_N \rangle$ with $f_i \stackrel{def}{=} \theta^{-1}(i)$ and observe that $\mathit{consts\text{-}of}_{\mathcal{T}_1}(\mathit{def\text{-}of}_{\mathcal{T}_1}(f_i)) \subseteq \Sigma^{log}_{op}(\mathcal{T}_1) \cup \{f_1, \ldots, f_{i-1}\}$ for $1 \leq i \leq N$. For if not, we would obtain some j with $i \leq j \leq N$ and $f_j \in \mathit{consts\text{-}of}_{\mathcal{T}_1}(\mathit{def\text{-}of}_{\mathcal{T}_1}(f_i))$ or, equivalently, $f_i \prec_{\Sigma_{op}(\mathcal{T}_1)} f_j$, which, in turn, would imply $j < i$, giving a contradiction. Thus, $\mathit{def\text{-}of}_{\mathcal{T}_1}(f_i) \in \mathit{Terms}(\Sigma^{log}_{op}(\mathcal{T}_1) \cup \{f_1, \ldots, f_{i-1}\})$ for all $1 \leq i \leq N$. This property of the sequence s allows us to iterate the operation map and target theory extension procedure on $f_1, \ldots, f_N$, starting with $\mathcal{T}_1$, $\mathcal{T}_2$, $\sigma(\Sigma^{log}_{op}(\mathcal{T}_1))$ and f_1.

Let the theory $\mathcal{T}_2^i$ and the signature morphism $\sigma(\Sigma_{op}^{log}(\mathcal{T}_1) \cup \{f_1, \dots, f_i\})$ denote the i-th intermediate result, with $0 \leq i \leq N$ and $\mathcal{T}_2^0 \stackrel{def}{=} \mathcal{T}_2$. So, the normalisation succeeds if the theory $\mathcal{T}_2^N$ and the signature morphism $\sigma(\Sigma_{op}^{log}(\mathcal{T}_1) \cup \{f_1, \dots, f_N\}) = \sigma(\Sigma_{op}(\mathcal{T}_1))$ are reached. In this case the procedure results in $\mathcal{T}_2^N$ and the complete σ. Otherwise, there has to be some j with $1 \leq j \leq N$ for which the operation map and target theory extension procedure produces the output `not normalisable` for the arguments $\mathcal{T}_1$, $\mathcal{T}_2^{j-1}$, $\sigma(\Sigma_{op}^{log}(\mathcal{T}_1) \cup \{f_1, \dots, f_{j-1}\})$ and f_j.

3.5.1 Matching of definitions

The normalisation procedure can be refined in the sense that for some derived constants in the source signature the extension of the target theory by the corresponding new derived constant can be avoided. This is the case when a suitable derived constant is already present in the target. Suitable means here that it satisfies the definitional condition (3.14).

More precisely, let $1 \leq j \leq N$, and $\mathcal{T}_2^{j-1}$, $\sigma(\Sigma)$, where $\Sigma \stackrel{def}{=} \Sigma_{op}^{log}(\mathcal{T}_1) \cup \{f_1, \dots, f_{j-1}\}$, f_j be the input in the j-th iteration step of the normalisation procedure above. Then we can initially search for a derived constant $f' \in \Sigma_{op}(\mathcal{T}_2)$ satisfying

$$\textit{def-of}_{\mathcal{T}_2}(f') \stackrel{\alpha}{=}_\theta \overline{\sigma}(\Sigma)(\textit{def-of}_{\mathcal{T}_1}(f_j)). \tag{3.18}$$

If such f' exists then an application of the operation map and target theory extension procedure can be skipped, because we can define $\mathcal{T}_2^j \stackrel{def}{=} \mathcal{T}_2^{j-1}$ and obtain $\sigma(\Sigma \cup \{f_j\})$ simply by an extension of the operation map of $\sigma(\Sigma)$ by $f_j \mapsto f'$.

Finally, it remains to show that the assignment $f_j \mapsto f'$ satisfies also the operation map condition (3.9). So, from $\textit{def-of}_{\mathcal{T}_1}(f_j)$ $\textit{of-type}_{\mathcal{T}_1}$ $\textit{type-of}_{\mathcal{T}_1}(f_j)$, we obtain

$$\overline{\sigma}(\Sigma)(\textit{def-of}_{\mathcal{T}_1}(f_j)) \ \textit{of-type}_{\mathcal{T}_2} \ \overline{\sigma}_{type}^{\sigma_{class}}(\textit{type-of}_{\mathcal{T}_1}(f_j)).$$

Further, from $\textit{def-of}_{\mathcal{T}_2}(f')$ $\textit{of-type}_{\mathcal{T}_2}$ $\textit{type-of}_{\mathcal{T}_2}(f')$ and (3.18) we have

$$\overline{\sigma}(\Sigma)(\textit{def-of}_{\mathcal{T}_1}(f_j)) \ \textit{of-type}_{\mathcal{T}_2} \ \theta(\textit{type-of}_{\mathcal{T}_2}(f'))$$

provided by the α-conversion property (2.5), where $\theta(\textit{type-of}_{\mathcal{T}_2}(f'))$ is the type with type variables renamed according to the bijection θ. Indeed, since f' is a derived constant, *typevars*(*def-of*$_{\mathcal{T}_2}(f')$) = *typevars*(*type-of*$_{\mathcal{T}_2}(f')$) holds, and hence θ is a bijection on *typevars*(*type-of*$_{\mathcal{T}_2}(f')$).

Taking further into account the uniqueness property of the of-type relation (Lemma 2.6), we have $\overline{\sigma}^{\sigma_{class}}_{type}(\textit{type-of}_{\mathcal{T}_1}(f_j)) = \theta(\textit{type-of}_{\mathcal{T}_2}(f'))$, and thus

$$\textit{type-of}_{\mathcal{T}_2}(f') \stackrel{\alpha}{=} \overline{\sigma}^{\sigma_{class}}_{type}(\textit{type-of}_{\mathcal{T}_1}(f_j)).$$

Although the normalisation procedure works without matching, this approach is very important in practice, because it avoids unnecessary blow-up of theories by derived constants, defined by the same term (up to the α-equivalence). Furthermore, another motivation is the following. If the normalisation introduces for a derived constant f in the source signature a fresh derived constant f' in the target, ignoring some existing derived constant $\hat{f}$ in the target such that $f \mapsto \hat{f}$ can be matched, then for any already proved property $P\,\hat{f}$ the property $P\,f'$ requires a proof. This is, of course, possible and simple (unfold the definition of $\hat{f}$ and fold f'), but doing this for hundreds of definitions is quite tedious, and can be simply avoided using the matching technique.

3.6 Composition of signature morphisms

In this section the composition of signature morphisms will be defined by compositions of the particular class, type and operations maps. So, let $\sigma^1 : \mathcal{T}_1 \longrightarrow \mathcal{T}_2$ and $\sigma^2 : \mathcal{T}_3 \longrightarrow \mathcal{T}_4$, such that $\mathcal{T}_2 \in Anc(\mathcal{T}_3)$. Then the signature morphism $\sigma \stackrel{def}{=} \langle \mathcal{T}_1, \mathcal{T}_4, \sigma_{class}, \sigma_{type}, \Sigma_{op}(\mathcal{T}_1), \sigma_{op}\rangle$ is called the *composition* of σ^1 and σ^2, denoted by $\sigma^2 \circ \sigma^1$, where σ_{class}, σ_{type}, and σ_{op} denote the compositions of the particular maps of σ^1 and σ^2, defined in the next three sections.

3.6.1 Composition of class maps

Let $\Sigma_{class}(\mathcal{T}_1) \stackrel{\sigma^1_{class}}{\longmapsto} \Sigma_{class}(\mathcal{T}_2)$ and $\Sigma_{class}(\mathcal{T}_3) \stackrel{\sigma^2_{class}}{\longmapsto} \Sigma_{class}(\mathcal{T}_4)$ be the class maps of σ^1 and σ^2, respectively. Then the class map of σ is defined as the functional composition $\sigma_{class} \stackrel{def}{=} \sigma^2_{class} \circ \sigma^1_{class}$, such that $\sigma_{class} : \Sigma_{class}(\mathcal{T}_1) \to \Sigma_{class}(\mathcal{T}_4)$ is well-defined since $\Sigma_{class}(\mathcal{T}_2) \subseteq \Sigma_{class}(\mathcal{T}_3)$, provided by the restriction $\mathcal{T}_2 \in Anc(\mathcal{T}_3)$. The class map condition, i.e. that $c \prec^*_{\Sigma_{class}(\mathcal{T}_1)} c'$ implies $\sigma_{class}(c) \prec^*_{\Sigma_{class}(\mathcal{T}_4)} \sigma_{class}(c')$, holds obviously, since the functional composition preserves monotonicity. So, we obtain $\Sigma_{class}(\mathcal{T}_1) \stackrel{\sigma_{class}}{\longmapsto} \Sigma_{class}(\mathcal{T}_4)$, and $\overline{\sigma}_{class} = \overline{\sigma}^2_{class} \circ \overline{\sigma}^1_{class}$, since $(f \circ g)\langle S\rangle = f\langle g\langle S\rangle\rangle$ holds.

3.6.2 Composition of type maps

The composition of type maps is not as straightforward. The reason is, of course, that a type map assigns type schemes to type constructors, such that we cannot use functional composition here. Instead, the core of the type map composition builds the application of a type scheme to a sequence of type schemes:

Definition 18 (application of a type scheme to a sequence of type schemes) Let Σ_{type} be a type signature, $\varsigma \in \widehat{Types}(\Sigma_{type})$ and $Bound(\varsigma) \leq b$. Furthermore, let $\langle \varsigma'_1, \ldots, \varsigma'_b \rangle \in \widehat{Types}(\Sigma_{type})^b$ be a sequence of type schemes. The *application* $\varsigma \triangleright \langle \varsigma'_1, \ldots, \varsigma'_b \rangle \in \widehat{Types}(\Sigma_{type})$ is defined recursively on the structure of ς:

(i) $`i \triangleright \langle \varsigma'_1, \ldots, \varsigma'_b \rangle = \varsigma'_i$;

(ii) $(\varsigma_1, \ldots, \varsigma_n) \cdot C \triangleright \langle \varsigma'_1, \ldots, \varsigma'_b \rangle = (\varsigma_1 \triangleright \langle \varsigma'_1, \ldots, \varsigma'_b \rangle, \ldots, \varsigma_n \triangleright \langle \varsigma'_1, \ldots, \varsigma'_b \rangle) \cdot C$

The most basic consequence of the definition is that $\langle `1, \ldots, `n \rangle$ is an identity element for any ς with $Bound(\varsigma) \leq n$, as the following lemma shows.

Lemma 3.9 *For any $\varsigma \in \widehat{Types}(\Sigma_{type})$ with $Bound(\varsigma) \leq n$*

$$\varsigma \triangleright \langle `1, \ldots, `n \rangle = \varsigma$$

holds.

Proof. (by structural induction on ς)
(i) Let $\varsigma = `i$ with $`i \in \mathbb{N}_+$. Then $\varsigma \triangleright \langle `1, \ldots, `n \rangle = `i$.
(ii) Let $\varsigma = (\varsigma_1, \ldots, \varsigma_m) \cdot C$. Then

$$\begin{array}{ll} (\varsigma_1, \ldots, \varsigma_m) \cdot C \triangleright \langle `1, \ldots, `n \rangle & = \\ (\varsigma_1 \triangleright \langle `1, \ldots, `n \rangle, \ldots, \varsigma_m \triangleright \langle `1, \ldots, `n \rangle) \cdot C & = \\ & \text{by ind. hyp.} \\ \varsigma & \end{array}$$

□

Some of the further basic properties of the application are emphasised by the following two lemmas.

Lemma 3.10 *Suppose $Bound(\varsigma'_i) \leq n$ for all $1 \leq i \leq b$ and $Bound(\varsigma) \leq b$. Then $Bound(\varsigma \triangleright \langle \varsigma'_1, \ldots, \varsigma'_b \rangle) \leq n$ holds.*

Proof. (by structural induction on ς)
(i) Let $\varsigma = `i$ with $`i \in \mathbb{N}_+$. Then $Bound(\varsigma \triangleright \langle \varsigma'_1, \ldots, \varsigma'_b \rangle) = Bound(\varsigma'_i) \leq n$.
(ii) Let $\varsigma = (\varsigma_1, \ldots, \varsigma_m) \cdot C$. Then

$$\begin{array}{ll} Bound(\varsigma \triangleright \langle \varsigma'_1, \ldots, \varsigma'_b \rangle) & = \\ Bound((\varsigma_1 \triangleright \langle \varsigma'_1, \ldots, \varsigma'_b \rangle, \ldots, \varsigma_m \triangleright \langle \varsigma'_1, \ldots, \varsigma'_b \rangle) \cdot C) & = \\ \mathrm{Max}_{1 \leq i \leq m}(Bound(\varsigma_i \triangleright \langle \varsigma'_1, \ldots, \varsigma'_b \rangle)). & \end{array}$$

By the induction hypothesis, this maximum is then less or equal n. □

Lemma 3.11 *Suppose $Bound(\varsigma'_i) \leq n$ for all $1 \leq i \leq b$ and $Bound(\varsigma) \leq b$. Then $(\varsigma \triangleright \langle \varsigma'_1, \ldots, \varsigma'_b \rangle) \bullet \langle \kappa_1, \ldots, \kappa_n \rangle = \varsigma \bullet \langle \varsigma'_1 \bullet \langle \kappa_1, \ldots, \kappa_n \rangle, \ldots, \varsigma'_b \bullet \langle \kappa_1, \ldots, \kappa_n \rangle \rangle$.*

Proof. (by structural induction on ς)
(i) Let $\varsigma = `i$ with $`i \in \mathbb{N}_+$. Then both *lhs* and *rhs* evaluate to $\varsigma'_i \bullet \langle \kappa_1, \ldots, \kappa_n \rangle$.

(ii) Let $\varsigma = (\varsigma_1, \ldots, \varsigma_m) \cdot C$. Then we can reason

$$\begin{array}{ll} (\varsigma \triangleright \langle \varsigma'_1, \ldots, \varsigma'_b \rangle) \bullet \langle \kappa_1, \ldots, \kappa_n \rangle & = \\ ((\varsigma_1 \triangleright \langle \varsigma'_1, \ldots, \varsigma'_b \rangle, \ldots, \varsigma_m \triangleright \langle \varsigma'_1, \ldots, \varsigma'_b \rangle) \cdot C) \bullet \langle \kappa_1, \ldots, \kappa_n \rangle & = \\ & \text{by def. of } \bullet \\ ((\varsigma_1 \triangleright \langle \varsigma'_1, \ldots, \varsigma'_b \rangle) \bullet \langle \kappa_1, \ldots, \kappa_n \rangle, \ldots, & \\ (\varsigma_m \triangleright \langle \varsigma'_1, \ldots, \varsigma'_b \rangle) \bullet \langle \kappa_1, \ldots, \kappa_n \rangle) C & = \\ & \text{by ind. hyp.} \\ (\varsigma_1 \bullet \langle \varsigma'_1 \bullet \langle \kappa_1, \ldots, \kappa_n \rangle, \ldots, \varsigma'_b \bullet \langle \kappa_1, \ldots, \kappa_n \rangle \rangle, \ldots, & \\ \varsigma_m \bullet \langle \varsigma'_1 \bullet \langle \kappa_1, \ldots, \kappa_n \rangle, \ldots, \varsigma'_b \bullet \langle \kappa_1, \ldots, \kappa_n \rangle \rangle) C & = \\ & \text{by def. of } \bullet \\ ((\varsigma_1, \ldots, \varsigma_m) \cdot C) \bullet \langle \varsigma'_1 \bullet \langle \kappa_1, \ldots, \kappa_n \rangle, \ldots, \varsigma'_b \bullet \langle \kappa_1, \ldots, \kappa_n \rangle \rangle & \end{array}$$

□

The application of type schemes to type schemes is used to define the operation, called *multiplication*. It assigns to any type scheme $\varsigma \in \widehat{Types}(\Sigma_{type}(\mathcal{T}))$ and any type map $\phi : \Sigma_{type}(\mathcal{T}) \to \widehat{Types}(\Sigma_{type}(\mathcal{T}'))$ the type scheme $\varsigma * \phi \in \widehat{Types}(\Sigma_{type}(\mathcal{T}'))$ as follows.

Definition 19 (multiplication)
Let $\varsigma \in \widehat{Types}(\Sigma_{type}(\mathcal{T}))$, and $\Sigma_{type}(\mathcal{T}) \overset{\phi}{\longmapsto}_{\varphi} \Sigma_{type}(\mathcal{T}')$ be a type map w.r.t. φ. Then the *multiplication* $\varsigma * \phi \in \widehat{Types}(\Sigma_{type}(\mathcal{T}'))$ is defined recursively on the structure of $\varsigma \in \widehat{Types}(\Sigma_{type}(\mathcal{T}))$ as follows:

(i) $`i * \phi = `i$;

(ii) $(\varsigma_1, \ldots, \varsigma_n) \cdot C * \phi = \phi(C) \triangleright \langle \varsigma_1 * \phi, \ldots, \varsigma_n * \phi \rangle$

where $\phi(C) \triangleright \langle \varsigma_1 * \phi, \ldots, \varsigma_n * \phi \rangle$ is well-defined since $C \in \Sigma_{type}(\mathcal{T})$ and ϕ is a type map, such that $\phi(C) \in \widehat{Types}(\Sigma_{type}(\mathcal{T}'))$ and $Bound(\phi(C)) \leq n$ hold.

The next two lemmas 'lift' the results of Lemma 3.10 and Lemma 3.11 to similar results for the multiplication.

Lemma 3.12 *Let* $\Sigma_{type}(\mathcal{T}) \stackrel{\phi}{\longmapsto}_{\varphi} \Sigma_{type}(\mathcal{T}')$. *Then* $Bound(\varsigma) \leq n$ *implies* $Bound(\varsigma * \phi) \leq n$ *for any* $\varsigma \in \widehat{Types}(\Sigma_{type}(\mathcal{T}))$.

Proof. (by structural induction on ς)
(i) Let $\varsigma = `i$ with $i \in \mathbb{N}_+$. Then $Bound(`i * \phi) = Bound(`i) \leq n$ holds by definition.
(ii) Let $\varsigma = (\varsigma_1, \ldots, \varsigma_m) \cdot C$ with $C \in \Sigma_{type}(\mathcal{T})$. Then

$$Bound(\varsigma * \phi) = Bound(\phi(C) \triangleright \langle \varsigma_1 * \phi, \ldots, \varsigma_m * \phi \rangle).$$

By the induction hypothesis $Bound(\varsigma_i * \phi) \leq n$ holds for all $1 \leq i \leq m$, and thus, by Lemma 3.10, we have $Bound(\varsigma * \phi) \leq n$ since $Bound(\phi(C)) \leq m$. □

Lemma 3.13 *Let* $\Sigma_{type}(\mathcal{T}) \stackrel{\phi}{\longmapsto}_{\varphi} \Sigma_{type}(\mathcal{T}')$. *Then* $Bound(\varsigma) \leq n$ *implies* $\overline{\phi}^{\varphi}(\varsigma \bullet \langle \kappa_1, \ldots, \kappa_n \rangle) = (\varsigma * \phi) \bullet \langle \overline{\phi}^{\varphi}(\kappa_1), \ldots, \overline{\phi}^{\varphi}(\kappa_n) \rangle$ *for any* $\varsigma \in \widehat{Types}(\Sigma_{type}(\mathcal{T}))$ *and* $\kappa_i \in Types_{\vartheta}(\Sigma_{type}(\mathcal{T}))$ *for* $1 \leq i \leq n$.

Proof. (by structural induction on ς)
(i) Let $\varsigma = `i$ with $`i \in \mathbb{N}_+$. Then $i \leq n$ holds by assumption, and so both *lhs* and *rhs* reduce to $\overline{\phi}^{\varphi}(\kappa_i)$.
(ii) Let $\varsigma = (\varsigma_1, \ldots, \varsigma_m) \cdot C$. From $Bound(\varsigma) \leq n$ we can deduce $Bound(\varsigma_i) \leq n$ for all $1 \leq i \leq m$. Furthermore, by Lemma 3.12 we obtain $Bound(\varsigma_i * \phi) \leq n$ for all $1 \leq i \leq m$. Then we can reason

$$
\begin{array}{ll}
(((\varsigma_1, \ldots, \varsigma_m) \cdot C) * \phi) \bullet \langle \overline{\phi}^{\varphi}(\kappa_1), \ldots, \overline{\phi}^{\varphi}(\kappa_n) \rangle & = \\
(\phi(C) \triangleright \langle \varsigma_1 * \phi, \ldots, \varsigma_m * \phi \rangle) \bullet \langle \overline{\phi}^{\varphi}(\kappa_1), \ldots, \overline{\phi}^{\varphi}(\kappa_n) \rangle & = \\
& \text{by Lemma 3.11 and} \\
& Bound(\phi(C)) \leq m \\
\phi(C) \bullet \langle (\varsigma_1 * \phi) \bullet \langle \overline{\phi}^{\varphi}(\kappa_1), \ldots, \overline{\phi}^{\varphi}(\kappa_n) \rangle, \ldots, & \\
\quad (\varsigma_m * \phi) \bullet \langle \overline{\phi}^{\varphi}(\kappa_1), \ldots, \overline{\phi}^{\varphi}(\kappa_n) \rangle \rangle & =
\end{array}
$$

$$
\begin{aligned}
& && \text{by ind. hyp.}\\
&\phi(C) \bullet \langle \overline{\phi}^{\varphi}(\varsigma_1 \bullet \langle \kappa_1, \ldots, \kappa_n \rangle), \ldots, && \\
&\qquad \overline{\phi}^{\varphi}(\varsigma_m \bullet \langle \kappa_1, \ldots, \kappa_n \rangle)\rangle && = \\
& && \text{by definition of } \overline{\phi}^{\varphi}\\
&\overline{\phi}^{\varphi}((\varsigma_1 \bullet \langle \kappa_1, \ldots, \kappa_n \rangle, \ldots, \varsigma_m \bullet \langle \kappa_1, \ldots, \kappa_n \rangle) C) && = \\
&\overline{\phi}^{\varphi}(\varsigma \bullet \langle \kappa_1, \ldots, \kappa_n \rangle).
\end{aligned}
$$

□

The composition of type maps will be now defined by means of the multiplication. So, let $\Sigma_{type}(\mathcal{T}_1) \overset{\sigma^1_{type}}{\longmapsto}_{\sigma^1_{class}} \Sigma_{type}(\mathcal{T}_2)$ and $\Sigma_{type}(\mathcal{T}_3) \overset{\sigma^2_{type}}{\longmapsto}_{\sigma^2_{class}} \Sigma_{type}(\mathcal{T}_4)$ be the type maps of σ^1 and σ^2, respectively. Their composition $(\sigma^2_{type} \circ \sigma^1_{type}) : \Sigma_{type}(\mathcal{T}_1) \to \widehat{Types}(\Sigma_{type}(\mathcal{T}_4))$ is defined by

$$(\sigma^2_{type} \circ \sigma^1_{type})(C) \overset{def}{=} \sigma^1_{type}(C) * \sigma^2_{type}$$

for any $C \in \Sigma_{type}(\mathcal{T}_1)$. Here we use, that $\mathcal{T}_2 \in Anc(\mathcal{T}_3)$, such that $\Sigma_{type}(\mathcal{T}_2) \subseteq \Sigma_{type}(\mathcal{T}_3)$, and so $\widehat{Types}(\Sigma_{type}(\mathcal{T}_2)) \subseteq \widehat{Types}(\Sigma_{type}(\mathcal{T}_3))$. The next goal is, of course, to show that $\sigma_{type} \overset{def}{=} \sigma^2_{type} \circ \sigma^1_{type}$ is indeed a type map.

Lemma 3.14 *$\Sigma_{type}(\mathcal{T}_1) \overset{\sigma_{type}}{\longmapsto}_{\sigma_{class}} \Sigma_{type}(\mathcal{T}_4)$ holds, where σ_{class} refers to the composed class map, constructed in the previous step.*

Proof. We start with the type map conditions as stated in Definition 13. To this end, let $C \in \Sigma_{type}(\mathcal{T}_1)$ with $rank_{\mathcal{T}_1}(C) = n$. Firstly, we have to show $Bound(\sigma_{type}(C)) \leq n$, which is equivalent to $Bound(\sigma^1_{type}(C) * \sigma^2_{type}) \leq n$. This holds by Lemma 3.12, since σ^1_{type}, σ^2_{type} are type maps, and hence $Bound(\sigma^1_{type}(C)) \leq n$ holds.

Secondly, we have to prove the condition (3.2). This can be done, proving the equivalent condition (3.3), as Lemma 3.3 shows. So, let $\alpha_1, \ldots, \alpha_n$ be distinct type variables and $(S_1, \ldots, S_n, S_{n+1}) \in arities\text{-}of_{\mathcal{T}_1}(C)$. Further, let $\kappa_i \overset{def}{=} \alpha_i ::_{sort} \overline{\sigma}_{class}(S_i)$ for all $1 \leq i \leq n$. We have then to show

$$(\sigma^1_{type}(C) * \sigma^2_{type}) \bullet \langle \kappa_1, \ldots, \kappa_n \rangle \text{ of sort}_{\mathcal{T}_4}\ \overline{\sigma}_{class}(S_{n+1}). \tag{3.19}$$

Firstly, using that $\sigma_{class} = \overline{\sigma}^2_{class} \circ \overline{\sigma}^1_{class}$, we observe that

$$\kappa_i = \alpha_i ::_{sort} \overline{\sigma}^2_{class}(\overline{\sigma}^1_{class}(S_i)) = \overline{\sigma}^{\sigma^2_{class}}_{type}(\alpha_i ::_{sort} \overline{\sigma}^1_{class}(S_i))$$

holds. Using this and Lemma 3.13 we obtain

$$\begin{aligned}
&(\sigma^1_{type}(C) * \sigma^2_{type}) \bullet \langle \kappa_1, \ldots, \kappa_n \rangle && = \\
&\overline{\sigma}^{\sigma^2_{class}}_{type}(\sigma^1_{type}(C) \bullet \langle \alpha_1 ::_{sort} \overline{\sigma}^1_{class}(S_1), \ldots, \alpha_n ::_{sort} \overline{\sigma}^1_{class}(S_n)\rangle) && = \\
&\overline{\sigma}^{\sigma^2_{class}}_{type}(\overline{\sigma}^{\sigma^1_{class}}_{type}((\alpha_1 ::_{sort} S_1, \ldots, \alpha_n ::_{sort} S_n)C))
\end{aligned}$$

such that (3.19) is implied by $(\alpha_1 ::_{sort} S_1, \ldots, \alpha_n ::_{sort} S_n)C$ *of-sort*$_{\mathcal{T}_1}$ S_{n+1}, which is true, by the definition of the of-sort relation.

Finally, the meta-logical requirements can be verified directly using the definitions:

1. $\sigma_{type}(\texttt{prop}) = \sigma^1_{type}(\texttt{prop}) * \sigma^2_{type} = \texttt{prop} * \sigma^2_{type} = \sigma^2_{type}(\texttt{prop}) = \texttt{prop}$;
2. $\sigma_{type}(\Rightarrow) = \sigma^1_{type}(\Rightarrow) * \sigma^2_{type} = ((\text{`}1, \text{`}2) \cdot \Rightarrow) * \sigma^2_{type} =$ $\sigma^2_{type}(\Rightarrow) \triangleright \langle \text{`}1 * \sigma^2_{type}, \text{`}2 * \sigma^2_{type}\rangle = ((\text{`}1, \text{`}2) \cdot \Rightarrow) \triangleright \langle \text{`}1, \text{`}2\rangle = (\text{`}1, \text{`}2) \cdot \Rightarrow$.

□

The next step, regarding the composition of type maps, proves that the composition respects the type map extension in the following sense:

Lemma 3.15 *For all sort assignments $\vartheta : \mathcal{X}_{type} \to Sorts(\Sigma_{class}(\mathcal{T}_1))$ and $\kappa \in Types_{\vartheta}(\Sigma_{type}(\mathcal{T}_1))$, $\overline{\sigma}^{\sigma_{class}}_{type}(\kappa) = (\overline{\sigma}^{\sigma^2_{class}}_{type} \circ \overline{\sigma}^{\sigma^1_{class}}_{type})(\kappa)$ holds, where $\overline{\sigma}^{\sigma_{class}}_{type}$ denotes the extension of the composition of σ^1_{type} and σ^2_{type}.*

Proof. (by structural induction on κ)
(i) Let $\kappa = \alpha ::_{sort} \vartheta(\alpha)$. Then

$$\begin{aligned}
&\overline{\sigma}^{\sigma_{class}}_{type}(\kappa) && = \\
&\alpha ::_{sort} \overline{\sigma}_{class}(\vartheta(\alpha)) && = \\
&\alpha ::_{sort} \overline{\sigma}^2_{class}(\overline{\sigma}^1_{class}(\vartheta(\alpha))) && = \\
&(\overline{\sigma}^{\sigma^2_{class}}_{type} \circ \overline{\sigma}^{\sigma^1_{class}}_{type})(\kappa).
\end{aligned}$$

(ii) Let $\kappa = (\kappa_1, \ldots, \kappa_n)C$ where $C \in \Sigma_{type}(\mathcal{T}_1)$ and $rank_{\mathcal{T}_1}(C) = n$. Then

$$\begin{aligned}
&\overline{\sigma}^{\sigma_{class}}_{type}((\kappa_1, \ldots, \kappa_n)C) && = \\
&\sigma_{type}(C) \bullet \langle \overline{\sigma}^{\sigma_{class}}_{type}(\kappa_1), \ldots, \overline{\sigma}^{\sigma_{class}}_{type}(\kappa_n)\rangle && = \\
&(\sigma^1_{type}(C) * \sigma^2_{type}) \bullet \langle \overline{\sigma}^{\sigma_{class}}_{type}(\kappa_1), \ldots, \overline{\sigma}^{\sigma_{class}}_{type}(\kappa_n)\rangle && = \\
& && \text{by ind. hyp.} \\
&(\sigma^1_{type}(C) * \sigma^2_{type}) \bullet \langle (\overline{\sigma}^{\sigma^2_{class}}_{type} \circ \overline{\sigma}^{\sigma^1_{class}}_{type})(\kappa_1), \ldots, \\
&\qquad (\overline{\sigma}^{\sigma^2_{class}}_{type} \circ \overline{\sigma}^{\sigma^1_{class}}_{type})(\kappa_n)\rangle && = \\
& && \text{by Lemma 3.13}
\end{aligned}$$

$$\begin{array}{ll}
\overline{\sigma}^{\sigma^2_{class}}_{type}(\sigma^1_{type}(C) \bullet \langle \overline{\sigma}^{\sigma^1_{class}}_{type}(\kappa_1), \ldots, \overline{\sigma}^{\sigma^1_{class}}_{type}(\kappa_n)\rangle) & = \\
\overline{\sigma}^{\sigma^2_{class}}_{type}(\overline{\sigma}^{\sigma^1_{class}}_{type}((\kappa_1, \ldots, \kappa_n)C)) & = \\
(\overline{\sigma}^{\sigma^2_{class}}_{type} \circ \overline{\sigma}^{\sigma^1_{class}}_{type})(\kappa). &
\end{array}$$

□

Associativity of the composition

Finally, we show that the composition of type maps also satisfies the associativity law. As a preliminary, we prove the following two proposition, starting with the associativity of the $\triangleright$-operation.

Lemma 3.16 *Let $\varsigma \in \widehat{Types}(\Sigma_{type}(\mathcal{T}))$ with $Bound(\varsigma) \leq m$. Further, let $\langle \varsigma'_1, \ldots, \varsigma'_m \rangle \in \widehat{Types}(\Sigma_{type}(\mathcal{T}))^m$ such that $Bound(\varsigma'_i) \leq k$ holds for all $1 \leq i \leq m$, and $\varsigma'' \in \widehat{Types}(\Sigma_{type}(\mathcal{T}))^k$. Then*

$$(\varsigma \triangleright \langle \varsigma'_1, \ldots, \varsigma'_m \rangle) \triangleright \varsigma'' = \varsigma \triangleright \langle \varsigma'_1 \triangleright \varsigma'', \ldots, \varsigma'_m \triangleright \varsigma'' \rangle \tag{3.20}$$

holds.

Proof. (by structural induction on ς)
(i) Let $\varsigma = {}^{\backprime}i$. Then $i \leq m$ and, both *lhs* and *rhs* of (3.20) reduce to $\varsigma'_i \triangleright \varsigma''$.
(ii) Let $\varsigma = (\varsigma_1, \ldots, \varsigma_n) \cdot C$ with $C \in \Sigma_{type}(\mathcal{T})$, $rank_{\mathcal{T}}(C) = n$. Then

$$\begin{array}{ll}
(\varsigma \triangleright \langle \varsigma'_1, \ldots, \varsigma'_m \rangle) \triangleright \varsigma'' & = \\
((\varsigma_1 \triangleright \langle \varsigma'_1, \ldots, \varsigma'_m \rangle, \ldots, \varsigma_n \triangleright \langle \varsigma'_1, \ldots, \varsigma'_m \rangle) \cdot C) \triangleright \varsigma'' & = \\
((\varsigma_1 \triangleright \langle \varsigma'_1, \ldots, \varsigma'_m \rangle) \triangleright \varsigma'', \ldots, (\varsigma_n \triangleright \langle \varsigma'_1, \ldots, \varsigma'_m \rangle) \triangleright \varsigma'') \cdot C & = \\
 & \text{by ind. hyp.} \\
(\varsigma_1 \triangleright \langle \varsigma'_1 \triangleright \varsigma'', \ldots, \varsigma'_m \triangleright \varsigma'' \rangle, \ldots, \varsigma_n \triangleright \langle \varsigma'_1 \triangleright \varsigma'', \ldots, \varsigma'_m \triangleright \varsigma'' \rangle) \cdot C & = \\
((\varsigma_1, \ldots, \varsigma_n) \cdot C) \triangleright \langle \varsigma'_1 \triangleright \varsigma'', \ldots, \varsigma'_m \triangleright \varsigma'' \rangle &
\end{array}$$

□

This result is used in the next lemma regarding multiplication.

Lemma 3.17 *Let $\Sigma_{type}(\mathcal{T}) \stackrel{\phi}{\longmapsto}_{\varphi} \Sigma_{type}(\mathcal{T}')$ be a type map. Further, let $\varsigma \in \widehat{Types}(\Sigma_{type}(\mathcal{T}))$ with $Bound(\varsigma) \leq b$ and $\langle \varsigma'_1, \ldots, \varsigma'_b \rangle \in \widehat{Types}(\Sigma_{type}(\mathcal{T}))^b$ be a sequence of type schemes. Then*

$$(\varsigma \triangleright \langle \varsigma'_1, \ldots, \varsigma'_b \rangle) * \phi = (\varsigma * \phi) \triangleright \langle \varsigma'_1 * \phi, \ldots, \varsigma'_b * \phi \rangle \tag{3.21}$$

holds.

Proof. (by structural induction on ς)
(i) Let $\varsigma = {}`i$. Then $i \leq b$, and, both *lhs* and *rhs* of (3.21) reduce to $\varsigma'_i * \phi$.

(ii) Let $\varsigma = (\varsigma_1, \ldots, \varsigma_n) \cdot C$ with $C \in \Sigma_{type}(\mathcal{T})$, $rank_{\mathcal{T}}(C) = n$. If $n = 0$ then $\varsigma * \phi = \phi(C)$, and, since $Bound(\phi(C)) \leq n$, both *lhs* and *rhs* of (3.21) reduce to $\phi(C)$. Otherwise, by Lemma 3.12, we have $Bound(\varsigma_i * \phi) \leq b$ for all $1 \leq i \leq n$, and can reason:

$$
\begin{array}{ll}
(\varsigma \triangleright \langle \varsigma'_1, \ldots, \varsigma'_b \rangle) * \phi & = \\
((\varsigma_1 \triangleright \langle \varsigma'_1, \ldots, \varsigma'_b \rangle, \ldots, \varsigma_n \triangleright \langle \varsigma'_1, \ldots, \varsigma'_b \rangle) \cdot C) * \phi & = \\
\phi(C) \triangleright \langle (\varsigma_1 \triangleright \langle \varsigma'_1, \ldots, \varsigma'_b \rangle) * \phi, \ldots, (\varsigma_n \triangleright \langle \varsigma'_1, \ldots, \varsigma'_b \rangle) * \phi \rangle & = \\
& \text{by ind. hyp.} \\
\phi(C) \triangleright \langle (\varsigma_1 * \phi) \triangleright \langle \varsigma'_1 * \phi, \ldots, \varsigma'_b * \phi \rangle, \ldots, & \\
\qquad (\varsigma_n * \phi) \triangleright \langle \varsigma'_1 * \phi, \ldots, \varsigma'_b * \phi \rangle \rangle & = \\
& \text{by Lemma 3.16} \\
(\phi(C) \triangleright \langle \varsigma_1 * \phi, \ldots, \varsigma_n * \phi \rangle) \triangleright \langle \varsigma'_1 * \phi, \ldots, \varsigma'_b * \phi \rangle & = \\
((\varsigma_1, \ldots, \varsigma_n) \cdot C * \phi) \triangleright \langle \varsigma'_1 * \phi, \ldots, \varsigma'_b * \phi \rangle & = \\
(\varsigma * \phi) \triangleright \langle \varsigma'_1 * \phi, \ldots, \varsigma'_b * \phi \rangle &
\end{array}
$$

□

Now, we can turn towards the actual problem: associativity of the composition of type maps. To this end let $\sigma^1_{type}, \sigma^2_{type}, \sigma^3_{type}$ be composable type maps with $\Sigma_{type}(\mathcal{T}_1) \overset{\sigma^1_{type}}{\longmapsto}_{\sigma^1_{class}} \Sigma_{type}(\mathcal{T}_2)$, $\Sigma_{type}(\mathcal{T}_3) \overset{\sigma^2_{type}}{\longmapsto}_{\sigma^2_{class}} \Sigma_{type}(\mathcal{T}_4)$, and $C_1 \in \Sigma_{type}(\mathcal{T}_1)$ an arbitrary type constructor with $rank_{\mathcal{T}_1}(C_1) = n$. Then, we need to show

$$(\sigma^1_{type}(C_1) * \sigma^2_{type}) * \sigma^3_{type} = \sigma^1_{type}(C_1) * (\sigma^3_{type} \circ \sigma^2_{type}) \tag{3.22}$$

This will be shown by a more general claim:

$$(\varsigma * \sigma^2_{type}) * \sigma^3_{type} = \varsigma * (\sigma^3_{type} \circ \sigma^2_{type}) \tag{3.23}$$

for any $\varsigma \in \widehat{Types}(\Sigma_{type}(\mathcal{T}_3))$ with $Bound(\varsigma) \leq n$. The proof proceeds by structural induction on ς.

(i) Let $\varsigma = {}`i$ with some $1 \leq i \leq n$. Then *lhs* and *rhs* of (3.23) simplify to $`i$.

(ii) Let $\varsigma = (\varsigma_1, \ldots, \varsigma_m) \cdot C_2$ with $m \in \mathbb{N}$. Then we can reason:

$$
\begin{array}{ll}
(\sigma^2_{type}(C_2) \rhd \langle \varsigma_1 * \sigma^2_{type}, \ldots, \varsigma_m * \sigma^2_{type} \rangle) * \sigma^3_{type} & = \\
& \text{by Lemma 3.17} \\
(\sigma^2_{type}(C_2) * \sigma^3_{type}) \rhd \langle (\varsigma_1 * \sigma^2_{type}) * \sigma^3_{type}, \ldots, & \\
\qquad (\varsigma_m * \sigma^2_{type}) * \sigma^3_{type} \rangle & \\
& \text{by ind. hyp.} \\
(\sigma^2_{type}(C_2) * \sigma^3_{type}) \rhd \langle \varsigma_1 * (\sigma^3_{type} \circ \sigma^2_{type}), \ldots, & \\
\qquad \varsigma_m * (\sigma^3_{type} \circ \sigma^2_{type}) \rangle & = \\
& \text{by def. of composition} \\
(\sigma^3_{type} \circ \sigma^2_{type})(C_2) \rhd \langle \varsigma_1 * (\sigma^3_{type} \circ \sigma^2_{type}), \ldots, & \\
\qquad \varsigma_m * (\sigma^3_{type} \circ \sigma^2_{type}) \rangle & = \\
(\varsigma_1, \ldots, \varsigma_m) \cdot C_2 * (\sigma^3_{type} \circ \sigma^2_{type}). &
\end{array}
$$

So, from (3.23) we have (3.22) instantiating ς by $\sigma^1_{type}(C_1)$, which is correct, since $\sigma^1_{type}(C_1) \in \widehat{Types}(\Sigma_{type}(\mathcal{T}_3))$ due to the restriction $\mathcal{T}_2 \in Anc(\mathcal{T}_3)$.

3.6.3 Composition of operation maps

Let $\Sigma_{op}(\mathcal{T}_1) \overset{\sigma^1_{op}}{\longmapsto}_{\sigma^1_{class}, \sigma^1_{type}} \Sigma_{op}(\mathcal{T}_2)$ and $\Sigma_{op}(\mathcal{T}_3) \overset{\sigma^2_{op}}{\longmapsto}_{\sigma^2_{class}, \sigma^2_{type}} \Sigma_{op}(\mathcal{T}_4)$ denote the particular operation maps. Then the operation map of σ is defined as the functional composition $\sigma_{op} \overset{def}{=} \sigma^2_{op} \circ \sigma^1_{op}$, such that $\sigma_{op} : \Sigma_{op}(\mathcal{T}_1) \to \Sigma_{op}(\mathcal{T}_4)$ is well-defined since $\Sigma_{op}(\mathcal{T}_2) \subseteq \Sigma_{op}(\mathcal{T}_3)$ by the restriction $\mathcal{T}_2 \in Anc(\mathcal{T}_3)$. The next lemma shows that σ_{op} is indeed an operation map.

Lemma 3.18 $\Sigma_{op}(\mathcal{T}_1) \overset{\sigma_{op}}{\longmapsto}_{\sigma_{class}, \sigma_{type}} \Sigma_{op}(\mathcal{T}_4)$ *holds, where* σ_{class} *and* σ_{type} *refer to the composed maps, constructed in the previous steps.*

Proof. By Definition 15, we first need to verify the operation map condition (3.9). To this end, let $f \in \Sigma_{op}(\mathcal{T}_1)$ be a constant. We have to show $\textit{type-of}_{\mathcal{T}_4}(\sigma_{op}(f)) \overset{\alpha}{=} \overline{\sigma}^{\sigma_{class}}_{type}(\textit{type-of}_{\mathcal{T}_1}(f))$. Since σ^1_{op} is an operation map, we obtain $\textit{type-of}_{\mathcal{T}_2}(\sigma^1_{op}(f)) \overset{\alpha}{=} \overline{\sigma}^{\sigma^1_{class}}_{type}(\textit{type-of}_{\mathcal{T}_1}(f))$. Further, $\sigma^1_{op}(f) \in \Sigma_{op}(\mathcal{T}_3)$ and $\textit{type-of}_{\mathcal{T}_2}(\sigma^1_{op}(f)) = \textit{type-of}_{\mathcal{T}_3}(\sigma^1_{op}(f))$ by the restriction $\mathcal{T}_2 \in Anc(\mathcal{T}_3)$. Hence,

$$
\begin{array}{ll}
\textit{type-of}_{\mathcal{T}_4}(\sigma_{op}(f)) & = \\
\textit{type-of}_{\mathcal{T}_4}(\sigma^2_{op}(\sigma^1_{op}(f))) & \overset{\alpha}{=} \\
& \text{by the operation map condition for } \sigma^2_{op} \\
\overline{\sigma}^{\sigma^2_{class}}_{type}(\textit{type-of}_{\mathcal{T}_3}(\sigma^1_{op}(f))) & =
\end{array}
$$

$$\begin{array}{ll}
\overline{\sigma}_{type}^{\sigma^2_{class}}(\mathit{type\text{-}of}_{\mathcal{T}_2}(\sigma^1_{op}(f))) & \stackrel{\alpha}{=} \\
& \text{by Lemma 3.5 and} \\
& \text{the operation map condition for } \sigma^1_{op} \\
\overline{\sigma}_{type}^{\sigma^2_{class}}(\overline{\sigma}_{type}^{\sigma^1_{class}}(\mathit{type\text{-}of}_{\mathcal{T}_1}(f))) & = \\
& \text{by Lemma 3.15} \\
\overline{\sigma}_{type}^{\sigma_{class}}(\mathit{type\text{-}of}_{\mathcal{T}_1}(f)). &
\end{array}$$

Finally, the meta-logical requirements $\sigma_{op}(\equiv) = \equiv$, $\sigma_{op}(\Longrightarrow) = \Longrightarrow$, $\sigma_{op}(\bigwedge) = \bigwedge$, hold obviously for σ_{op} by the definition of the composition using the properties of σ^1_{op} and σ^2_{op}. □

As with class and type map compositions, we show that the composition of the operation maps respects the operation map extension.

Lemma 3.19 *For all sort assignments* $\vartheta : \mathcal{X}_{type} \to \mathit{Sorts}(\Sigma_{class}(\mathcal{T}_1))$*, type assignments* $\varrho_\vartheta : \mathcal{X}_{term} \to \mathit{Types}_\vartheta(\Sigma_{type}(\mathcal{T}_1))$*, and* $t \in \mathit{Terms}_{\varrho_\vartheta}(\Sigma_{op}(\mathcal{T}_1))$*,* $\overline{\sigma}_{op}(t) = (\overline{\sigma}^2_{op} \circ \overline{\sigma}^1_{op})(t)$ *holds.*

Proof. (by structural induction on t)
(i) Let $t = x ::_{type} \varrho_\vartheta(x)$. Then

$$\begin{array}{ll}
\overline{\sigma}_{op}(t) & = \\
x ::_{type} \overline{\sigma}_{type}^{\sigma_{class}}(\varrho_\vartheta(x)) & = \\
& \text{by Lemma 3.15} \\
x ::_{type} \overline{\sigma}_{type}^{\sigma^2_{class}}(\overline{\sigma}_{type}^{\sigma^1_{class}}(\varrho_\vartheta(x))) & = \\
(\overline{\sigma}^2_{op} \circ \overline{\sigma}^1_{op})(t). &
\end{array}$$

(ii) Let $t = f ::_{type} \kappa$ with $f \in \Sigma_{op}(\mathcal{T}_1)$ and $\kappa \in \mathit{Types}_\vartheta(\Sigma_{type}(\mathcal{T}_1))$. Then

$$\begin{array}{ll}
\overline{\sigma}_{op}(t) & = \\
\sigma_{op}(f) ::_{type} \overline{\sigma}_{type}^{\sigma_{class}}(\kappa) & = \\
& \text{by Lemma 3.15} \\
\sigma^2_{op}(\sigma^1_{op}(f)) ::_{type} \overline{\sigma}_{type}^{\sigma^2_{class}}(\overline{\sigma}_{type}^{\sigma^1_{class}}(\kappa)) & = \\
\overline{\sigma}^2_{op}(\sigma^1_{op}(f) ::_{type} \overline{\sigma}_{type}^{\sigma^1_{class}}(\kappa)) & = \\
(\overline{\sigma}^2_{op} \circ \overline{\sigma}^1_{op})(t). &
\end{array}$$

(iii) Let $t = t_1\, t_2$ with $t_1, t_2 \in \mathit{Terms}_{\varrho_\vartheta}(\Sigma_{op}(\mathcal{T}_1))$. Then

$$\begin{array}{ll}
\overline{\sigma}_{op}(t) & = \\
\overline{\sigma}_{op}(t_1)\, \overline{\sigma}_{op}(t_2) & = \\
 & \text{by ind. hyp.} \\
(\overline{\sigma}^2_{op} \circ \overline{\sigma}^1_{op})(t_1)\, (\overline{\sigma}^2_{op} \circ \overline{\sigma}^1_{op})(t_2) & = \\
\overline{\sigma}^2_{op}(\overline{\sigma}^1_{op}(t_1)\, \overline{\sigma}^1_{op}(t_2)) & = \\
(\overline{\sigma}^2_{op} \circ \overline{\sigma}^1_{op})(t). &
\end{array}$$

(iv) Let $t = \lambda(x ::_{type} \varrho_\vartheta(x)).t_1$ with $t_1 \in \mathit{Terms}_{\varrho_\vartheta}(\Sigma_{op}(\mathcal{T}_1))$ and $x \in \mathcal{X}_{term}$. Then

$$\begin{array}{ll}
\overline{\sigma}_{op}(t) & = \\
\lambda(x ::_{type} \overline{\sigma}^{\sigma_{class}}_{type}(\varrho_\vartheta(x))).\overline{\sigma}_{op}(t_1) & = \\
 & \text{by ind. hyp. and} \\
 & \text{Lemma 3.15} \\
\lambda(x ::_{type} (\overline{\sigma}^{\sigma^2_{class}}_{type} \circ \overline{\sigma}^{\sigma^1_{class}}_{type})(\varrho_\vartheta(x))).(\overline{\sigma}^2_{op} \circ \sigma^1_{op})(t_1) & = \\
\overline{\sigma}^2_{op}(\lambda(x ::_{type} \overline{\sigma}^{\sigma^1_{class}}_{type}(\varrho_\vartheta(x))).\overline{\sigma}^1_{op}(t_1)) & = \\
(\overline{\sigma}^2_{op} \circ \overline{\sigma}^1_{op})(t). &
\end{array}$$

□

Verifying the definitional conditions for composition. Finally, we need to check the definitional conditions, as stated in Definition 16, in order to justify that $\sigma \stackrel{def}{=} \langle \mathcal{T}_1, \mathcal{T}_4, \sigma_{class}, \sigma_{type}, \Sigma_{op}(\mathcal{T}_1), \sigma_{op} \rangle$ is well-defined. So, firstly, for any $f \in \Sigma^{der}_{op}(\mathcal{T}_1)$ we have $\sigma_{op}(f) = \sigma^2_{op}(\sigma^1_{op}(f)) \in \Sigma^{der}_{op}(\mathcal{T}_4)$, because $\sigma^1_{op}(f) \in \Sigma^{der}_{op}(\mathcal{T}_2)$, and, hence, $\sigma^1_{op}(f) \in \Sigma^{der}_{op}(\mathcal{T}_3)$ since $\mathcal{T}_2 \in \mathit{Anc}(\mathcal{T}_3)$. Secondly, the condition (3.14) can be proved as follows:

$$\begin{array}{ll}
\mathit{def\text{-}of}_{\mathcal{T}_4}(\sigma_{op}(f)) & = \\
\mathit{def\text{-}of}_{\mathcal{T}_4}(\sigma^2_{op}(\sigma^1_{op}(f))) & \stackrel{\alpha}{=} \\
 & \text{by the condition (3.14) for } \sigma^2_{op} \\
\overline{\sigma}^2_{op}(\mathit{def\text{-}of}_{\mathcal{T}_3}(\sigma^1_{op}(f))) & = \\
\overline{\sigma}^2_{op}(\mathit{def\text{-}of}_{\mathcal{T}_2}(\sigma^1_{op}(f))) & \stackrel{\alpha}{=} \\
 & \text{by Lemma 3.12 and} \\
 & \text{the condition (3.14) for } \sigma^1_{op} \\
\overline{\sigma}^2_{op}(\overline{\sigma}^1_{op}(\mathit{def\text{-}of}_{\mathcal{T}_1}(f))) & = \\
 & \text{by Lemma 3.19} \\
\overline{\sigma}_{op}(\mathit{def\text{-}of}_{\mathcal{T}_1}(f)). &
\end{array}$$

This completes the description of the composition of signature morphisms.

3.7 Construction of signature morphisms in Isabelle

This section describes how the construction of signature morphisms has been implemented in Isabelle. The description starts with the presentation of the top-level syntax for declarations of signature morphisms. Then, the semantics of such a declaration will be given by a description of signature morphism it constitutes (if any).

Basically, the syntax of signature morphism declarations is closely related to the definition of signature morphisms in the sense, that it comprises the declarations of the source, target, class, type and operations maps. On the other hand, the construction pursues a special strategy. Firstly, it takes into account the normalisation procedure (Section 3.5). This provides the possibility to declare only a base signature morphism, which is then internally extended to a complete signature morphism if such exists, using the normalisation procedure. Secondly, in order to construct a signature morphism from $\mathcal{T}_1$ to $\mathcal{T}_2$, its class, type and operation maps have to be declared only on the *domain* classes, type constructors and logical constants, while all remaining global elements are then mapped to themselves (or in the case of type constructors to the corresponding type scheme). This strategy is based on the observation that the common ancestors of two theories form some kind of their common global context, and mapping of the global elements by identity fits well into the definition of signature morphisms, since it can be seen as an extension of the meta-logical requirements from meta-logical elements to all global elements. This strategy is 'user-friendly' in the sense that one does not have to care about assignments for global elements (the number of these can be impressive), but can concentrate on the essential problem in the construction of signature morphisms, namely the assignments for domain elements.

On the other hand, this strategy does not allow to construct signature morphisms, which assign for instance a constant $g \neq f$ to some global constant f. However, this is not a proper restriction, since any global element can be explicitly made local by changing the setup of the particular theory development.

3.7.1 The sigmorph command

Signature morphisms are constructed at the top-level of Isabelle via the integrated command **sigmorph**. Two ways are provided:

1. By source and target theories, and declarations $\mathcal{D}_{class}$, $\mathcal{D}_{type}$, $\mathcal{D}_{op}$ of class,

type and operation maps, respectively, using the following form:
sigmorph $\sigma^{\mathcal{I}}$: *Source* $\longrightarrow$ *Target*
class_map: $\mathcal{D}_{class}$
type_map: $\mathcal{D}_{type}$
op_map: $\mathcal{D}_{op}$

Basically, in this context mapping declarations like $\mathcal{D}_{class}, \mathcal{D}_{type}$ and $\mathcal{D}_{op}$ have the following common syntactic form: $[\ l_1 \mapsto r_1, \ldots, l_n \mapsto r_n\ ]$, i.e. a list of assignments. The particular semantical restrictions on l_i, r_i are described in the following sections.

2. By composition of existing signature morphisms σ_1 and σ_2, where the target of σ_1 is an ancestor of the source of σ_2, as described in Section 3.6. At the top-level of Isabelle, this is represented by the *compositional* declaration: **sigmorph** σ **by_composition** $\sigma_2 \circ \sigma_1$

Since the semantics of the compositional declaration corresponds exactly to the composition of signature morphisms, given in Section 3.6, only the first case has to be explained. To this end, let $\mathcal{T}_1$ and $\mathcal{T}_2$ denote the in the first case declared source and target theories, respectively. The next three sections describe the syntax and semantics of the declarations $\mathcal{D}_{class}$, $\mathcal{D}_{type}$ and $\mathcal{D}_{op}$.

3.7.2 Class map construction

A class map declaration $\mathcal{D}_{class}$ consists of a list of assignments $(c \mapsto c')$, where $c \in \Sigma_{class}(\mathcal{T}_1)$, $c' \in \Sigma_{class}(\mathcal{T}_2)$. If $c \in Dom_{class}(\mathcal{T}_1, \mathcal{T}_2)$ then an assignment $(c \mapsto c')$ has to be declared, such that, for any other declaration $(c \mapsto c'')$, $c' = c''$ holds. In the following, a procedure is described, which simultaneously checks a given declaration $\mathcal{D}_{class}$, and constructs a class map $\sigma^{\mathcal{I}}_{class}$ from it, such that $\Sigma_{class}(\mathcal{T}_1) \overset{\sigma^{\mathcal{I}}_{class}}{\longmapsto} \Sigma_{class}(\mathcal{T}_2)$ holds.

Let $D_0 \overset{def}{=} Global_{class}(\mathcal{T}_1, \mathcal{T}_2)$. Since $\prec_{\Sigma_{class}(\mathcal{T}_1)}$ respects the ancestor relation, $Super_{\Sigma_{class}(\mathcal{T}_1)}(c) \subseteq D_0$ holds for any $c \in D_0$. So, defining $\sigma^0_{class}(c) \overset{def}{=} c$ for any $c \in D_0$, $\Sigma_{class}(\mathcal{T}_1) \overset{\sigma^0_{class}}{\longmapsto}_{D_0} \Sigma_{class}(\mathcal{T}_2)$ obviously holds by Definition 9, i.e. σ^0_{class} is a class map w.r.t. D_0.

Further, let $N \overset{def}{=} |Dom_{class}(\mathcal{T}_1, \mathcal{T}_2)|$. If $N = 0$, i.e. there are no type classes in the domain, then we can define $\sigma^{\mathcal{I}}_{class} \overset{def}{=} \sigma^0_{class}$. Now, let $N > 0$ and $\theta : Dom_{class}(\mathcal{T}_1, \mathcal{T}_2) \to \{1, \ldots, N\}$ be a bijection, such that for any $c, c' \in Dom_{class}(\mathcal{T}_1, \mathcal{T}_2)$, $c \prec_{\Sigma_{class}(\mathcal{T}_1)} c'$ implies $\theta(c') < \theta(c)$. Such θ exists, due to the acyclic property of $\prec_{\Sigma_{class}(\mathcal{T}_1)}$. For θ and arbitrary $c, c' \in Dom_{class}(\mathcal{T}_1, \mathcal{T}_2)$ we can conclude:

1. if $c \prec^{+}_{\Sigma_{class}(\mathcal{T}_1)} c'$ then $\theta(c') < \theta(c)$, and, hence,
2. if $c' \in Super_{\Sigma_{class}(\mathcal{T}_1)}(c)$ then $\theta(c') \leq \theta(c)$.

Provided by θ, we obtain the sequence $s \stackrel{def}{=} \langle c_1, \ldots, c_N \rangle$ of the domain classes, where $c_i \stackrel{def}{=} \theta^{-1}(i)$. Let $D_i \stackrel{def}{=} D_0 \cup \{c_1, \ldots, c_i\}$ for $1 \leq i \leq N$, such that $D_0 \subset \ldots \subset D_N = \Sigma_{class}(\mathcal{T}_1)$ holds. The construction of the sequence s ensures

$$Super_{\Sigma_{class}(\mathcal{T}_1)}(c) \subseteq D_i \text{ for all } 1 \leq i \leq N \text{ and } c \in D_i \tag{3.24}$$

For if not, there would exist $1 \leq i \leq N$, $c \in D_i$ and $c' \in Super_{\Sigma_{class}(\mathcal{T}_1)}(c)$ with $c' \notin D_i$. By the definition of D_i, this implies $c' = c_k$ with $i < k \leq N$, and, hence, $c' \in D_k$. On the other hand, $c \in D_i$ implies that

1. either $c \in D_0$, i.e. is a global class; in this case $Super_{\Sigma_{class}(\mathcal{T}_1)}(c) \subseteq D_0 \subset D_i$ holds, and, hence, $c' \in D_i$, which is a contradiction;
2. or there exists l with $1 \leq l \leq i$ such that $c = c_l$; then from $c_k \in Super_{\Sigma_{class}(\mathcal{T}_1)}(c_l)$ we can conclude $\theta(c_k) \leq \theta(c_l)$; since $c_k = \theta^{-1}(k)$ and $c_l = \theta^{-1}(l)$, we have $k \leq l$, which is again a contradiction.

The basic step, extending a class map by a type class assignment, can be abstracted by the following

Procedure *extension of a class map by a class assignment*
Input:

1. two theories $\mathcal{T}_1$, $\mathcal{T}_2$;
2. a set of classes $D \subset \Sigma_{class}(\mathcal{T}_1)$ with $Super_{\Sigma_{class}(\mathcal{T}_1)}(c) \subseteq D$ for all $c \in D$;
3. a class map φ with $\Sigma_{class}(\mathcal{T}_1) \stackrel{\varphi}{\longmapsto}_D \Sigma_{class}(\mathcal{T}_2)$;
4. a class assignment $(c \mapsto c')$ with $c \in \Sigma_{class}(\mathcal{T}_1)$, $c \notin D$, $Super_{\Sigma_{class}(\mathcal{T}_1)}(c) \subseteq D \cup \{c\}$, and $c' \in \Sigma_{class}(\mathcal{T}_2)$;

Output: the class assignment $(c \mapsto c')$ is either

- accepted, yielding the extended class map φ' with $D \cup \{c\} \stackrel{\varphi'}{\longmapsto}_{\prec_{\Sigma_{class}(\mathcal{T}_1)}} \Sigma_{class}(\mathcal{T}_2)$ and $\varphi'(c) = c'$, $\varphi'(x) = \varphi(x)$ for all $x \neq c$;

- or rejected;

Description: Let $D' \stackrel{def}{=} D \cup \{c\}$. The mapping $\varphi' : D' \to \Sigma_{class}(\mathcal{T}_2)$ extends φ by: $\varphi'(c) \stackrel{def}{=} c'$ and $\varphi'(x) \stackrel{def}{=} \varphi(x)$ for all $x \neq c$. Whether

$$\Sigma_{class}(\mathcal{T}_1) \stackrel{\varphi'}{\longmapsto}_{D'} \Sigma_{class}(\mathcal{T}_2) \tag{3.25}$$

holds, depends on the given class assignment $(c \mapsto c')$. There are the following cases:

(i) the condition

$$\overline{\varphi'}(Super_{\Sigma_{class}(\mathcal{T}_1)}(c))\downarrow^{\mathcal{T}_2}_{sort} = \{c'\} \tag{3.26}$$

holds, then (3.25) holds, and $(c \mapsto c')$ is accepted;

(ii) the condition (3.26) does not hold, so $(c \mapsto c')$ is rejected.

In order to justify the usage of the condition (3.26), we first note, that since $c' \in \overline{\varphi'}(Super_{\Sigma_{class}(\mathcal{T}_1)}(c))$ holds by construction, (3.26) is equivalent to the condition

$$\overline{\varphi'}(Super_{\Sigma_{class}(\mathcal{T}_1)}(c)) \subseteq Super_{\Sigma_{class}(\mathcal{T}_2)}(c')$$

by Lemma 2.1 and Lemma 2.2. Furthermore, this is equivalent to

$$\Sigma_{class}(\mathcal{T}_1) \stackrel{\varphi'}{\longmapsto}_{D'} \Sigma_{class}(\mathcal{T}_2)$$

by Lemma 3.1, since $Super_{\Sigma_{class}(\mathcal{T}_1)}(c) \subseteq D'$ holds for any $c \in D'$, provided by the preconditions of the procedure.

Now, in the current situation this procedure is applicable to $\mathcal{T}_1$, $\mathcal{T}_2$, D_0, σ^0_{class}, and the declared class assignment $(c_1 \mapsto c'_1)$, since these values satisfy the preconditions. Let D_{k-1}, σ^{k-1}_{class}, $(c_k \mapsto c'_k)$ with $1 \leq k \leq N$ be the arguments of the extension procedure above satisfying the preconditions w.r.t. $\mathcal{T}_1$, $\mathcal{T}_2$. If the procedure succeeds then the resulting values D_k and σ^k_{class} also satisfy the preconditions. Altogether, the iteration of the procedure on the sequence $\langle c_1, \ldots, c_N \rangle$ yields the class map $\sigma^{\mathcal{I}}_{class} \stackrel{def}{=} \sigma^N_{class}$ satisfying $\Sigma_{class}(\mathcal{T}_1) \stackrel{\sigma^{\mathcal{I}}_{class}}{\longmapsto} \Sigma_{class}(\mathcal{T}_2)$, if it succeeds N-times. Otherwise, there has to be some declared class assignment

$(c_k \mapsto c'_k)$ with $1 \leq k \leq N$, which has been rejected by the procedure. In this case the entire class map declaration $\mathcal{D}_{class}$ is rejected, because it does not constitute a correct class map.

Finally, note that the construction of $\sigma^{\mathcal{I}}_{class}$ ensures that the sort map $\overline{\sigma}^{\mathcal{I}}_{class}$ maps global sorts by identity, i.e. $\overline{\sigma}^{\mathcal{I}}_{class}(S) = S$ for any $S \in Sorts(\Sigma_{class}(\mathcal{T}))$ with $\mathcal{T} \in Global_{thys}(\mathcal{T}_1, \mathcal{T}_2)$.

3.7.3 Type map construction

A type map declaration $\mathcal{D}_{type}$ consists of a list of assignment declarations $(p \mapsto s)$ where

1. $p \in \widehat{Types}_{\mathcal{X}_{type}}(\Sigma_{type}(\mathcal{T}_1))$ is a special type scheme with type variables as indices, called *type pattern*, of the form $(`\alpha_1, \ldots, `\alpha_n) \cdot C$ (or just $(\alpha_1, \ldots, \alpha_n)C$, for short), where $C \in \Sigma_{type}(\mathcal{T}_1)$ with $rank_{\mathcal{T}_1}(C) = n$, and $\alpha_i \in \mathcal{X}_{type}$ are distinct type variables. Such p is then called a type pattern for C;

2. $s \in \widehat{Types}_{\mathcal{X}_{type}}(\Sigma_{type}(\mathcal{T}_2))$, i.e. a type scheme over $\Sigma_{type}(\mathcal{T}_2)$ with type variables as indices. Moreover, the condition $Inds_{\mathcal{X}_{type}}(s) \subseteq \{\alpha_1, \ldots, \alpha_n\}$ has to be satisfied.

If p and s satisfy these conditions then $(p \mapsto s)$ is called a *type assignment for* the type constructor C. Moreover, for any $C \in Dom_{type}(\mathcal{T}_1, \mathcal{T}_2)$ a type map declaration has to contain exactly one type assignment for C.

Let $((\alpha_1, \ldots, \alpha_n)C \mapsto s)$ be a type assignment for C. Since α_i are distinct, we can define the mapping $\iota_C : \mathcal{X}_{type} \to \widehat{Types}(\Sigma_{type}(\mathcal{T}_2))$ by

$$\iota_C(\beta) \stackrel{def}{=} \begin{cases} `i & \text{if there exists } 1 \leq i \leq n \text{ with } \beta = \alpha_i, \\ `(n+1) & \text{otherwise.} \end{cases}$$

Let $\overline{\iota}_C : \widehat{Types}_{\mathcal{X}_{type}}(\Sigma_{type}(\mathcal{T}_2)) \to \widehat{Types}(\Sigma_{type}(\mathcal{T}_2))$ be the unique homomorphism extending ι_C to $\widehat{Types}_{\mathcal{X}_{type}}(\Sigma_{type}(\mathcal{T}_2))$. In particular, we obtain that $\overline{\iota}_C(s) \in \widehat{Types}(\Sigma_{type}(\mathcal{T}_2))$ holds, and, provided by the condition $Inds_{\mathcal{X}_{type}}(s) \subseteq \{\alpha_1, \ldots, \alpha_n\}$, $Bound(\overline{\iota}_C(s)) \leq n$ holds as well.

So, any type assignment declaration $((\alpha_1, \ldots, \alpha_n)C \mapsto s)$ determines the mapping $\iota_C : \mathcal{X}_{type} \to \widehat{Types}(\Sigma_{type}(\mathcal{T}_2))$, and, thus the assignment $(C \mapsto \varsigma)$ with $\varsigma \stackrel{def}{=} \overline{\iota}_C(s)$ such that the condition $Bound(\varsigma) \leq n$ holds.

By the way, we can also go other way around. So, any assignment $(C \mapsto \varsigma)$ where $rank_{\mathcal{T}_1}(C) = n$ and $Bound(\varsigma) \leq n$ together with $n+1$ distinct type variables $\alpha_1, \ldots, \alpha_{n+1}$ determine the map $\iota_C : \mathbb{N}_+ \to \widehat{Types}_{\mathcal{X}_{type}}(\Sigma_{type}(\mathcal{T}_2))$ with $\iota_C(i) \stackrel{def}{=} `\alpha_i$ for $1 \leq i \leq n$ and $\iota_C(x) \stackrel{def}{=} `\alpha_{n+1}$ otherwise. Then the extension $\overline{\iota}_C$ to $\widehat{Types}(\Sigma_{type}(\mathcal{T}_2))$ such that $\overline{\iota}_C(`i) = \iota_C(i)$ holds, maps ς to $\overline{\iota}_C(\varsigma) \in \widehat{Types}_{\mathcal{X}_{type}}(\Sigma_{type}(\mathcal{T}_2))$ which contains at most the type variables $\alpha_1, \ldots, \alpha_n$. That is, we obtain the type assignment $((\alpha_1, \ldots, \alpha_n)C \mapsto \overline{\iota}_C(\varsigma))$.

Finally, $((\alpha_1, \ldots, \alpha_n)C \mapsto s)$ is called *correct*, if the derived assignment $(C \mapsto \varsigma)$ is accepted by the procedure for type constructor assignment check, described in Section 3.2, for $\mathcal{T}_1$ and $\mathcal{T}_2$ and the class map $\sigma^{\mathcal{I}}_{class}$, constructed from $\mathcal{D}_{class}$ in the previous section. Consequently, $\mathcal{D}_{type}$ is called *correct* if it contains only correct type assignment declarations.

Altogether, the mapping $\sigma^{\mathcal{I}}_{type} : \Sigma_{type}(\mathcal{T}_1) \to \widehat{Types}(\Sigma_{type}(\mathcal{T}_2))$, satisfying $\Sigma_{type}(\mathcal{T}_1) \overset{\sigma^{\mathcal{I}}_{type}}{\longmapsto}_{\sigma^{\mathcal{I}}_{class}} \Sigma_{type}(\mathcal{T}_2)$, is constituted by a correct declaration $\mathcal{D}_{type}$ as follows:

1. $\sigma^{\mathcal{I}}_{type}(C) \stackrel{def}{=} (`1, \ldots, `n) \cdot C$, if $C \in Global_{type}(\mathcal{T}_1, \mathcal{T}_2)$ with $rank_{\mathcal{T}_1}(C) = n$;

2. $\sigma^{\mathcal{I}}_{type}(C) \stackrel{def}{=} \varsigma$, where $(C \mapsto \varsigma)$ is derived from the type assignment declaration for C, otherwise, i.e. if $C \in Dom_{type}(\mathcal{T}_1, \mathcal{T}_2)$.

The correctness of $\sigma^{\mathcal{I}}_{type}$ on type constructors in the domain is given by the correctness of $\mathcal{D}_{type}$. So it remains to justify the correctness of the definition of $\sigma^{\mathcal{I}}_{type}$ on global type constructors. To this end, let $C \in Global_{type}(\mathcal{T}_1, \mathcal{T}_2)$ be some global type constructor with the rank n. Firstly, we note $Inds((`1, \ldots, `n) \cdot C) = \{1, \ldots, n\}$ such that $Bound(\sigma^{\mathcal{I}}_{type}(C)) = n$ holds. Secondly, the condition (3.2) holds in this case as well, as the following argumentation shows.

Let $(S_1, \ldots, S_n, S_{n+1}) \in arities\text{-}of_{\mathcal{T}_1}(C)$ and $\kappa_1, \ldots, \kappa_n \in Types_{\vartheta}(\Sigma_{type}(\mathcal{T}_2))$ with κ_i $of\text{-}sort_{\mathcal{T}_2}$ $\overline{\sigma}^{\mathcal{I}}_{class}(S_i)$ for all $1 \leq i \leq n$. Since C is global and $arities\text{-}of_{\mathcal{T}_1}$ respects the ancestor relation, we have

$$(\overline{\sigma}^{\mathcal{I}}_{class}(S_1), \ldots, \overline{\sigma}^{\mathcal{I}}_{class}(S_n), \overline{\sigma}^{\mathcal{I}}_{class}(S_{n+1})) = (S_1, \ldots, S_n, S_{n+1}) \in arities\text{-}of_{\mathcal{T}_2}(C)$$

Using this we can show $\sigma^{\mathcal{I}}_{type}(C) \bullet \langle \kappa_1, \ldots, \kappa_n \rangle$ $of\text{-}sort_{\mathcal{T}_2}$ $\overline{\sigma}^{\mathcal{I}}_{class}(S_{n+1})$, which is then equivalent to $(\kappa_1, \ldots, \kappa_n)C$ $of\text{-}sort_{\mathcal{T}_2}$ S_{n+1} and follows from the assumptions on C, κ_i and by the definition of the of-sort relation.

Finally, regarding the extension $\overline{\sigma}^{\sigma^{\mathcal{I}}_{class}}_{type}$ of $\sigma^{\mathcal{I}}_{type}$, we can observe $\overline{\sigma}^{\sigma^{\mathcal{I}}_{class}}_{type}(\kappa) = \kappa$ for any $\kappa \in Types(\Sigma_{type}(\mathcal{T}))$ with $\mathcal{T} \in Global_{thys}(\mathcal{T}_1, \mathcal{T}_2)$, i.e. $\overline{\sigma}^{\sigma^{\mathcal{I}}_{class}}_{type}$ does not change global types.

3.7.4 Operation map construction

An operation map declaration $\mathcal{D}_{op}$ consists of a list of assignments $(f \mapsto f')$ where

1. $f \in Dom_{op}(\mathcal{T}_1, \mathcal{T}_2) \cap \Sigma_{op}^{log}(\mathcal{T}_1)$, i.e. a *logical* constant from the domain;
2. $f' \in \Sigma_{op}(\mathcal{T}_2)$, i.e. a (not necessary logical) constant from the operation signature of $\mathcal{T}_2$;
3. $\textit{type-of}_{\mathcal{T}_2}(f') \stackrel{\alpha}{=} \overline{\sigma}_{type}^{\sigma_{class}^{\mathcal{I}}}(\textit{type-of}_{\mathcal{T}_1}(f))$, i.e. the condition (3.9) holds.

If f and f' satisfy these conditions then $(f \mapsto f')$ is called an *operation assignment for* the logical constant f. So, $\mathcal{D}_{type}$ is called *correct* if it contains exactly one operation assignment for any $f \in Dom_{op}(\mathcal{T}_1, \mathcal{T}_2) \cap \Sigma_{op}^{log}(\mathcal{T}_1)$.

Any correct operation map declaration $\mathcal{D}_{op}$ determines the mapping $\sigma_{op}^{\mathcal{I}} : \Sigma_{op}^{log}(\mathcal{T}_1) \to \Sigma_{op}(\mathcal{T})$, which maps global logical constants by identity, while logical constants in the domain are mapped according to the declaration. Provided by this construction, $\Sigma_{op}^{log}(\mathcal{T}_1) \stackrel{\sigma_{op}^{\mathcal{I}}}{\longmapsto}_{\sigma_{class}^{\mathcal{I}}, \sigma_{type}^{\mathcal{I}}} \Sigma_{op}(\mathcal{T}_2)$ holds, i.e. we obtain a base signature morphism $\sigma^{\mathcal{I}}(\Sigma_{op}^{log}(\mathcal{T}_1)) : \mathcal{T}_1 \longrightarrow \mathcal{T}_2$.

Now, if the normalisation procedure, described in Section 3.5, applied to $\mathcal{T}_1$, $\mathcal{T}_2$ and $\sigma^{\mathcal{I}}(\Sigma_{op}^{log}(\mathcal{T}_1))$, succeeds then we obtain the theory $\mathcal{T}_2'$, (possibly) extended by new definitions from the domain, and the signature morphism $\sigma^{\mathcal{I}} : \mathcal{T}_1 \longrightarrow \mathcal{T}_2'$. Important fact is, that all global derived constants are *matched* by themselves, which is provided by the properties of $\overline{\sigma}_{class}^{\mathcal{I}}$ and $\overline{\sigma}_{type}^{\sigma_{class}^{\mathcal{I}}}$ on global sorts and types, respectively. Consequently, for the homomorphic extension of the signature morphism $\sigma^{\mathcal{I}}$ we obtain $\overline{\sigma}^{\mathcal{I}}(t) = t$ for any $t \in Terms(\Sigma_{op}(\mathcal{T}))$ with $\mathcal{T} \in Global_{thy}(\mathcal{T}_1, \mathcal{T}_2') = Global_{thy}(\mathcal{T}_1, \mathcal{T}_2)$.

Otherwise, if the normalisation procedure fails then it means, that the base signature morphism $\sigma^{\mathcal{I}}(\Sigma_{op}^{log}(\mathcal{T}_1)) : \mathcal{T}_1 \longrightarrow \mathcal{T}_2$ is not normalisable according to Definition 17, i.e. there is no complete signature morphism $\sigma^{\mathcal{I}} : \mathcal{T}_1 \longrightarrow \mathcal{T}_2'$ with $\mathcal{T}_2'$ extending $\mathcal{T}_2$. Then either the declaration $\mathcal{D}_{type}$, defining the type map $\sigma_{type}^{\mathcal{I}} : \Sigma_{type}(\mathcal{T}_1) \to \widehat{Types}(\Sigma_{type}(\mathcal{T}_2))$, or the definition in $Dom_{thy}(\mathcal{T}_1, \mathcal{T}_2)$ have to be changed in order to allow normalisation.

3.7.5 Example: sets as monads

Taking our example signature morphism $\sigma_{\mathsf{PSet\_Monad}} : \mathsf{Monad} \longrightarrow \mathsf{PSet\_Monad}$, defined in Section 3.4.1, we can represent it using the introduced facilities by, e.g.

sigmorph *pset-as-monad* : *Monad* $\longrightarrow$ *PSet_Monad*
type_map: $[\alpha\ T \quad \mapsto \quad \alpha \Rightarrow bool]$
op_map: $[T \quad \mapsto \quad T^{PSet},$
$\eta \quad \mapsto \quad \eta^{PSet},$
$\mu \quad \mapsto \quad \mu^{PSet}]$

or taking alternatively the operation map declaration
op_map: $[\eta \quad \mapsto \quad \eta^{PSet},$
$\mu \quad \mapsto \quad \mu^{PSet},$
$T \quad \mapsto \quad T^{PSet}]$

since the order of assignments does not matter.

3.8 Conclusion

In this chapter the notion of signature morphisms in a logical framework, specified in Chapter 2, has been introduced. Further, an associative composition operation for signature morphisms as well as the homomorphic extension of a signature morphism have been defined. These will allow a categorical view on theories and morphisms in the next chapter. Apart from this, procedures for construction of a signature morphism from given class, type, and operation assignments declaration have been described and shown, how this works in the concrete case of the theorem prover Isabelle. The presented normalisation procedure together with the matching approach for definitions allows us to reduce the problem of construction of an operation map to assignments for logical constants in the given source signature.

Chapter 4

Theory Morphisms

In this chapter the notion of theory morphisms will be introduced. Basically, any theory morphism extends some signature morphism σ, additionally relating global contexts by a mapping (called theorem map) from the global context of the source theory of σ to the global context of the target theory of σ, satisfying certain rules called theorem map conditions. Since the set of definitions of a theory is a part of its global context, the notions of theory and signature morphisms intersect on these, because signature morphisms already treat this special subset of axioms by the definitional condition (3.14). So, theory morphisms will use a slightly weaker version of (3.14) in order to relate global contexts by a theorem map.

As shown in the previous chapter, the homomorphic extension of a signature morphism maps propositions in the source to propositions in the target. It is, however, not always the case that a signature morphism maps derivable proposition in the source to derivable propositions in the target. The most obvious example is given when the source theory is inconsistent whereas the target theory consistent, such that any proposition in the source is derivable, but its image under the homomorphic extension has not to be derivable in the target. So, one of the main motivations for the introduction of theory morphisms in this context is that they will provide a sufficient condition for this property of signature morphisms. Further, since derivable propositions are characterised by proof terms, in order to show that a proposition is derivable in a theory the existence of an appropriate admissible proof term has to be shown. This, in turn, will be done in this chapter in a constructive way, based on the so-called proof term translation, which can be seen as the extension of a theory morphism to proof terms. Another important motivation for the introduction of theory morphisms is that the concept of parameterised theories and, especially, of their instantiation is based on this notion. The instantiation of parameterised theories can be seen as a special case of pushout construction, and forms the basic device for transformational development.

4.1 Theory morphisms and proof term translation

First of all, theory morphisms are defined as follows.

Definition 20 (theory morphisms)
A *theory morphism* $\tau(D) = \langle \sigma, D, \xi \rangle$ comprises

1. A signature morphism $\sigma : \mathcal{T}_1 \longrightarrow \mathcal{T}_2$ with $\mathcal{G}(\mathcal{T}_1) = \langle D_1, \mathit{thm\text{-}of}_{\mathcal{T}_1} \rangle$ and $\mathcal{G}(\mathcal{T}_2) = \langle D_2, \mathit{thm\text{-}of}_{\mathcal{T}_2} \rangle$;

2. A set D of proposition labels, such that $AX(\mathcal{T}_1) \subseteq D \subseteq D_1$ holds;

3. A mapping $\xi : D \rightarrow D_2$, such that for all $d \in D$ the following two *theorem map conditions* hold:

$$\mathit{thm\text{-}of}_{\mathcal{T}_2}(\xi(d)) \stackrel{\alpha}{=} \overline{\sigma}(\mathit{thm\text{-}of}_{\mathcal{T}_1}(d)) \tag{4.1}$$

   and

$$\frac{d \in \mathit{Thms}(\mathcal{T}_1)}{\xi(d) \in \mathit{Thms}(\mathcal{T}_2)} \tag{4.2}$$

Similarly to the previous constructions, *meta-logical requirements* on ξ say that $\xi(a) = a$ holds for any $a \in AX(\mathsf{Pure})$.

Further, σ is called the *underlying signature morphism of* τ, ξ is called the *theorem map* of τ, and we also write $\tau(D) : \mathcal{T}_1 \longrightarrow \mathcal{T}_2$ in order to emphasise the source and the target theories of a theory morphism. Moreover, similarly to signature morphisms, $\tau(AX(\mathcal{T}_1))$ is called a *base* theory morphism, while $\tau(D_1)$ a *complete* theory morphism. The notation $\tau : \mathcal{T}_1 \longrightarrow \mathcal{T}_2$ will be used as a shortening for $\tau(D_1) : \mathcal{T}_1 \longrightarrow \mathcal{T}_2$. The theories $\mathcal{T}_1$ and $\mathcal{T}_2$ are also called the *source* and the *target* of the theory morphism.

An important observation connecting signature and theory morphisms is that the definitional condition (3.14) on signature morphisms is a special case of the theorem map condition (4.1). Let $\sigma : \mathcal{T}_1 \longrightarrow \mathcal{T}_2$ be a signature morphism with the operation map σ_{op} and $\mathcal{G}(\mathcal{T}_2) = \langle D_2, \mathit{thm\text{-}of}_{\mathcal{T}_2}\rangle$. Further, let $f \in \Sigma_{op}^{der}(\mathcal{T}_1)$ be a derived constant in the source signature. By the definition of signature morphisms $\sigma_{op}(f)$ is a derived constant as well, i.e. $\sigma_{op}(f) \in \Sigma_{op}^{der}(\mathcal{T}_2)$ holds. So, let $f_{def} \in AX(\mathcal{T}_1)$ be the axiom defining f and $\sigma_{op}(f)_{def} \in AX(\mathcal{T}_2)$ be the axiom defining $\sigma_{op}(f)$. By the restriction on definitions, $\mathit{prop\text{-}of}_{\mathcal{T}_1}(f_{def})$ has the form $f \equiv \mathit{def\text{-}of}_{\mathcal{T}_1}(f)$ and $\mathit{prop\text{-}of}_{\mathcal{T}_2}(\sigma_{op}(f)_{def})$ has the form $\sigma_{op}(f) \equiv \mathit{def\text{-}of}_{\mathcal{T}_2}(\sigma_{op}(f))$. Thus, we can reason:

$$\begin{array}{ll}
\overline{\sigma}(\mathit{thm\text{-}of}_{\mathcal{T}_1}(f_{def})) & = \\
\overline{\sigma}(\mathit{prop\text{-}of}_{\mathcal{T}_1}(f_{def})) & = \\
\overline{\sigma}(f \;\equiv\; \mathit{def\text{-}of}_{\mathcal{T}_1}(f)) & = \\
 & \text{by the meta-logical requirements} \\
\sigma_{op}(f) \;\equiv\; \overline{\sigma}(\mathit{def\text{-}of}_{\mathcal{T}_1}(f)) & \stackrel{\alpha}{=} \\
 & \text{by the definitional condition (3.14)} \\
\sigma_{op}(f) \;\equiv\; \mathit{def\text{-}of}_{\mathcal{T}_2}(\sigma_{op}(f)) & = \\
\mathit{prop\text{-}of}_{\mathcal{T}_2}(\sigma_{op}(f)_{def}) & = \\
\mathit{thm\text{-}of}_{\mathcal{T}_2}(\sigma_{op}(f)_{def}) &
\end{array}$$

So, the theorem map assignment $f_{def} \mapsto \sigma_{op}(f)_{def}$ is derivable from the operation map assignment $f \mapsto \sigma_{op}(f)$, i.e. it satisfies the theorem map conditions (4.1) and (4.2). As a nice result we obtain that already a mapping $\xi : AX(\mathcal{T}_1) \setminus \mathit{Defs}(\mathcal{T}_1) \to D_2$, satisfying the theorem map conditions, constitutes a base theory morphism $\tau(D) = \langle \sigma, AX(\mathcal{T}_1), \xi' \rangle$, where ξ' extends ξ to definitions in accordance with the operation map of σ. In other words, the problem of construction of a theorem map reduces to mapping of 'proper' axioms in $AX(\mathcal{T}_1) \setminus \mathit{Defs}(\mathcal{T}_1)$.

Matching. As with signature morphisms and definitions (Section 3.5.1), axioms and theorems can be matched in a similar manner. More precisely, let $\sigma : \mathcal{T}_1 \longrightarrow \mathcal{T}_2$ be a signature morphism, and $\mathcal{G}(\mathcal{T}_1) = \langle D_1, \mathit{thm\text{-}of}_{\mathcal{T}_1}\rangle$, $\mathcal{G}(\mathcal{T}_2) = \langle D_2, \mathit{thm\text{-}of}_{\mathcal{T}_2}\rangle$. Then for $d_1 \in D_1$ we can search for $d_2 \in D_2$ satisfying

$$\mathit{thm\text{-}of}_{\mathcal{T}_2}(d_2) \stackrel{\alpha}{=} \overline{\sigma}(\mathit{thm\text{-}of}_{\mathcal{T}_1}(d_1))$$

i.e. some d_2 satisfying exactly the theorem map condition (4.1) w.r.t. d_1. If such d_2 can be found then the theorem map of any theory morphism extending σ, can assign d_2 to d_1 (in the case when $d_1 \in \mathit{Thms}(\mathcal{T}_1)$, $d_2 \in \mathit{Thms}(\mathcal{T}_2)$ need to be additionally satisfied as required by (4.2)). However, if such $d_2 \in D_2$

exists it is clearly not always unique, i.e. there might be $d_3 \in D_2$ with $d_2 \neq d_3$ also satisfying the theorem map conditions w.r.t. d_1. On the other hand, the theorem map conditions in this case ensure that the propositions of d_2 and d_3 are α-convertible, and therefore logically equivalent. We will use this approach later on, in the context of theory morphism construction in Isabelle (Section 4.3).

4.1.1 Example: sets as monads

Now we can continue the development of Section 3.4.1. Recall that we already have the signature morphism $\sigma_{\mathsf{PSet\_Monad}} : \mathit{Monad} \longrightarrow \mathit{PSet\_Monad}$. Thus, the goal is to construct the theory morphism

$$\tau_{\mathsf{PSet\_Monad}}(AX(\mathit{Monad})) = \langle \sigma_{\mathsf{PSet\_Monad}}, AX(\mathit{Monad}), \xi \rangle$$

i.e. we need to give an appropriate mapping ξ from $AX(\mathit{Monad})$ to the admissible global context of *PSet_Monad*. This will be done in two steps:

1. we extend *PSet_Monad* conservatively, now defining the constants introduced in Section 3.4.1;

2. by this extension, the admissible global context $\mathcal{G}(\mathit{PSet\_Monad})$ can be chosen in such a way, that $\mathit{Thms}(\mathit{PSet\_Monad})$ contains the images of the monad axioms under $\overline{\sigma}_{\mathsf{PSet\_Monad}}$; in other words, these propositions can be then proved in *PSet_Monad*.

A suitable conservative extension of *PSet_Monad* is given by the following three definitions:
defs

$$T^{PSet}_{def} \; : \; T^{PSet}\, f\, S \;\equiv\; \{y \mid \exists x \in S .\, f\, x = y\}$$

$$\eta^{PSet}_{def} \; : \; \eta^{PSet}\, x \;\equiv\; \{x\}$$

$$\mu^{PSet}_{def} \; : \; \mu^{PSet}\, S \;\equiv\; \{x \mid \exists X \in S .\, x \in X\}$$

Notice that $\{x \mid C\}$ is merely an abbreviation for $\lambda x.\, C$, $\{x\}$ – for $\lambda y.\, y = x$, and $x \in P$ – for $P\, x$.

In this extended theory *PSet_Monad* it is possible to give proofs for the propositions comprising the images of the monad axioms under $\overline{\sigma}_{\mathsf{PSet\_Monad}}$, e.g. for the propositions

1. $T^{\mathsf{PSet}}\, id = id$, and

2. $\bigwedge f\ g.\ T^{\mathsf{PSet}}(f \circ g) = T^{\mathsf{PSet}} f \circ T^{\mathsf{PSet}} g$

This works, because the definitions introduced above can be unfolded, reducing the problem to propositions built only of the basic operations of the higher order logic, like existential quantifier or equality. In particular, since $T^{\mathsf{PSet}} f\ S$ is defined to be the image of S under f, the proposition $\bigwedge f\ g.\ T^{\mathsf{PSet}}(f \circ g) = T^{\mathsf{PSet}} f \circ T^{\mathsf{PSet}} g$ is equivalent to the proposition that the image of any set under the composition of two functions can be equivalently expressed building two successive images of this set.

Altogether, the constructed theory morphism $\tau_{\mathsf{PSet\_Monad}}(AX(\mathsf{Monad}))$ reflects the consistency of the monad axiomatisation in the higher order logic on page 70, since the target theory *PSet_Monad* has been conservatively constructed in this object logic. This conjecture holds, because any derivable proposition $p \in Thms(\mathsf{Monad})$ constitutes the derivable proposition $\overline{\sigma}_{\mathsf{PSet\_Monad}}(p) \in Thms(\mathsf{PSet\_Monad})$. This result is based on by theory morphisms induced translation of proof terms, which is generally treated in the following sections.

4.1.2 Proof term translation

Now, theory morphisms will be used to define mappings of admissible proof terms in the source to admissible proof terms in the target. The next lemma is crucial as it forms the base for the translation of proof terms between theories, and justifies the claim that the existence of a theory morphism, extending a signature morphism σ, already implies that $\overline{\sigma}$ preserves derivability of propositions.

Lemma 4.1 *Let $\tau(D) = \langle \sigma, D, \xi \rangle$ be a theory morphism, where $\sigma : \mathcal{T}_1 \longrightarrow \mathcal{T}_2$, $\mathcal{G}(\mathcal{T}_1) = \langle D_1, \textit{thm-of}_{\mathcal{T}_1} \rangle$ and $\mathcal{G}(\mathcal{T}_2) = \langle D_2, \textit{thm-of}_{\mathcal{T}_2} \rangle$. Then for any proof term $\pi \in \textit{Proofs}^{Adm}_{\varrho_\vartheta}(\mathcal{T}_1)$ with $\textit{proof-names}(\pi) \subseteq D$ there exists a proof term $\pi' \in \textit{Proofs}^{Adm}_{\varrho'_{\vartheta'}}(\mathcal{T}_2)$, such that for all propositions $p \in \textit{Props}_{\varrho_\vartheta}(\mathcal{T}_1)$ and any context $H : \mathcal{X}_{proof} \rightharpoonup \textit{Props}_{\varrho_\vartheta}(\mathcal{T}_1)$*

$$\frac{H \vdash \pi\ \textit{proof-of}_{\mathcal{T}_1}\ p}{H' \vdash \pi'\ \textit{proof-of}_{\mathcal{T}_2}\ \overline{\sigma}(p)} \tag{4.3}$$

holds, where $\vartheta' \stackrel{def}{=} \overline{\sigma}_{class} \circ \vartheta$, $\varrho'_{\vartheta'} \stackrel{def}{=} \overline{\sigma}^{\sigma_{class}}_{type} \circ \varrho_\vartheta$, and for all X, $H'(X) \stackrel{def}{=} \overline{\sigma}(H(X))$ if $H(X)$ is defined and $H'(X) \stackrel{def}{=} \bot$, otherwise.

Proof. (by structural induction on π)
(i) If $\pi = X$ with $X \in \mathcal{X}_{proof}$ then $\pi' \stackrel{def}{=} X$. For the proof of (4.3) we can assume

$H \vdash \pi$ *proof-of*$_{\mathcal{T}_1}$ p, which implies $H(X) = p$. Hence, $H'(X) = \overline{\sigma}(p)$, such that $H' \vdash \pi'$ *proof-of*$_{\mathcal{T}_2}$ $\overline{\sigma}(p)$ holds.

(ii) Let $\pi = d_{\overline{[\alpha_i := \kappa_i]}_n}$ with $d \in \mathcal{G}(\mathcal{T}_1)$, *typevars*(*thm-of*$_{\mathcal{T}_1}(d)) \subseteq \{\alpha_1, \ldots, \alpha_n\}$ and $\kappa_1, \ldots, \kappa_n \in Types_{\vartheta}(\Sigma_{type}(\mathcal{T}_1))$. By the assumption *proof-names*$(\pi) \subseteq D$ we obtain $d \in D$. Further, since $\tau(D)$ is a theory morphism,

$$\overline{\sigma}(\textit{thm-of}_{\mathcal{T}_1}(d)) \overset{\alpha}{=}_{\theta} \textit{thm-of}_{\mathcal{T}_2}(\xi(d))$$

holds, provided by a bijection θ. Hence, by Lemma 2.5, we obtain a mapping θ' extending θ to $\{\alpha_1, \ldots, \alpha_n\}$, such that

$$\overline{\sigma}(\textit{thm-of}_{\mathcal{T}_1}(d))\overline{[\alpha_i := \overline{\sigma}^{\sigma_{class}}_{type}(\kappa_i)]}_n = \textit{thm-of}_{\mathcal{T}_2}(\xi(d))\overline{[\theta'(\alpha_i) := \overline{\sigma}^{\sigma_{class}}_{type}(\kappa_i)]}_n \quad (4.4)$$

holds. Let *typevars*$(\overline{\sigma}(\textit{thm-of}_{\mathcal{T}_1}(d))) = \{\alpha_{i_1}, \ldots, \alpha_{i_m}\}$ with $m \leq n$. Since θ and θ' agree on $\{\alpha_{i_1}, \ldots, \alpha_{i_m}\}$, (4.4) can be reduced to

$$\overline{\sigma}(\textit{thm-of}_{\mathcal{T}_1}(d))\overline{[\alpha_{i_k} := \overline{\sigma}^{\sigma_{class}}_{type}(\kappa_{i_k})]}_m = \textit{thm-of}_{\mathcal{T}_2}(\xi(d))\overline{[\theta(\alpha_{i_k}) := \overline{\sigma}^{\sigma_{class}}_{type}(\kappa_{i_k})]}_m$$

Using this and $\xi(d) \in D_2$, we define

$$\pi' \overset{def}{=} \xi(d)_{\overline{[\theta(\alpha_{i_k}) := \overline{\sigma}^{\sigma_{class}}_{type}(\kappa_{i_k})]}_m}.$$

For the proof of the condition (4.3), let $H \vdash \pi$ *proof-of*$_{\mathcal{T}_1}$ p. This implies $p = \textit{thm-of}_{\mathcal{T}_1}(d)\overline{[\alpha_i := \kappa_i]}_n$. Then we can reason

$$\begin{array}{ll} \overline{\sigma}(p) & = \\ \overline{\sigma}(\textit{thm-of}_{\mathcal{T}_1}(d)\overline{[\alpha_i := \kappa_i]}_n) & = \\ & \text{by (3.11)} \\ \overline{\sigma}(\textit{thm-of}_{\mathcal{T}_1}(d))\overline{[\alpha_i := \overline{\sigma}^{\sigma_{class}}_{type}(\kappa_i)]}_n & = \\ & \text{by (4.4)} \\ \textit{thm-of}_{\mathcal{T}_2}(\xi(d))\overline{[\theta'(\alpha_i) := \overline{\sigma}^{\sigma_{class}}_{type}(\kappa_i)]}_n & \end{array}$$

On the other hand,

$$H' \vdash \pi' \textit{ proof-of}_{\mathcal{T}_2} \textit{ thm-of}_{\mathcal{T}_2}(\xi(d))\overline{[\theta'(\alpha_i) := \overline{\sigma}^{\sigma_{class}}_{type}(\kappa_i)]}_n$$

holds as well, and thus $H' \vdash \pi'$ *proof-of*$_{\mathcal{T}_2}$ $\overline{\sigma}(p)$.

(iii) Let $\pi = \overline{\lambda}(x ::_{type} \varrho_\vartheta(x)).\pi_1$ with $\pi_1 \in Proofs^{Adm}_{\varrho_\vartheta}(\mathcal{T}_1)$ and $x \in \mathcal{X}_{term}$. Since $proof\text{-}names(\pi) \subseteq D$, we also have $proof\text{-}names(\pi_1) \subseteq D$ and obtain by the induction hypothesis a proof term $\pi'_1 \in Proofs^{Adm}_{\varrho'_{\vartheta'}}(\mathcal{T}_2)$ satisfying (4.3) w.r.t. π_1. Let $\pi' \stackrel{def}{=} \overline{\lambda}(x ::_{type} \varrho'_{\vartheta'}(x)).\pi'_1$. For the proof of (4.3), let $H \vdash \pi\ proof\text{-}of_{\mathcal{T}_1}\ p$. Then there is a proposition p_1, such that $H \vdash \pi_1\ proof\text{-}of_{\mathcal{T}_1}\ p_1$, x does not occur in H and $p = \bigwedge(x ::_{type} \varrho_\vartheta(x)).p_1$. Firstly, by the assumption on π_1, we obtain $H' \vdash \pi'_1\ proof\text{-}of_{\mathcal{T}_1}\ \overline{\sigma}(p_1)$. Secondly, since x does not occur in H and $\overline{\sigma}$ maintains term variables, x does not occur in H' as well. So, by the definition of derivability relation, we can conclude $H' \vdash \pi'\ proof\text{-}of_{\mathcal{T}_2}\ \bigwedge(x::_{type}\varrho'_{\vartheta'}(x)).\overline{\sigma}(p_1)$,, and hence also $H' \vdash \pi'\ proof\text{-}of_{\mathcal{T}_2}\ \overline{\sigma}(p)$, using the properties of the homomorphic extension of σ:

$$
\begin{array}{ll}
\overline{\sigma}(p) & = \\
\overline{\sigma}(\bigwedge(x ::_{type} \varrho_\vartheta(x)).p_1) & = \\
\bigwedge(x ::_{type} \overline{\sigma}^{\sigma_{class}}_{type}(\varrho_\vartheta(x))).\overline{\sigma}(p_1) & = \\
\bigwedge(x ::_{type} \varrho'_{\vartheta'}(x)).\overline{\sigma}(p_1). &
\end{array}
$$

(iv) Let $\pi = \overline{\overline{\lambda}}(X ::_{prop} p_0).\pi_1$ with $X \in \mathcal{X}_{proof}$, $p_0 \in Terms_{\varrho_\vartheta}(\Sigma_{op}(\mathcal{T}_1))$, $p_0 \in Props(\mathcal{T}_1)$, and $\pi_1 \in Proofs^{Adm}_{\varrho_\vartheta}(\mathcal{T}_1)$. By the induction hypothesis for π_1 we obtain a proof term $\pi'_1 \in Proofs^{Adm}_{\varrho'_{\vartheta'}}(\mathcal{T}_2)$ satisfying the claim. We define

$$\pi' \stackrel{def}{=} \overline{\overline{\lambda}}(X ::_{prop} \overline{\sigma}(p_0)).\pi'_1$$

Since $\overline{\sigma}(p_0) \in Terms_{\varrho'_{\vartheta'}}(\Sigma_{op}(\mathcal{T}_2))$ and $\overline{\sigma}$ preserves the of-type relation, i.e. $\overline{\sigma}(p_0) \in Props(\mathcal{T}_2)$ holds, $\pi' \in Proofs^{Adm}_{\varrho'_{\vartheta'}}(\mathcal{T}_2)$ holds as well. For the proof of (4.3) for π', let $H \vdash \pi\ proof\text{-}of_{\mathcal{T}_1}\ p$. Then $H(X) = \perp$ and there is a proposition $p_1 \in Terms_{\varrho_\vartheta}(\Sigma_{op}(\mathcal{T}_1))$ such that $H[X \mapsto p_0] \vdash \pi_1\ proof\text{-}of_{\mathcal{T}_1}\ p_1$ and $p = p_0 \Longrightarrow p_1$. So, by the assumptions on π_1 we have $H'[X \mapsto \overline{\sigma}(p_0)] \vdash \pi'_1\ proof\text{-}of_{\mathcal{T}_2}\ \overline{\sigma}(p_1)$, where $H'(X) = \perp$. By the definition of derivability relation, we can then conclude

$$H' \vdash \overline{\overline{\lambda}}(X ::_{prop} \overline{\sigma}(p_0)).\pi'_1\ proof\text{-}of_{\mathcal{T}_2}\ \overline{\sigma}(p_0) \Longrightarrow \overline{\sigma}(p_1)$$

and hence $H' \vdash \pi'\ proof\text{-}of_{\mathcal{T}_2}\ \overline{\sigma}(p)$, taking into account that $\overline{\sigma}(p) = \overline{\sigma}(p_0) \Longrightarrow \overline{\sigma}(p_1)$ holds.

(v) Let $\pi = \pi_1\, t$ with $t \in Terms_{\varrho_\vartheta}(\Sigma_{op}(\mathcal{T}_1))$ and $\pi_1 \in Proofs^{Adm}_{\varrho_\vartheta}(\mathcal{T}_1)$. By the induction hypothesis for π_1 we obtain a proof term $\pi'_1 \in Proofs^{Adm}_{\varrho'_{\vartheta'}}(\mathcal{T}_2)$ satisfying

(4.3). Let $\pi' \overset{def}{=} \pi_1' \,\overline{\sigma}(t)$, which is well-defined since $\overline{\sigma}(t) \in \mathit{Terms}_{\varrho'_{\vartheta'}}(\Sigma_{op}(\mathcal{T}_2))$. For the proof of (4.3) for π', let $H \vdash \pi$ *proof-of*$_{\mathcal{T}_1}$ p. Then, by the definition of derivability, there is a proposition $p_1 \in \mathit{Terms}_{\varrho_\vartheta}(\Sigma_{op}(\mathcal{T}_1))$ such that

$$H \vdash \pi_1 \text{ proof-of}_{\mathcal{T}_1} \bigwedge(x ::_{type} \varrho_\vartheta(x)).p_1$$

holds. Moreover, we have t *of-type*$_{\mathcal{T}_1}$ $\varrho_\vartheta(x)$ and $p = p_1[t/x]$. By the assumptions on π_1, we can conclude:

$$H' \vdash \pi_1' \text{ proof-of}_{\mathcal{T}_2} \bigwedge(x ::_{type} \varrho'_{\vartheta'}(x)).\overline{\sigma}(p_1).$$

Furthermore, since the homomorphic extension preserves of-type relation, we have $\overline{\sigma}(t)$ *of-type*$_{\mathcal{T}_2}$ $\varrho'_{\vartheta'}(x)$, and, hence:

$$H' \vdash \pi_1' \,\overline{\sigma}(t) \text{ proof-of}_{\mathcal{T}_2} \overline{\sigma}(p_1)[\overline{\sigma}(t)/x]$$

and, hence, using Lemma 3.13, $H' \vdash \pi'$ *proof-of*$_{\mathcal{T}_2}$ $\overline{\sigma}(p_1[t/x])$.

(vi) Let $\pi = \pi_1\,\pi_2$ with $\pi_1, \pi_2 \in \mathit{Proofs}^{Adm}_{\varrho_\vartheta}(\mathcal{T}_1)$. Since *proof-names*$(\pi) \subseteq D$, we also have *proof-names*$(\pi_1) \subseteq D$ and *proof-names*$(\pi_2) \subseteq D$ and, hence, obtain by the induction hypothesis proof terms $\pi_1', \pi_2' \in \mathit{Proofs}^{Adm}_{\varrho'_{\vartheta'}}(\mathcal{T}_2)$ satisfying (4.3). Then, $\pi_1'\,\pi_2' \in \mathit{Proofs}^{Adm}_{\varrho'_{\vartheta'}}(\mathcal{T}_2)$ holds, and we define $\pi' \overset{def}{=} \pi_1'\,\pi_2'$. For the proof of (4.3) for π', let $H \vdash \pi$ *proof-of*$_{\mathcal{T}_1}$ p. Then, by the definition of derivability there is a proposition $p_1 \in \mathit{Terms}_{\varrho_\vartheta}(\Sigma_{op}(\mathcal{T}_1))$ such that $H \vdash \pi_1$ *proof-of*$_{\mathcal{T}_1}$ $(p_1 \Longrightarrow p)$ and $H \vdash \pi_2$ *proof-of*$_{\mathcal{T}_1}$ p_1 hold. By the assumptions on π_1 and π_2, we also obtain $H' \vdash \pi_1'$ *proof-of*$_{\mathcal{T}_2}$ $\overline{\sigma}(p_1 \Longrightarrow p)$ and $H' \vdash \pi_2'$ *proof-of*$_{\mathcal{T}_2}$ $\overline{\sigma}(p_1)$. Since $\overline{\sigma}(p_1 \Longrightarrow p) = \overline{\sigma}(p_1) \Longrightarrow \overline{\sigma}(p)$, we can deduce $H' \vdash \pi_1'\,\pi_2'$ *proof-of*$_{\mathcal{T}_2}$ $\overline{\sigma}(p)$. □

So, for any theory morphism $\tau(D) : \mathcal{T}_1 \longrightarrow \mathcal{T}_2$ the proof establishes a recursive function $\overline{\tau}(D) : \mathit{Proofs}^{Adm}(\mathcal{T}_1) \to \mathit{Proofs}^{Adm}(\mathcal{T}_2)$, called the *extension* of $\tau(D)$, such that for any $\pi \in \mathit{Proofs}^{Adm}(\mathcal{T}_1)$ with *proof-names*$(\pi) \subseteq D$ the rule

$$\frac{\emptyset \vdash \pi \text{ proof-of}_{\mathcal{T}_1} p}{\emptyset \vdash \overline{\tau}(D)(\pi) \text{ proof-of}_{\mathcal{T}_2} \overline{\sigma}(p)} \qquad (4.5)$$

holds, where σ denotes the underlying signature morphism.

4.1.3 Theorem translation

Based on the theory morphism extension, a procedure for translation of theorems from the global context of the source to theorems in the target theory will be formulated in this section. The basic primitive step of this procedure is, in turn, abstracted by the following

Procedure *theorem translation step*
Input:

1. two theories $\mathcal{T}_1$, $\mathcal{T}_2$ with $\mathcal{G}(\mathcal{T}_1) = \langle D_1, \mathit{thm\text{-}of}_{\mathcal{T}_1}\rangle$, $\mathcal{G}(\mathcal{T}_2) = \langle D_2, \mathit{thm\text{-}of}_{\mathcal{T}_2}\rangle$;

2. a theory morphism $\tau(D) = \langle \sigma, D, \xi\rangle$ with $\sigma : \mathcal{T}_1 \longrightarrow \mathcal{T}_2$;

3. a theorem $d \in \mathit{Thms}(\mathcal{T}_1)$, such that for any d' with $d \prec_{\mathit{Thms}(\mathcal{T}_1)} d'$, $d' \in D$ holds;

Output: a theory $\mathcal{T}_2'$ (possibly) extending the global context of $\mathcal{T}_2$ and the theory theory morphism $\tau(D \cup \{d\}) : \mathcal{T}_1 \longrightarrow \mathcal{T}_2'$;

Description: Since $d \in \mathit{Thms}(\mathcal{T}_1)$, $d \in D_1$ and $d \notin AX(\mathcal{T}_1)$ holds. The trivial case is, when $d \in D$ holds, because the output is then just $\mathcal{T}_2$ and $\tau(D)$. Otherwise, using $d \in \mathit{Thms}(\mathcal{T}_1)$, we can take $\pi \stackrel{def}{=} \mathit{proof\text{-}of}_{\mathcal{T}_1}(d)$ such that $\emptyset \vdash \pi\ \mathit{proof\text{-}of}_{\mathcal{T}_1}\ \mathit{thm\text{-}of}_{\mathcal{T}_1}(d)$ holds. On the other hand, using the precondition, i.e. that $d' \in D$ holds for all d' with $d \prec_{\mathit{Thms}(\mathcal{T}_1)} d'$, we also obtain $\mathit{proof\text{-}names}(\pi) \subseteq D$. By the rule (4.5), $\emptyset \vdash \overline{\tau}(D)(\pi)\ \mathit{proof\text{-}of}_{\mathcal{T}_2}\ \overline{\sigma}(\mathit{thm\text{-}of}_{\mathcal{T}_1}(d))$ holds then as well, and allows us to define using some $\overline{d} \notin D_2$:

$$\mathcal{T}_2' \stackrel{def}{=} \mathcal{T}_2 +_{thm} (\overline{d}, \overline{\tau}(D)(\pi))$$

such that $\mathit{thm\text{-}of}_{\mathcal{T}_2'}(\overline{d}) = \overline{\sigma}(\mathit{thm\text{-}of}_{\mathcal{T}_1}(d))$ holds. Finally, let $\tau(D \cup \{d\}) \stackrel{def}{=} \langle \sigma', D \cup \{d\}, \xi[d \mapsto \overline{d}]\rangle$, where $\sigma' : \mathcal{T}_1 \longrightarrow \mathcal{T}_2'$ is basically the same signature morphism as σ (i.e. constituted by the same mappings) and $\xi[d \mapsto \overline{d}]$ extends ξ on d. So, by the construction $\tau(D \cup \{d\})$ is indeed a theory morphism.

Now, let $\mathcal{T}_1$ and $\mathcal{T}_2$ be theories, $\tau(D) : \mathcal{T}_1 \longrightarrow \mathcal{T}_2$ a theory morphism, and $d \in \mathit{Thms}(\mathcal{T}_1)$ with $d \notin D$ a theorem. Further, let

$$S \stackrel{def}{=} \{d' \mid d \prec^*_{\mathit{Thms}(\mathcal{T}_1)} d' \text{ and } d' \notin AX(\mathcal{T}_1)\},$$

i.e. the subset of theorems in $\mathcal{T}_1$, which are in the image of the reflexive-transitive closure of $\prec_{\mathit{Thms}(\mathcal{T}_1)}$ under $\{d\}$. Moreover, let $N \stackrel{def}{=} |S|$ and $\theta : S \to \{1, \ldots, N\}$ be a bijection, such that $d_1 \prec_{\mathit{Thms}(\mathcal{T}_1)} d_2$ implies $\theta(d_1) > \theta(d_2)$ for all $d_1, d_2 \in S$.

Provided by θ we can define the sequence $s \stackrel{def}{=} \langle d_1, \ldots d_N \rangle$, where $d_i \stackrel{def}{=} \theta^{-1}(i)$. This construction ensures that for any $1 \leq i \leq N$

$$\{d' \mid d_i \prec_{Thms(\mathcal{T}_1)} d'\} \subseteq AX(\mathcal{T}_1) \cup \{d_1, \ldots, d_{i-1}\}$$

holds. Let $D_i \stackrel{def}{=} D \cup \{d_1, \ldots, d_{i-1}\}$ such that, in particular, $D_1 = D$ holds. Since $AX(\mathcal{T}_1) \subseteq D$ holds by definition of theory morphisms, $\{d' \mid d_i \prec_{Thms(\mathcal{T}_1)} d'\} \subseteq D_i$ holds for any $1 \leq i \leq N$ as well.

Altogether, this means that the theorem translation step described above can be applied to a theory morphism $\tau(D_i) : \mathcal{T}_1 \longrightarrow \mathcal{T}_2^i$ and theorem d_i for any $1 \leq i \leq N$, and yields the possibly extended target theory $\mathcal{T}_2^{i+1}$ and the theory morphism $\tau(D_{i+1}) : \mathcal{T}_1 \longrightarrow \mathcal{T}_2^{i+1}$. So, setting $\mathcal{T}_2^1 \stackrel{def}{=} \mathcal{T}_2$ we have the theory morphism $\tau(D_1) : \mathcal{T}_1 \longrightarrow \mathcal{T}_2^1$, such that the theorem translation step can be iterated for $i = 1, \ldots, N$, and yields the theory morphism $\tau(D_{N+1}) : \mathcal{T}_1 \longrightarrow \mathcal{T}_2^{N+1}$. Now we have $d \in D_{N+1}$, or in other words the theorem d has been translated along the theory morphism $\tau(D)$.

The theorem translation provides the possibility to use complete theory morphisms for theoretical purposes as in the case of the definition of composition of theory morphisms in the next section. Though, in contrast to signature morphisms which are basically always normalised if possible, this kind of normalisation of theory morphisms is not very practical. In a concrete implementation, the approach 'translate theorems on-demand' seems to be more suitable than 'translate all theorems at once'.

4.2 Composition of theory morphisms

In this section composition of two theory morphisms will be defined as an extension of the composition of the underlying signature morphisms.

Let $\sigma^1 : \mathcal{T}_1 \longrightarrow \mathcal{T}_2$ and $\sigma^2 : \mathcal{T}_3 \longrightarrow \mathcal{T}_4$ be signature morphisms, with $\mathcal{T}_2 \in Anc(\mathcal{T}_3)$. Further, let $\tau^1 = \langle \sigma^1, D_1, \xi^1 \rangle$ and $\tau^2 = \langle \sigma^2, D_3, \xi^2 \rangle$ be theory morphisms extending, where $\mathcal{G}(\mathcal{T}_1) = \langle D_1, \mathit{thm\text{-}of}_{\mathcal{T}_1} \rangle$, $\mathcal{G}(\mathcal{T}_2) = \langle D_2, \mathit{thm\text{-}of}_{\mathcal{T}_2} \rangle$, $\mathcal{G}(\mathcal{T}_3) = \langle D_3, \mathit{thm\text{-}of}_{\mathcal{T}_3} \rangle$, $\mathcal{G}(\mathcal{T}_4) = \langle D_2, \mathit{thm\text{-}of}_{\mathcal{T}_2} \rangle$.

Then the *composition* of τ^1 and τ^2, denoted by $\tau^2 \circ \tau^1$, is the theory morphism $\tau : \mathcal{T}_1 \longrightarrow \mathcal{T}_4$, given by $\langle \sigma^2 \circ \sigma^1, D_1, \xi^2 \circ \xi^1 \rangle$. Firstly, we note that $\sigma^2 \circ \sigma^1$ and $\xi^2 \circ \xi^1$ are well-defined, since $\mathcal{T}_2 \in Anc(\mathcal{T}_3)$ and, hence $D_2 \subseteq D_3$. Secondly, we need to verify the theorem map conditions. For the proof of (4.1) let $d \in D_1$.

Then we can reason as follows:

$$
\begin{array}{ll}
\mathit{thm\text{-}of}_{\mathcal{T}_4}(\xi^2(\xi^1(d))) & \stackrel{\alpha}{=} \\
& \text{by (4.1) for } \tau^2 \\
\overline{\sigma}^2(\mathit{thm\text{-}of}_{\mathcal{T}_3}(\xi^1(d))) & = \\
& \text{since } \xi^1(d) \in D_2 \\
\overline{\sigma}^2(\mathit{thm\text{-}of}_{\mathcal{T}_2}(\xi^1(d))) & \stackrel{\alpha}{=} \\
& \text{by (4.1) for } \tau^1 \text{ and (3.12)} \\
\overline{\sigma}^2(\overline{\sigma}^1(\mathit{thm\text{-}of}_{\mathcal{T}_1}(d))) & = \\
& \text{by Lemma 3.19} \\
(\overline{\sigma^2 \circ \sigma^1})(\mathit{thm\text{-}of}_{\mathcal{T}_1}(d)). &
\end{array}
$$

The theorem map condition (4.2) holds obviously for the composition of theorem maps $\xi^2 \circ \xi^1$.

Similarly to the composition of signature morphisms, we finally prove, that the composition of theory morphisms preserves their extensions.

Lemma 4.2 *Let $\tau^1 : \mathcal{T}_1 \longrightarrow \mathcal{T}_2$ and $\tau^2 : \mathcal{T}_3 \longrightarrow \mathcal{T}_4$ be theory morphisms, with $\mathcal{T}_2 \in Anc(\mathcal{T}_3)$. Then for any proof term $\pi \in Proofs_{\varrho\vartheta}^{Adm}(\mathcal{T}_1)$*

$$
(\overline{\tau^2 \circ \tau^1})(\pi) = (\overline{\tau}^2 \circ \overline{\tau}^1)(\pi) \tag{4.6}
$$

holds.

Proof. (by structural induction on π)
(i) Let $\pi = X$ with $X \in \mathcal{X}_{proof}$. Then both *lhs* and *rhs* reduce to X.
(ii) Let $\pi = d_{\overline{[\alpha_i := \kappa_i]_n}}$ with $d \in \mathcal{G}(\mathcal{T}_1)$, *typevars*$(\mathit{thm\text{-}of}_{\mathcal{T}_1}(d)) \subseteq \{\alpha_1, \ldots, \alpha_n\}$ and $\kappa_1, \ldots, \kappa_n \in Types_{\vartheta}(\Sigma_{type}(\mathcal{T}_1))$. Since τ_1 and τ_2 are theory morphisms, we have firstly

$$
\overline{\sigma}^1(\mathit{thm\text{-}of}_{\mathcal{T}_1}(d)) \stackrel{\alpha}{=}_{\theta_1} \mathit{thm\text{-}of}_{\mathcal{T}_2}(\xi^1(d))
$$

and, secondly,

$$
\overline{\sigma}^2(\mathit{thm\text{-}of}_{\mathcal{T}_3}(\xi^1(d))) \stackrel{\alpha}{=}_{\theta_2} \mathit{thm\text{-}of}_{\mathcal{T}_4}(\xi^2(\xi^1(d)))
$$

Then

$$
\begin{array}{ll}
\overline{\sigma}^2(\overline{\sigma}^1(\mathit{thm\text{-}of}_{\mathcal{T}_1}(d))) & = \\
& \text{by (3.19)} \\
(\overline{\sigma^2 \circ \sigma^1})(\mathit{thm\text{-}of}_{\mathcal{T}_1}(d))) & \stackrel{\alpha}{=}_{\theta} \\
\mathit{thm\text{-}of}_{\mathcal{T}_4}(\xi^2(\xi^1(d))) &
\end{array}
$$

holds, where $\theta \stackrel{def}{=} \theta_2 \circ \theta_1$. Further, let

$$typevars(\overline{\sigma}^2(\overline{\sigma}^1(thm\text{-}of_{\mathcal{T}_1}(d)))) = \{\alpha_{i_1}, \ldots, \alpha_{i_m}\}$$

with $m \leq n$. Then we can reason

$$\begin{array}{ll}
(\overline{\tau^2 \circ \tau^1})(d_{\overline{[\alpha_i := \kappa_i]_n}}) & = \\
\xi^2(\xi^1(d))_{\overline{[\theta(\alpha_{i_k}) := (\overline{\sigma^{\sigma^2_{class}}_{type} \circ \sigma^{\sigma^1_{class}}_{type}})(\kappa_{i_k}))]_m}} & = \\
 & \text{by (3.15)} \\
\xi^2(\xi^1(d))_{\overline{[\theta(\alpha_{i_k}) := \overline{\sigma}^{\sigma^2_{class}}_{type}(\overline{\sigma}^{\sigma^1_{class}}_{type}(\kappa_{i_k}))]_m}} & = \\
\overline{\tau}^2(\overline{\tau}^1(d_{\overline{[\alpha_i := \kappa_i]_n}}))
\end{array}$$

(iii) Let $\pi = \overline{\lambda}(x ::_{type} \varrho_\vartheta(x)).\pi_1$ with $\pi_1 \in Proofs^{Adm}_{\varrho_\vartheta}(\mathcal{T}_1)$ and $x \in \mathcal{X}_{term}$. Then

$$\begin{array}{ll}
(\overline{\tau^2 \circ \tau^1})(\pi) & = \\
\overline{\lambda}(x ::_{type} \overline{\sigma}^{\sigma^2_{class}}_{type}(\overline{\sigma}^{\sigma^1_{class}}_{type}(\varrho_\vartheta(x)))).(\overline{\tau^2 \circ \tau^1})(\pi_1) & = \\
 & \text{by ind. hyp.} \\
\overline{\lambda}(x ::_{type} \overline{\sigma}^{\sigma^2_{class}}_{type}(\overline{\sigma}^{\sigma^1_{class}}_{type}(\varrho_\vartheta(x)))).\overline{\tau}^2(\overline{\tau}^1(\pi_1)) & = \\
\overline{\tau}^2(\overline{\lambda}(x ::_{type} \overline{\sigma}^{\sigma^1_{class}}_{type}(\varrho_\vartheta(x)))).\overline{\tau}^1(\pi_1) & = \\
\overline{\tau}^2(\overline{\tau}^1(\pi))
\end{array}$$

(iv) Let $\pi = \overline{\overline{\lambda}}(X ::_{prop} p).\pi_1$ with $X \in \mathcal{X}_{proof}$, $p \in Terms_{\varrho_\vartheta}(\Sigma_{op}(\mathcal{T}_1))$, $p \in Props(\mathcal{T}_1)$, and $\pi_1 \in Proofs^{Adm}_{\varrho_\vartheta}(\mathcal{T}_1)$. Then

$$\begin{array}{ll}
(\overline{\tau^2 \circ \tau^1})(\pi) & = \\
\overline{\overline{\lambda}}(X ::_{prop} (\overline{\sigma^2 \circ \sigma^1})(p)).(\overline{\tau^2 \circ \tau^1})(\pi_1) & = \\
 & \text{by ind. hyp. and (3.19)} \\
\overline{\overline{\lambda}}(X ::_{prop} \overline{\sigma}^2(\overline{\sigma}^1(p))).\overline{\tau}^2(\overline{\tau}^1(\pi_1)) & = \\
\overline{\tau}^2(\overline{\tau}^1(\pi))
\end{array}$$

(v) Let $\pi = \pi_1\ t$ with $t \in Terms_{\varrho_\vartheta}(\Sigma_{op}(\mathcal{T}_1))$ and $\pi_1 \in Proofs^{Adm}_{\varrho_\vartheta}(\mathcal{T}_1)$. Then

$$\begin{array}{ll}
(\overline{\tau^2 \circ \tau^1})(\pi) & = \\
(\overline{\tau^2 \circ \tau^1})(\pi_1)\ (\overline{\sigma^2 \circ \sigma^1})(t) & = \\
 & \text{by ind. hyp. and (3.19)} \\
\overline{\tau}^2(\overline{\tau}^1(\pi_1))\ \overline{\sigma}^2(\overline{\sigma}^1(t)) & = \\
\overline{\tau}^2(\overline{\tau}^1(\pi_1)\ \overline{\sigma}^1(t)) & = \\
\overline{\tau}^2(\overline{\tau}^1(\pi))
\end{array}$$

(vi) Let $\pi = \pi_1\,\pi_2$ with $\pi_1, \pi_2 \in \mathit{Proofs}^{Adm}_{\varrho\vartheta}(\mathcal{T}_1)$. Then

$$\begin{array}{ll}
(\overline{\tau^2 \circ \tau^1})(\pi) & = \\
(\overline{\tau^2 \circ \tau^1})(\pi_1)\,(\overline{\tau^2 \circ \tau^1})(\pi_2) & = \\
 & \text{by ind. hyp.} \\
\overline{\tau}^2(\overline{\tau}^1(\pi_1))\,\overline{\tau}^2(\overline{\tau}^1(\pi_2)) & = \\
\overline{\tau}^2(\overline{\tau}^1(\pi_1)\,\overline{\tau}^1(\pi_2)) & = \\
\overline{\tau}^2(\overline{\tau}^1(\pi)) &
\end{array}$$

□

4.3 Theory morphisms in Isabelle

As defined in Section 4.1, theory morphisms extend signature morphisms by a theorem mapping. So this section seamlessly continues Section 3.7, where the syntax and semantics of the construction of signature morphisms in Isabelle have been introduced.

One of the central properties of the theorem translation by theory morphisms is that this procedure does not depend on which subset of the global context of the source theory the theorem map of the given morphism is defined as long this subset contains the axioms of the source. This already allows to reduce the efforts in the construction of theory morphisms to the set of axioms. Furthermore, assignments for definitions, among these axioms, are provided by the underlying signature morphisms via definitional conditions. This, in turn, reduces the efforts in the construction to the remaining set of 'proper' axioms. On this note, for the construction of theory morphisms the name *axiom map* is more suitable than the name *theorem map*, and will be also used below.

Moreover, since we consider signature morphisms which are constructed using the strategy 'global elements are mapped by identity', the efforts in the construction of theory morphisms are finally reduced to the set of 'proper' axioms in the domain, because global axioms are considered to be mapped by any axiom map to themselves.

4.3.1 The thymorph command

Theory morphisms are constructed at the top-level of Isabelle via the integrated command **thymorph**. Two basic ways of construction are provided:

1. By a given signature morphism $\sigma^{\mathcal{I}} : \mathcal{T}_1 \longrightarrow \mathcal{T}_2$ and an axiom map declaration $\mathcal{D}_{axiom}$, using the following form:
   thymorph $\tau^{\mathcal{I}}$ **by_sigmorph** $\sigma^{\mathcal{I}}$
   axiom_map: $\mathcal{D}_{axiom}$

2. By composition of existing theory morphisms τ_1 and τ_2, where the target of τ_1 is an ancestor of the source of τ_2, as described in Section 4.2. At the top-level of Isabelle, this is represented by the *compositional* declaration:
   thymorph τ **by_composition** $\tau_2 \circ \tau_1$

Moreover, another possibility to construct a theory morphism is a derived one: by source and target theories, and declarations $\mathcal{D}_{class}$, $\mathcal{D}_{type}$, $\mathcal{D}_{op}$, $\mathcal{D}_{axiom}$ of class, type, operation, and axiom maps, respectively, using the following form:
thymorph $\tau^{\mathcal{I}}$: *Source* $\longrightarrow$ *Target*
class_map: $\mathcal{D}_{class}$
type_map: $\mathcal{D}_{type}$
op_map: $\mathcal{D}_{op}$
axiom_map: $\mathcal{D}_{axiom}$
which proceeds as follows. Firstly, the declaration of the underlying signature morphism is given by
sigmorph $\sigma^{\mathcal{I}}$: *Source* $\longrightarrow$ *Target*
class_map: $\mathcal{D}_{class}$
type_map: $\mathcal{D}_{type}$
op_map: $\mathcal{D}_{op}$
such that the declaration of $\tau^{\mathcal{I}}$ is then given by
thymorph $\tau^{\mathcal{I}}$ **by_sigmorph** $\sigma^{\mathcal{I}}$
axiom_map: $\mathcal{D}_{axiom}$

Axiom map. Let $\sigma^{\mathcal{I}} : \mathcal{T}_1 \longrightarrow \mathcal{T}_2$ be a given signature morphism, which should be extended to a theory morphism provided by a given axiom map declaration $\mathcal{D}_{axiom}$. Also, let $\mathcal{G}(\mathcal{T}_2) = \langle D_2, \mathit{thm\text{-}of}_{\mathcal{T}_2} \rangle$. Similarly to the signature morphism declaration, $\mathcal{D}_{axiom}$ consists of a list of assignments $(a \mapsto d)$, where $a \in AX(\mathcal{T}_1)$ and $d \in D_2$. Such $(a \mapsto d)$ is also called an *axiom assignment for a*. Furthermore, an axiom assignment $(a \mapsto d)$ is called *correct* if

$$\mathit{thm\text{-}of}_{\mathcal{T}_2}(d) \stackrel{\alpha}{=} \overline{\sigma}^{\mathcal{I}}(\mathit{prop\text{-}of}_{\mathcal{T}_1}(a)) \tag{4.7}$$

holds, which is almost exactly the theorem map condition (4.1). Moreover, let $Dom^{\sharp}_{axiom}(\mathcal{T}_1, \mathcal{T}_2)$ denote the set of domain axioms, which are not associated with a derived constant as its definition (i.e. 'proper' axioms in the domain of the underlying signature morphism), or, formally

$$Dom^{\sharp}_{axiom}(\mathcal{T}_1, \mathcal{T}_2) \stackrel{def}{=} Dom_{axiom}(\mathcal{T}_1, \mathcal{T}_2) \setminus Defs(\mathcal{T}_1)$$

So, $\mathcal{D}_{axiom}$ is called *correct*, if for any $a \in Dom^{\sharp}_{axiom}(\mathcal{T}_1, \mathcal{T}_2)$ it contains exactly one axiom assignment which is correct, i.e. satisfies (4.7).

From a correct axiom map declaration $\mathcal{D}_{axiom}$, a base theory morphism $\tau^{\mathcal{I}}$ is constructed by

$$\tau^{\mathcal{I}} \stackrel{def}{=} \langle \sigma^{\mathcal{I}}, AX(\mathcal{T}_1), \xi^{\mathcal{I}} \rangle$$

where the mapping $\xi^{\mathcal{I}} : AX(\mathcal{T}_1) \rightarrow D_2$ is defined below. Let $a \in AX(\mathcal{T}_1)$, then there are the following cases:

1. If $a \in Global_{axiom}(\mathcal{T}_1, \mathcal{T}_2)$ then there is a theory $\mathcal{T}$ with $\mathcal{T} \in Anc(\mathcal{T}_i)$ for $i = 1, 2$, such that $a \in AX(\mathcal{T})$. So, $a \in D_2$ holds as well, and we set $\xi^{\mathcal{I}}(a) \stackrel{def}{=} a$. In order to justify the correctness, the theorem map condition (4.1) has to be verified (the condition (4.2) holds obviously). This is done as follows:

$$\begin{array}{ll} \textit{thm-of}_{\mathcal{T}_2}(\xi^{\mathcal{I}}(a)) & = \\ \textit{thm-of}_{\mathcal{T}_2}(a) & = \\ \textit{prop-of}_{\mathcal{T}_2}(a) & = \\ & \text{since } a \in AX(\mathcal{T}) \\ \textit{prop-of}_{\mathcal{T}}(a) & = \\ & \text{since } \textit{prop-of}_{\mathcal{T}}(a) \in Terms(\Sigma_{op}(\mathcal{T})) \\ \overline{\sigma}^{\mathcal{I}}(\textit{prop-of}_{\mathcal{T}}(a)) & = \\ \overline{\sigma}^{\mathcal{I}}(\textit{prop-of}_{\mathcal{T}_1}(a)) & \end{array}$$

2. If $a \in Dom_{axiom}(\mathcal{T}_1, \mathcal{T}_2)$ and $a \in Defs(\mathcal{T}_1)$ then there is a constant $f \in \Sigma^{der}_{op}(\mathcal{T}_1)$ with $\textit{prop-of}_{\mathcal{T}_1}(a) = f \equiv \textit{def-of}_{\mathcal{T}_1}(f)$. Let $f' \stackrel{def}{=} \sigma^{\mathcal{I}}_{op}(f)$. So, by the definition of signature morphisms $f' \in \Sigma^{der}_{op}(\mathcal{T}_2)$ holds as well, and hence there is $f'_{def} \in AX(\mathcal{T}_2)$ which defines f'. Altogether, we set: $\xi^{\mathcal{I}}(a) \stackrel{def}{=} f'_{def}$. This assignment satisfies the theorem map conditions which is ensured by the definitional conditions satisfied by $\sigma^{\mathcal{I}}$, as shown at the beginning of Section 4.1.

3. If $a \in Dom_{axiom}(\mathcal{T}_1, \mathcal{T}_2)$ and $a \notin Defs(\mathcal{T}_1)$, i.e. $a \in Dom^{\sharp}_{axiom}(\mathcal{T}_1, \mathcal{T}_2)$, then $\mathcal{D}_{axiom}$ contains an assignment $(a \mapsto d)$, and we define $\xi^{\mathcal{I}}(a) \stackrel{def}{=} d$, such that the theorem map conditions are satisfied by the correctness of $\mathcal{D}_{axiom}$, i.e. due to (4.7).

Furthermore, the matching technique, described at the end of Section 4.1 on page 107, forms the core of a procedure for completion of a given axiom map declaration $\mathcal{D}_{axiom}$, searching for correct assignments for domain axioms.

4.3.2 Example: sets as monads

Using the example declaration *pset-as-monad* from Section 3.7.5 on page 102, representing the signature morphism $\sigma_{\mathsf{PSet\_Monad}} : \mathsf{Monad} \longrightarrow \mathsf{PSet\_Monad}$, its extension to the theory morphism

$$\tau_{\mathsf{PSet\_Monad}}(AX(\mathsf{Monad})) = \langle \sigma_{\mathsf{PSet\_Monad}}, AX(\mathsf{Monad}), \xi \rangle$$

from Section 4.1.1 can be now represented in Isabelle by
thymorph *powerset-as-monad* **by_sigmorph** *pset-as-monad*
axiom_map: [F1 $\mapsto$ $\mathrm{F1}^{\mathsf{PSet}}$,
F2 $\mapsto$ $\mathrm{F2}^{\mathsf{PSet}}$,
N1 $\mapsto$ $\mathrm{N1}^{\mathsf{PSet}}$,
...]

Here, the labels $\mathrm{F1}^{\mathsf{PSet}}, \mathrm{F2}^{\mathsf{PSet}}, \ldots$ refer to those theorems in the context of *PSet_Monad* which comprise the images of the domain axioms F1, F2, ... (see Section 3.2.2) under the homomorphic extension of *pset-as-monad*, as already mentioned in Section 4.1.1. So, e.g. $\textit{thm-of}_{\mathsf{PSet\_Monad}}(\mathrm{F1}^{\mathsf{PSet}})$ gives the derived proposition $T^{\mathsf{PSet}}\ id = id$, and $\textit{thm-of}_{\mathsf{PSet\_Monad}}(\mathrm{F1}^{\mathsf{PSet}})$ – the derived proposition $\bigwedge f\ g.\ T^{\mathsf{PSet}}(f \circ g) = T^{\mathsf{PSet}}\ f \circ T^{\mathsf{PSet}}\ g$.

Taking also into account the matching of domain axioms, the declaration
thymorph *powerset-as-monad* **by_sigmorph** *pset-as-monad*
represents in this case the same theory morphism as the declaration above, since the domain axioms will be automatically matched by $\mathrm{F1}^{\mathsf{PSet}}, \mathrm{F2}^{\mathsf{PSet}}, \ldots$, derived in *PSet_Monad*. Generally, if some explicit axiom assignments are omitted then the automatically found assignments may also differ from 'intended', due to the possible ambiguity mentioned in Section 4.1.

4.3.3 The translate_thm command

The theorem translation procedure, described in Section 4.1.3, is represented at the top-level of Isabelle by

translate_thm d **as** $\overline{d}$ **along** τ
renames: $\mathcal{D}_{renames}$
where

- $\tau(D) : \mathcal{T}_1 \longrightarrow \mathcal{T}_2$ is a theory morphism,
- if $\mathcal{G}(\mathcal{T}_1) = \langle D_1, \mathit{thm\text{-}of}_{\mathcal{T}_1}\rangle$ and $\mathcal{G}(\mathcal{T}_2) = \langle D_2, \mathit{thm\text{-}of}_{\mathcal{T}_2}\rangle$ then $d \in D_1$ denotes the theorem to be translated, while $\overline{d} \notin D_2$ is a new label which is assigned to the translated proposition $\mathit{thm\text{-}of}_{\mathcal{T}_1}(d)$,
- $\mathcal{D}_{renames}$ is a list of assignments $(d_1 \mapsto d_2)$ which can be used to give an explicit new label d_2 to any theorem $d_1 \in D_1$ with $d \prec_{Thms(\mathcal{T}_1)} d_1$ and $\xi_\tau(d_1)$ undefined, such that d_1 has to be translated before d.

4.4 Inclusions, identities and categories

The presented way of construction of morphisms in Isabelle in the manner 'global elements are mapped by identity' has already hinted at this, but now we specify it exactly.

Definition 21 (inclusion)
Let $\mathcal{T}_1, \mathcal{T}_2 \in \mathcal{Th}$. A theory morphism $\iota : \mathcal{T}_1 \longrightarrow \mathcal{T}_2$ is called an *inclusion*, if

1. the class map $\iota_{class} : \Sigma_{class}(\mathcal{T}_1) \to \Sigma_{class}(\mathcal{T}_2)$ is an identity map;
2. the type map $\iota_{type} : \Sigma_{type}(\mathcal{T}_1) \to \widehat{Types}(\Sigma_{type}(\mathcal{T}_2))$ satisfies $\iota_{type}(C) = (\text{'}1, \ldots, \text{'}n) \cdot C$ for any $C \in \Sigma_{type}(\mathcal{T}_1)$ with $rank_{\mathcal{T}_1}(C) = n$;
3. the operation map $\iota_{op} : \Sigma_{op}(\mathcal{T}_1) \to \Sigma_{op}(\mathcal{T}_2)$ is an identity map;
4. the theorem map $\iota_{thm} : D_1 \to D_2$ is an identity map, where $\mathcal{G}(\mathcal{T}_1) = \langle D_1, \mathit{thm\text{-}of}_{\mathcal{T}_1}\rangle$ and $\mathcal{G}(\mathcal{T}_2) = \langle D_2, \mathit{thm\text{-}of}_{\mathcal{T}_2}\rangle$.

The notation $\iota : \mathcal{T}_1 \hookrightarrow \mathcal{T}_2$ will be used for inclusions as well.

By the definition, there can be at most one inclusion $\mathcal{T} \hookrightarrow \mathcal{T}'$ for any two theories $\mathcal{T}, \mathcal{T}'$. On the other hand, the morphism construction in Isabelle shows that global elements can be mapped exactly as the definition above requires. Furthermore, the basic observation is that in the special case, when the source theory is an ancestor of the target theory, *all* elements in the source are global, or in other words the particular class, type, operation, and axiom domains are empty. This can be summarised by the following general causality.

Fact 4.4.0.1 *Any two theories $\mathcal{T}$ and $\mathcal{T}'$ satisfying $\mathcal{T} \in Anc(\mathcal{T}')$ determine exactly one inclusion $\iota : \mathcal{T} \hookrightarrow \mathcal{T}'$.*

The converse, however, is in general not true: an inclusion $\iota : \mathcal{T} \hookrightarrow \mathcal{T}'$ does not always imply $\mathcal{T} \in Anc(\mathcal{T}')$, because $\mathcal{T}'$ might be almost the same theory as $\mathcal{T}$ apart of, e.g. some new axioms such that $Ax(\mathcal{T}) \subset Ax(\mathcal{T}')$.

Let $\iota : \mathcal{T}_1 \hookrightarrow \mathcal{T}_2$ be an inclusion and $\tau : \mathcal{T}_3 \longrightarrow \mathcal{T}_4$ a theory morphism with $\mathcal{T}_2 \in Anc(\mathcal{T}_3)$. Then the theory morphism $(\tau \circ \iota) : \mathcal{T}_1 \longrightarrow \mathcal{T}_4$ is a *restriction* of τ to $\mathcal{T}_1$. Further, if $\tau : \mathcal{T}_3 \longrightarrow \mathcal{T}_4$ a theory morphism with $\mathcal{T}_4 \in Anc(\mathcal{T}_1)$ then the theory morphism $(\iota \circ \tau) : \mathcal{T}_3 \longrightarrow \mathcal{T}_2$ is an *extension* of τ to $\mathcal{T}_2$.

In turn, the special kind of inclusions, which source and target theories are the same, are as usually called *identities*. So, any theory $\mathcal{T}$ has its identity $\iota_{\mathcal{T}} : \mathcal{T} \hookrightarrow \mathcal{T}$, such that for any two theory morphisms $\tau_1 : \mathcal{T}_1 \longrightarrow \mathcal{T}$ and $\tau_2 : \mathcal{T} \longrightarrow \mathcal{T}_2$, the left and right unit laws

1. $\iota_{\mathcal{T}} \circ \tau_1 = \tau_1$, and
2. $\tau_2 \circ \iota_{\mathcal{T}} = \tau_2$

hold. Taking also into account associativity of the composition, the set of theories $\mathcal{T}h$ as objects together with theory morphisms as arrows forms the category **Th**. Furthermore, there are the following functors into **Set**:

1. F_{sort} : **Th** $\longrightarrow$ **Set**, assigning to any theory $\mathcal{T}$ the set $Sorts(\Sigma_{class}(\mathcal{T}))$ and to any theory morphism $\tau : \mathcal{T}_1 \longrightarrow \mathcal{T}_2$ the mapping $\overline{\sigma}_{class}$, where σ_{class} is the class map of the underlying signature morphism;
2. F_{type} : **Th** $\longrightarrow$ **Set**, assigning to any theory $\mathcal{T}$ the set $Types(\Sigma_{type}(\mathcal{T}))$ and to any theory morphism $\tau : \mathcal{T}_1 \longrightarrow \mathcal{T}_2$ the mapping $\overline{\sigma}_{type}^{\sigma_{class}}$, where σ_{type} is the type map of the underlying signature morphism;
3. F_{term} : **Th** $\longrightarrow$ **Set**, assigning to any theory $\mathcal{T}$ the set $Terms(\Sigma_{op}(\mathcal{T}))^\sharp$ of well-typed terms (i.e. from which a type via *of-type*$_{\mathcal{T}}$ is derivable), and to any theory morphism $\tau : \mathcal{T}_1 \longrightarrow \mathcal{T}_2$ the homomorphic extension $\overline{\sigma}$ of the underlying signature morphism σ;
4. F_{prop} : **Th** $\longrightarrow$ **Set**, assigning to any theory $\mathcal{T}$ the set of propositions $Props(\mathcal{T})$ and to any theory morphism $\tau : \mathcal{T}_1 \longrightarrow \mathcal{T}_2$ the homomorphic extension $\overline{\sigma}$ of the underlying signature morphism σ;
5. F_{proof} : **Th** $\longrightarrow$ **Set**, assigning to any theory $\mathcal{T}$ the set $Proofs^{Adm}(\mathcal{T})^\sharp$ of admissible proof terms (i.e. from which a proposition in $Props(\mathcal{T})$ via

proof-of$_{\mathcal{T}}$ is derivable), and to any theory morphism $\tau : \mathcal{T}_1 \longrightarrow \mathcal{T}_2$ its extension $\overline{\tau}$.

Moreover, Lemma 3.6 and Lemma 4.1 establish the natural transformations

$$\begin{array}{ll} \textit{of-type} & : F_{term} \stackrel{\cdot}{\rightarrow} F_{type} \\ \textit{proof-of} & : F_{proof} \stackrel{\cdot}{\rightarrow} F_{prop} \end{array}$$

whereas F_{prop} and F_{term} are connected by inclusion. This basically abstracts most of the so far presented properties of theories and morphisms.

Finally, isomorphic theories can be now defined in the standard way: $\mathcal{T}_1 \stackrel{\sim}{=} \mathcal{T}_2$ holds iff there are theory morphisms $\tau_1 : \mathcal{T}_1 \longrightarrow \mathcal{T}_2$ and $\tau_2 : \mathcal{T}_2 \longrightarrow \mathcal{T}_1$, such that $\tau_2 \circ \tau_1 = \iota_{\mathcal{T}_1}$ and $\tau_1 \circ \tau_2 = \iota_{\mathcal{T}_2}$ holds.

4.5 Parameterised theories and instantiation

At this point, the framework and morphisms are sufficiently described, and we can turn towards instantiation of theories – the basic concept providing re-use of abstract developments and transformation *in-the-large*. In this context, instantiation can be also understood as a sort of application of theories to theories, in some sense similar to application of an ML-functor to an ML-structure. More precisely, in order to apply a theory it should be parameterised in some sense. So, a theory $\mathcal{B}$ is called *parameterised* by a theory $\mathcal{P}$ if an inclusion $\iota : \mathcal{P} \hookrightarrow \mathcal{B}$ exists. This fact is then denoted by $\langle \mathcal{P}, \mathcal{B} \rangle$, and $\mathcal{P}$ is called the *parameter* part, $\mathcal{B}$ the *body* part of the parametrisation $\langle \mathcal{P}, \mathcal{B} \rangle$. According to this definition, parametrisation is just another name for inclusion, but nevertheless allows a special view on the structure of theories.

A theory $\mathcal{B}$ is applicable to a theory $\mathcal{I}$ provided by a parametrisation $\langle \mathcal{P}, \mathcal{B} \rangle$ and a theory morphism $\tau : \mathcal{P} \longrightarrow \mathcal{I}$. The theory $\mathcal{I}$ is also called the *instantiating* theory. The result of such an application is a theory $\mathcal{T}_{inst}$ extending $\mathcal{I}$, called *instantiation*, and a theory morphism $\tau' : \mathcal{B} \longrightarrow \mathcal{T}_{inst}$ extending τ. The instantiation procedure, formulated in the next section, gives an exact description of how $\mathcal{T}_{inst}$ and τ' can be constructed.

Parametrisations allow us to encode general design tactics and transformation rules using the following scheme: the parameter part (axiomatically) specifies some structures, while the body part contains deductions of properties, constructions of new instances, etc. Using the application, all the deductions in the body part are translated into the usually more concrete context of the instantiating theory. This approach has been already sketched by the stack-example in the introduction, where the parametrisation $\langle$*QuotientType_Parameters*, *QuotientType*$\rangle$ (Figure 1.2) has been involved. So, *QuotientType_Parameters* has been used to specify a type parameter with an equivalence relation on it, whereas *QuotientType* has been used to introduce the corresponding quotient type together with some derived constants and to deduce their properties. This parametrisation has been applied to *StackImpl* using the theory morphism *parameters*.

Furthermore, a parametrisation $\langle \mathcal{P}, \mathcal{B} \rangle$ is called *well-formed* if $AX(\mathcal{B}) \setminus AX(\mathcal{P}) \subseteq \mathit{Defs}(\mathcal{B})$, i.e. if any axiom, introduced in some $\mathcal{T} \in Anc(\mathcal{B})$ with $\mathcal{T} \notin Anc(\mathcal{P})$, forms a conservative extension. If $\langle \mathcal{P}, \mathcal{B} \rangle$ is well-formed then an instantiation $\mathcal{T}_{inst}$ by a theory morphism $\tau : \mathcal{P} \longrightarrow \mathcal{I}$, if such $\mathcal{T}_{inst}$ exists, will be a conservative extension of the instantiating theory $\mathcal{I}$. This is one of the central properties of the instantiation: referring again to Figure 1.1 from the introduction, where the small triangle inside the axiomatic development can be now seen as the body part of a well-formed parametrisation, this property justifies that the instantiation, obtained by an application to a theory from the conservative development sub-tree, seamlessly extends this development. So, e.g. the parametrisation $\langle$*QuotientType_Parameters*, *QuotientType*$\rangle$ is well-formed, because the axioms introduced in *QuotientType* comprise only definitions and a single type definition axiom for the quotient type, which is a conservative extension as already mentioned in the introduction. Hence, the application of $\langle$*QuotientType_Parameters*, *QuotientType*$\rangle$ to the theory *StackImpl* yields the theory *StackImpl'*, which extends *StackImpl* in a conservative way.

4.5.1 Instantiation

The application of a parameterised theory is defined by means of instantiation, defined below.

Definition 22 (instantiation)
Let $\langle \mathcal{P}, \mathcal{B} \rangle$ be a parametrisation and $\tau : \mathcal{P} \longrightarrow \mathcal{I}$ a theory morphism. The *instantiation of $\mathcal{B}$ by τ* is the (unique up to isomorphism) pushout $(\mathcal{T}_{inst}, \tau_1 : \mathcal{B} \longrightarrow \mathcal{T}_{inst}, \tau_2 : \mathcal{I} \longrightarrow \mathcal{T}_{inst})$ of the span $\mathcal{B} \overset{\iota}{\hookleftarrow} \mathcal{P} \overset{\tau}{\longrightarrow} \mathcal{I}$, which is also visualised by the diagram (4.8).

Computing instantiation

Procedure *instantiation*
Input:

1. a parametrisation $\langle \mathcal{P}, \mathcal{B} \rangle$;

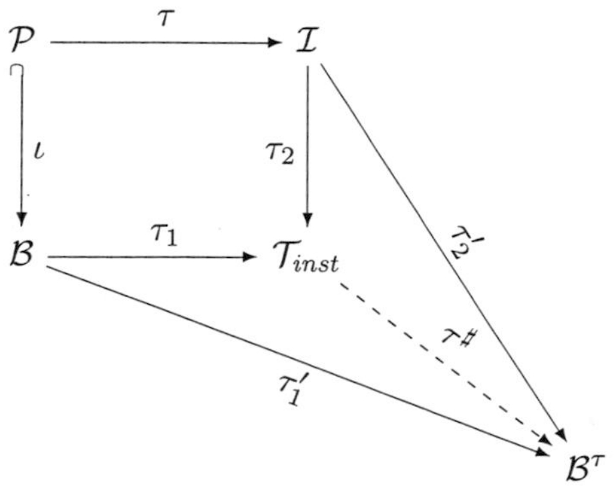

(4.8)

2. a theory morphism $\tau = \langle \sigma, D, \xi \rangle$ with the source in $\mathcal{P}$, the target in $\mathcal{I}$, the class map σ_{class}, the type map σ_{type} and the operation map σ_{op};

Output: either a theory $\mathcal{T}_{inst}$ extending $\mathcal{I}$, together with a theory morphism $\tau' : \mathcal{B} \longrightarrow \mathcal{T}_{inst}$, or the output: `no instantiation`;
Description: the instantiation procedure of $\mathcal{B}$ by τ proceeds by the following steps, considering type classes, type constructors, logical constants, definitions, axioms, and theorems.

1. Let $\mathcal{T}_0$ be the theory with $Par(\mathcal{T}_0) = \{\mathcal{I}\}$, $\mathcal{G}(\mathcal{T}_0) = \mathcal{G}(\mathcal{I})$ and empty sets of type classes, type constructors, constants and axioms;

2. Let $S \stackrel{def}{=} \Sigma_{class}(\mathcal{B}) \setminus \Sigma_{class}(\mathcal{P})$ and $c_1, \ldots, c_{|S|}$ be a sequence of the type classes in S such that $Super_{\Sigma_{class}(\mathcal{B})}(c_i) \subseteq \{c_1, \ldots, c_i\} \cup \Sigma_{class}(\mathcal{P})$, for all $1 \leq i \leq N$. Starting with $\mathcal{T}_0^1 \stackrel{def}{=} \mathcal{T}_0$ and $\varphi^1 \stackrel{def}{=} \sigma_{class}$, where σ_{class} is the class map of σ, we iterate for all $i = 1, \ldots, |S|$ the steps

$$\mathcal{T}_0^{i+1} \stackrel{def}{=} \mathcal{T}_0^i +_{class} (c_i', \langle \varphi^i(c_{i,1}), \ldots, \varphi^i(c_{i,n}) \rangle)$$

and

$$\varphi^{i+1} \stackrel{def}{=} \varphi^i[c_i \mapsto c_i']$$

where $c'_i \notin \Sigma_{class}(\mathcal{T}_0^i)$, $\{c_{i,1}, \ldots, c_{i,n}\} = Super_{\Sigma_{class}(\mathcal{B})}(c_i)$. Any assignment $(c_i \mapsto c'_i)$ satisfies the class map condition, as stated in Definition 9, since

$$\overline{\varphi}^i(Super_{\Sigma_{class}(\mathcal{B})}(c_i)) = Super_{\Sigma_{class}(\mathcal{T}_0^{i+1})}(c'_i)$$

holds. So, we obtain $\Sigma_{class}(\mathcal{B}) \overset{\sigma'_{class}}{\longmapsto} \Sigma_{class}(\mathcal{T}_1)$, where $\mathcal{T}_1 \overset{def}{=} \mathcal{T}_0^{|S|+1}$ and $\sigma'_{class} \overset{def}{=} \varphi^{|S|+1}$.

3. Let $S \overset{def}{=} \Sigma_{type}(\mathcal{B}) \setminus \Sigma_{type}(\mathcal{P})$ and $C_1, \ldots, C_{|S|}$ be a sequence of the type constructors in S. Starting with $\mathcal{T}_1^1 \overset{def}{=} \mathcal{T}_1$ and $\phi^1 \overset{def}{=} \sigma_{type}$, where σ_{type} is the type map of σ, we iterate for all $i = 1, \ldots, |S|$ the steps

$$\begin{aligned}\mathcal{T}_1^{i+1} \overset{def}{=} (\mathcal{T}_1^i \quad & +_{tcons}\ (C'_i,\ n)) \\ & +_{arities}\ (C'_i, \langle(\overline{\sigma}'_{class}(S_{1,1}), \ldots, \overline{\sigma}'_{class}(S_{1,n+1})), \\ & \qquad \ldots, \\ & \qquad (\overline{\sigma}'_{class}(S_{a_i,1}), \ldots, \overline{\sigma}'_{class}(S_{a_i,n+1}))\rangle)\end{aligned}$$

and

$$\phi^{i+1} \overset{def}{=} \phi^i[C_i \mapsto (\text{`}1, \ldots, \text{`}n) \cdot C'_i]$$

where $n = rank_{\mathcal{B}}(C)$, $C'_i \notin \Sigma_{type}(\mathcal{T}_1^i)$,

$$(S_{1,1}, \ldots, S_{1,n+1}), \ldots, (S_{a_i,1}, \ldots, S_{a_i,n+1})$$

is a sequence of arities of C_i in $\mathcal{B}$, and $\overline{\sigma}'_{class}$ denotes the extension of σ'_{class} defined in the previous step.

Provided by this construction, any assignment $(C_i \mapsto (\text{`}1, \ldots, \text{`}n) \cdot C'_i)$ satisfies the type map conditions as stated in Definition 13. So, we obtain $\mathcal{T}_2 \overset{def}{=} \mathcal{T}_1^{|S|+1}$ and $\sigma'_{type} \overset{def}{=} \phi^{|S|+1}$, such that $\Sigma_{type}(\mathcal{B}) \overset{\sigma'_{type}}{\longmapsto}_{\sigma'_{class}} \Sigma_{type}(\mathcal{T}_2)$ holds.

4. Let $S \overset{def}{=} \Sigma_{op}^{log}(\mathcal{B}) \setminus \Sigma_{op}^{log}(\mathcal{P})$ and $f_1, \ldots, f_{|S|}$ be a sequence of the logical constants in S. Starting with $\mathcal{T}_2^1 \overset{def}{=} \mathcal{T}_1$ and $\psi^1 \overset{def}{=} \sigma_{op}$, where σ_{op} is the operation map of σ, we iterate for all $i = 1, \ldots, |S|$ the steps

$$\mathcal{T}_2^{i+1} \overset{def}{=} \mathcal{T}_2^i +_{op} (f'_i, \overline{\sigma}'^{\sigma'_{class}}_{type}(\textit{type-of}_{\mathcal{B}}(f_i)))$$

and

$$\psi^{i+1} \overset{def}{=} \psi^i[f_i \mapsto f'_i]$$

where $f'_i \notin \Sigma_{op}(\mathcal{T}_2^i)$ and $\overline{\sigma}^{\sigma'_{class}}_{type}$ denotes the extension of the type map σ'_{type} defined in the previous step. Provided by this construction, any assignment $(f_i \mapsto f'_i)$ satisfies the operation map condition, as stated in Definition 15. Furthermore, *lhs* and *rhs* in (3.9) are even equal. So, we obtain $\mathcal{T}_3 \stackrel{def}{=} \mathcal{T}_2^{|S|+1}$ and $\sigma'_{op} \stackrel{def}{=} \psi^{|S|+1}$, such that

$$(\Sigma_{op}(\mathcal{P}) \cup \Sigma_{op}^{log}(\mathcal{B})) \stackrel{\sigma'_{op}}{\longmapsto}_{\sigma'_{class}, \sigma'_{type}} \Sigma_{op}(\mathcal{T}_3)$$

holds.

5. The subtle problem regarding definitions and derived constants is that there can be some constant $f \in \Sigma_{op}^{log}(\mathcal{P})$ with $f \in \Sigma_{op}^{der}(\mathcal{B})$, i.e. a logical constant in parameters for which a defining axiom is introduced in the body of the parametrisation. This is the reason which forbids us to derive a signature morphism $\sigma'(\Sigma_{op}(\mathcal{P}) \cup \Sigma_{op}^{log}(\mathcal{B}))$ from σ'_{class}, σ'_{type}, and σ'_{op}. Even though this situation is quite unusual in practice, it has to be treated here. To this end, we will directly consider the definitions in $Defs(\mathcal{B}) \setminus Defs(\mathcal{P})$ in contrast to the normalisation procedure (Section 3.5), where we have accessed definitions indirectly via the derived constants in the signature.

   So, let $S \stackrel{def}{=} Defs(\mathcal{B}) \setminus Defs(\mathcal{P})$ and $d_1, \ldots, d_{|S|}$ be a sequence of definitions in S, ordered by $\prec^{Defs(\mathcal{B})}_{AX(\mathcal{B})}$ such that $d_i \prec^{Defs(\mathcal{B})}_{AX(\mathcal{B})} d_j$ implies $j < i$ for any $1 \leq i, j \leq |S|$, which relies on the global assumption that $\prec^{Defs(\mathcal{B})}_{AX(\mathcal{B})}$ is acyclic.

   Let $prop\text{-}of_{\mathcal{B}}(d_1) = f_1 \equiv t_1$. Provided by the order on d_i,

$$t_1 \in Terms(\Sigma_{op}(\mathcal{P}) \cup \Sigma_{op}^{log}(\mathcal{B}))$$

   holds. Further, let $\mathcal{T}_3^1 \stackrel{def}{=} \mathcal{T}_3$, $\psi_1 \stackrel{def}{=} \sigma'_{op}$, and $prop\text{-}of_{\mathcal{B}}(d_i) = f_i \equiv t_i$ with $1 \leq i \leq |S|$. Moreover, we assume $t_i \in Terms(\Sigma_{op}(\mathcal{P}) \cup \Sigma_{op}^{log}(\mathcal{B}) \cup \{f_1, \ldots, f_{i-1}\})$ and

$$(\Sigma_{op}(\mathcal{P}) \cup \Sigma_{op}^{log}(\mathcal{B}) \cup \{f_1, \ldots, f_{i-1}\})) \stackrel{\psi_i}{\longmapsto}_{\sigma'_{class}, \sigma'_{type}} \Sigma_{op}(\mathcal{T}_3^i)$$

   such that

$$t'_i \stackrel{def}{=} \overline{\psi}_i^{\sigma'_{class}, \sigma'_{type}}(t_i)$$

   is a well-defined term with $t'_i \in Terms(\Sigma_{op}(\mathcal{T}_3^i))$.

Since t_i *of-type*$_{\mathcal{B}}$ *type-of*$_{\mathcal{B}}(f_i)$ holds,

$$t'_i \ \textit{of-type}_{\mathcal{T}_3^i} \ \overline{\sigma}_{type}^{\sigma'_{class}}(\textit{type-of}_{\mathcal{B}}(f_i)) \tag{4.9}$$

holds as well, by Lemma 3.6. Similarly to the normalisation we have to check the type variable condition:

$$\textit{typevars}(t'_i) \subseteq \textit{typevars}(\overline{\sigma}_{type}^{\sigma'_{class}}(\textit{type-of}_{\mathcal{B}}(f_i))) \tag{4.10}$$

If it does not hold, return the output `no instantiation`. Otherwise, we consider the cases on the in $\mathcal{B}$ derived constant f_i, which might be logical in $\mathcal{P}$:

(a) if $f_i \in \Sigma_{op}^{log}(\mathcal{P})$ then there are the following cases for $\sigma_{op}(f_i) \in \Sigma_{op}(\mathcal{I})$:

(I) $\sigma_{op}(f_i) \in \Sigma_{op}^{log}(\mathcal{T}_3^i)$; if $\sigma_{op}(f_i)$ is a final constant, return the output `no instantiation`; otherwise, since $\sigma_{op}(f_i) \equiv t'_i$ forms a pre-definition in $\mathcal{T}_3^i$, we define

$$\mathcal{T}_3^{i+1} \stackrel{def}{=} \mathcal{T}_3^i +_{def} (d'_i, \sigma_{op}(f_i) \equiv t'_i)$$

and $\psi_{i+1} \stackrel{def}{=} \psi_i$. Note, that in contrast to normalisation, in this situation the pre-definition $\sigma_{op}(f) \equiv t'_i$ could introduce cycles., such that $\mathcal{T}_3^{i+1}$ is only well-defined if

$$\prec^{Defs(\mathcal{T}_3^i) \cup \{d'_i\}}_{AX(\mathcal{T}_3^i +_{axiom} (d'_i, \sigma_{op}(f_i) \equiv t'_i))}$$

is acyclic. Otherwise, the output `no instantiation` is produced;

(II) $\sigma_{op}(f_i) \in \Sigma_{op}^{der}(\mathcal{T}_3^i)$, i.e. there has to be some $j < i$ with $\sigma_{op}(f_j) = \sigma_{op}(f_i)$ such that the definition of f_j has been already translated to $\mathcal{T}_3^j$; so we check whether $t'_i \stackrel{\alpha}{=} \textit{def-of}_{\mathcal{T}_3^i}(\sigma_{op}(f_i))$ holds; if yes then define $\mathcal{T}_3^{i+1} \stackrel{def}{=} \mathcal{T}_3^i$ and $\psi_{i+1} \stackrel{def}{=} \psi_i$, otherwise return the output `no instantiation`;

(b) if $f_i \in \Sigma_{op}^{der}(\mathcal{P})$ then $\sigma_{op}(f_i) \in \Sigma_{op}^{der}(\mathcal{I})$ holds by definitional conditions for signature morphisms, and as above we check whether $t'_i \stackrel{\alpha}{=} \textit{def-of}_{\mathcal{I}}(\sigma_{op}(f_i))$ holds; if yes then define $\mathcal{T}_3^{i+1} \stackrel{def}{=} \mathcal{T}_3^i$ and $\psi_{i+1} \stackrel{def}{=} \psi_i$, otherwise return the output `no instantiation`;

So, it remains the case $f_i \in \Sigma_{op}^{der}(\mathcal{B}) \setminus \Sigma_{op}(\mathcal{P})$. Now we can proceed as with normalisation. We define

$$\mathcal{T}_3^{i+1} \stackrel{def}{=} (\mathcal{T}_3^i +_{op} (f_i', \overline{\sigma}_{type}^{\sigma_{class}}(type\text{-}of_{\mathcal{B}}(f_i)))) +_{def} (d_i', f_i' \equiv t_i')$$

which is well-defined, provided f_i' is a fresh operation symbol, i.e. $f_i' \notin \Sigma_{op}(\mathcal{T}_3)$ (which is assumed here). Furthermore, we update the operation map ψ_i by

$$\psi_{i+1} \stackrel{def}{=} \psi_i[f_i \mapsto f_i']$$

Altogether, let $\mathcal{T}_4 \stackrel{def}{=} \mathcal{T}_3^{|S|+1}$ and $\sigma_{op}'' \stackrel{def}{=} \psi_{|S|+1}$, if $\mathcal{T}_3^{|S|+1}$ and $\psi_{|S|+1}$ exist. Firstly we have

$$\Sigma_{op}(\mathcal{B}) \xmapsto{\sigma_{op}''}_{\sigma_{class}', \sigma_{type}'} \Sigma_{op}(\mathcal{T}_4)$$

Secondly, we can also conclude, that now the definitional conditions for signature morphisms (Definition 16) are satisfied by σ_{op}''. That is, if $\mathcal{T}_4$ and σ_{op}'' are well-defined then σ_{class}', σ_{type}' and σ_{op}'' give us the complete signature morphism $\sigma' : \mathcal{B} \longrightarrow \mathcal{T}_4$. Otherwise, i.e. if $\mathcal{T}_4$ and σ_{op}'' are undefined, it follows from the construction above, that no complete signature morphism $\sigma' : \mathcal{B} \longrightarrow \mathcal{T}_4$ extending σ exists.

6. Let $S \stackrel{def}{=} AX(\mathcal{B}) \setminus (AX(\mathcal{P}) \cup Defs(\mathcal{B}))$ and $a_1, \ldots, a_{|S|}$ be a sequence of the 'proper' axioms in S. Starting with $\mathcal{T}_4^1 \stackrel{def}{=} \mathcal{T}_4$ and $\xi^1 \stackrel{def}{=} \xi$, we iterate for all $i = 1, \ldots, |S|$ the steps

$$\mathcal{T}_4^{i+1} \stackrel{def}{=} \mathcal{T}_4^i +_{axiom} (a_i', \overline{\sigma}'(prop\text{-}of_{\mathcal{B}}(a_i)))$$

and

$$\xi^{i+1} \stackrel{def}{=} \xi^i[a_i \mapsto a_i']$$

where $\mathcal{G}(\mathcal{T}_4^i) = \langle D_i, thm\text{-}of_{\mathcal{T}_4^i} \rangle$, $a_i' \notin D_i$. Provided by this construction, any assignment $(a_i \mapsto a_i')$ satisfies the theorem map conditions, as stated in Definition 20. Defining $\mathcal{T}_5 \stackrel{def}{=} \mathcal{T}_4^{|S|+1}$ and $\xi' \stackrel{def}{=} \xi^{|S|+1}$, we obtain the theory morphism $\tau'(AX(\mathcal{B})) : \mathcal{B} \longrightarrow \mathcal{T}_5$ with the underlying signature morphism σ' and the theorem map $\overline{\xi}'$, extending ξ' to $Defs(\mathcal{B})$ in accordance with the translated derived constants in the previous step.

Finally, $\mathcal{T}_5$ is extended to $\mathcal{T}_{inst}$ and $\tau'(AX(\mathcal{B}))$ is extended to the theory morphism $\tau' : \mathcal{B} \longrightarrow \mathcal{T}_{inst}$, translating any theorem $d \in Thms(\mathcal{B}) \setminus Thms(\mathcal{P})$ to $\mathcal{T}_5$.

If the given parametrisation $\langle \mathcal{P}, \mathcal{B} \rangle$ is well-formed, the set $AX(\mathcal{B}) \setminus (AX(\mathcal{P}) \cup Defs(\mathcal{B}))$ in the last step is empty, and hence the theory $\mathcal{T}_4$ is extended only by theorems. Otherwise, this step is in general not safe, in the sense that $AX(\mathcal{T}_{inst}) \setminus (AX(\mathcal{I}) \cup Defs(\mathcal{T}_{inst})) \neq \emptyset$ holds, i.e. $\mathcal{I}$ has been extended by 'proper' axioms. A possible treatment of this problem is presented in the next section in the context of a concrete implementation in Isabelle.

The following diagram

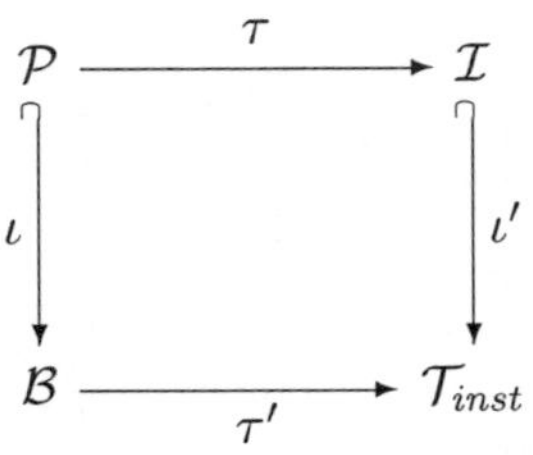

(4.11)

summarises the construction, done by the instantiation procedure. That is, if it succeeds then the resulting theory $\mathcal{T}_{inst}$ extending $\mathcal{I}$, and the theory morphism τ' make the diagram commute. This result follows directly from the construction of τ' which extends τ. A somewhat more complicated task is to prove the universality of the construction. Referring again to the diagram (4.8), where τ_1 is now replaced by τ' and τ_2 by ι', unique existence of the morphism $\tau^\sharp$ has to be shown.

Universal property of the construction

Regarding the presented procedure, in order to establish that $(\mathcal{T}_{inst}, \tau', \iota')$ is indeed an instantiation of $\mathcal{B}$ by τ, we also need to argue that this result is universal, i.e. for any other triple $(\mathcal{B}^\tau, \tau_1 : \mathcal{B} \longrightarrow \mathcal{B}^\tau, \tau_2 : \mathcal{I} \longrightarrow \mathcal{B}^\tau)$, which makes the diagram (4.11) commute, there is the unique theory morphism $\tau^\sharp : \mathcal{T}_{inst} \longrightarrow \mathcal{B}^\tau$ such that

$$\tau_1 = \tau^\sharp \circ \tau' \qquad (4.12)$$

$$\tau_2 = \tau^\sharp \circ \iota' \qquad (4.13)$$

hold. The following constructions and argumentations are based on the property of the instantiation procedure that $Anc(\mathcal{T}_{inst}) \setminus Anc(\mathcal{I})$ maintains the structure of $Anc(\mathcal{B}) \setminus Anc(\mathcal{P})$, and that τ' extends τ only by injective assignments.

Class map of $\tau^\sharp$. Let $c \in \Sigma_{class}(\mathcal{T}_{inst})$ be a type class. Then there are the following cases:

1. $c \in \Sigma_{class}(\mathcal{T}_{inst}) \setminus \Sigma_{class}(\mathcal{I})$. Then, there is unique $c' \in \Sigma_{class}(\mathcal{B}) \setminus \Sigma_{class}(\mathcal{P})$ such that $c = \sigma'_{class}(c')$ holds, where σ'_{class} is the class map of τ' In this case, we define

   $$\sigma^\sharp_{class}(c) \stackrel{def}{=} \sigma^1_{class}(c')$$

   where σ^1_{class} is the class map of τ_1.

2. $c \in \Sigma_{class}(\mathcal{I})$. In this case, we simply define

   $$\sigma^\sharp_{class}(c) \stackrel{def}{=} \sigma^2_{class}(c)$$

   where σ^2_{class} is the class map of τ_2.

In order to show that $\Sigma_{class}(\mathcal{T}_{inst}) \stackrel{\sigma^\sharp_{class}}{\longmapsto} \Sigma_{class}(\mathcal{B}^\tau)$ holds, we prove

$$\overline{\sigma}^\sharp_{class}(Super_{\Sigma_{class}(\mathcal{T}_{inst})}(c)) \subseteq Super_{\Sigma_{class}(\mathcal{B}^\tau)}(\sigma^\sharp_{class}(c))$$

for any $c \in \Sigma_{class}(\mathcal{T}_{inst})$. If $c \in \Sigma_{class}(\mathcal{I})$ then $Super_{\Sigma_{class}(\mathcal{T}_{inst})}(c) \subseteq \Sigma_{class}(\mathcal{I})$, and hence

$$\begin{array}{ll} \overline{\sigma}^\sharp_{class}(Super_{\Sigma_{class}(\mathcal{T}_{inst})}(c)) & = \\ \overline{\sigma}^2_{class}(Super_{\Sigma_{class}(\mathcal{T}_{inst})}(c)) & \subseteq \\ & \text{since } \overline{\sigma}^2_{class} \text{ is a class map} \\ Super_{\Sigma_{class}(\mathcal{B}^\tau)}(\sigma^2_{class}(c)) & = \\ Super_{\Sigma_{class}(\mathcal{B}^\tau)}(\sigma^\sharp_{class}(c)) & \end{array}$$

holds. Otherwise, let $c \in \Sigma_{class}(\mathcal{T}_{inst}) \setminus \Sigma_{class}(\mathcal{I})$. Then we have to show

$$\overline{\sigma}^\sharp_{class}(Super_{\Sigma_{class}(\mathcal{T}_{inst})}(\sigma'_{class}(c'))) \subseteq Super_{\Sigma_{class}(\mathcal{B}^\tau)}(\sigma^\sharp_{class}(\sigma'_{class}(c')))$$

with the unique $c' \in \Sigma_{class}(\mathcal{B}) \setminus \Sigma_{class}(\mathcal{P})$. Moreover, by the instantiation procedure

$$Super_{\Sigma_{class}(\mathcal{T}_{inst})}(\sigma'_{class}(c')) = \overline{\sigma}'_{class}(Super_{\Sigma_{class}(\mathcal{B})}(c'))$$

holds, i.e. we can show

$$\overline{\sigma}^\sharp_{class}(\overline{\sigma}'_{class}(Super_{\Sigma_{class}(\mathcal{B})}(c'))) \subseteq Super_{\Sigma_{class}(\mathcal{B}^\tau)}(\sigma^1_{class}(c')).$$

To this end, let $\overline{c}' \in Super_{\Sigma_{class}(\mathcal{B})}(c')$.

1. If $\overline{c}' \in \Sigma_{class}(\mathcal{P})$ then $\sigma^{\sharp}_{class}(\sigma'_{class}(\overline{c}')) = \sigma^{\sharp}_{class}(\sigma_{class}(\overline{c}'))$ holds. By the definition of $\sigma^{\sharp}_{class}$ and $\sigma_{class}(\overline{c}') \in \Sigma_{class}(\mathcal{I})$
$$\sigma^{\sharp}_{class}(\sigma_{class}(\overline{c}')) = \sigma^{2}_{class}(\sigma_{class}(\overline{c}')) = \sigma^{1}_{class}(\overline{c}')$$
holds as well. Altogether, since σ^{1}_{class} is a class map we obtain
$$\sigma^{\sharp}_{class}(\sigma'_{class}(\overline{c}')) \in Super_{\Sigma_{class}(\mathcal{B}^{\mathcal{T}})}(\sigma^{1}_{class}(c'))).$$

2. If $\overline{c}' \in \Sigma_{class}(\mathcal{B}) \setminus \Sigma_{class}(\mathcal{P})$ then $\sigma^{\sharp}_{class}(\sigma'_{class}(\overline{c}')) = \sigma^{1}_{class}(\overline{c}')$ by the definition of $\sigma^{\sharp}_{class}$. Again, since σ^{1}_{class} is a class map, we obtain the result
$$\sigma^{\sharp}_{class}(\sigma'_{class}(\overline{c}')) \in Super_{\Sigma_{class}(\mathcal{B}^{\mathcal{T}})}(\sigma^{1}_{class}(c'))).$$

The proof of the condition (4.12) for class maps proceeds in a similar manner. Let $c \in \Sigma_{class}(\mathcal{B})$. Then, if $c \in \Sigma_{class}(\mathcal{P})$ holds we can conclude

$$\begin{array}{ll} (\sigma^{\sharp}_{class} \circ \sigma'_{class})(c) & = \\ \sigma^{\sharp}_{class}(\sigma_{class}(c)) & = \\ & \text{by the definition of } \sigma^{\sharp}_{class} \text{ for } \sigma_{class}(c) \in \Sigma_{class}(\mathcal{I}) \\ \sigma^{2}_{class}(\sigma_{class}(c)) & = \\ \sigma^{1}_{class}(c). & \end{array}$$

Otherwise, let $c \in \Sigma_{class}(\mathcal{B}) \setminus \Sigma_{class}(\mathcal{P})$. Then $(\sigma^{\sharp}_{class} \circ \sigma'_{class})(c) = \sigma^{1}_{class}(c)$ holds by the definition of $\sigma^{\sharp}_{class}$. The condition (4.13) for class maps follows directly from the definition of $\sigma^{\sharp}_{class}$.

Finally, we need to argue that $\sigma^{\sharp}_{class}$ is unique. To this end, let $\sigma^{\flat}_{class}$ be some class map satisfying (4.12) and (4.13). Further, let $c \in \Sigma_{class}(\mathcal{T}_{inst})$. Again, the instantiation procedure provides that if $c \in \Sigma_{class}(\mathcal{T}_{inst}) \setminus \Sigma_{class}(\mathcal{I})$ then there is unique $c' \in \Sigma_{class}(\mathcal{B}) \setminus \Sigma_{class}(\mathcal{P})$ with $\sigma'_{class}(c') = c$. So, $\sigma^{\flat}_{class}(\sigma'_{class}(c')) = \sigma^{1}_{class}(c')$. On the other hand, if $c \in \Sigma_{class}(\mathcal{I})$ then $\sigma^{\flat}_{class}(c) = \sigma^{2}_{class}(c) = \sigma^{\sharp}_{class}(c)$ holds. This establishes $\sigma^{\flat}_{class} = \sigma^{\sharp}_{class}$.

Type map of $\tau^{\sharp}$. Let $C \in \Sigma_{type}(\mathcal{T}_{inst})$ with $rank_{\mathcal{T}_{inst}}(C) = n$ be a type constructor. Then there are the following cases:

1. $C \in \Sigma_{type}(\mathcal{T}_{inst}) \setminus \Sigma_{type}(\mathcal{I})$. Then the instantiation procedure provides, that there is unique $C' \in \Sigma_{type}(\mathcal{B}) \setminus \Sigma_{class}(\mathcal{P})$ such that $rank_{\mathcal{B}}(C') = n$ and $\sigma'_{type}(C') = (`1, \ldots, `n) \cdot C$. In this case, we define

$$\sigma^{\sharp}_{type}(C) \stackrel{def}{=} \sigma^{1}_{type}(C')$$

where σ^{1}_{type} is the type map of τ_1.

2. $C \in \Sigma_{type}(\mathcal{I})$. In this case, we simply define

$$\sigma^{\sharp}_{type}(C) \stackrel{def}{=} \sigma^{2}_{type}(C)$$

In order to show that $\Sigma_{type}(\mathcal{T}_{inst}) \overset{\sigma^{\sharp}_{type}}{\longmapsto}_{\sigma^{\sharp}_{class}} \Sigma_{type}(\mathcal{B}^{\tau})$ holds, the type map conditions need to be verified. Firstly, $Bound(\sigma^{\sharp}_{type}(C)) \leq n$ holds, since σ^{1}_{type} and σ^{2}_{type} are type maps.

In order to verify (3.2), let $(S_1, \ldots, S_{n+1}) \in arities\text{-}of_{\mathcal{T}_{inst}}(C)$.

1. If $C \in \Sigma_{type}(\mathcal{T}_{inst}) \setminus \Sigma_{type}(\mathcal{I})$, the instantiation procedure provides, that there is an arity $(S'_1, \ldots, S'_{n+1}) \in arities\text{-}of_{\mathcal{B}}(C')$ with $S_i = \overline{\sigma}'_{class}(S'_i)$ for $1 \leq i \leq n+1$. Further, let $\kappa_1, \ldots, \kappa_n \in Types_{\vartheta}(\Sigma_{type}(\mathcal{B}^{\tau}))$ with κ_i *of-sort*$_{\mathcal{B}^{\tau}}$ $\overline{\sigma}^{\sharp}_{class}(S_i)$ for $1 \leq i \leq n$. We have then to show

$$\sigma^{\sharp}_{type}(C) \bullet \langle \kappa_1, \ldots, \kappa_n \rangle \; of\text{-}sort_{\mathcal{B}^{\tau}} \; \overline{\sigma}^{\sharp}_{class}(S_{n+1})$$

which is equivalent to

$$\sigma^{1}_{type}(C') \bullet \langle \kappa_1, \ldots, \kappa_n \rangle \; of\text{-}sort_{\mathcal{B}^{\tau}} \; \overline{\sigma}^{\sharp}_{class}(\overline{\sigma}'_{class}(S'_{n+1})).$$

Using $\overline{\sigma}^{\sharp}_{class} \circ \overline{\sigma}'_{class} = \overline{(\sigma^{\sharp}_{class} \circ \sigma'_{class})} = \overline{\sigma}^{1}_{class}$, this is equivalent to

$$\sigma^{1}_{type}(C') \bullet \langle \kappa_1, \ldots, \kappa_n \rangle \; of\text{-}sort_{\mathcal{B}^{\tau}} \; \overline{\sigma}^{1}_{class}(S'_{n+1})$$

which is true, since σ^{1}_{type} satisfies (3.2).

2. On the other hand, if $C \in \Sigma_{type}(\mathcal{I})$ then $S_i \in Sorts(\Sigma_{class}(\mathcal{I}))$ holds for all $1 \leq i \leq n+1$, and hence $\overline{\sigma}^{\sharp}_{class}(S_i) = \overline{\sigma}^{2}_{class}(S_i)$ holds as well. So, let $\kappa_1, \ldots, \kappa_n \in Types_{\vartheta}(\Sigma_{type}(\mathcal{B}^{\tau}))$ with κ_i *of-sort*$_{\mathcal{B}^{\tau}}$ $\overline{\sigma}^{2}_{class}(S_i)$ for $1 \leq i \leq n$. Then

$$\sigma^{\sharp}_{type}(C) \bullet \langle \kappa_1, \ldots, \kappa_n \rangle \; of\text{-}sort_{\mathcal{B}^{\tau}} \; \overline{\sigma}^{2}_{class}(S_{n+1})$$

holds, since $\sigma^{\sharp}_{type}(C) = \sigma^{2}_{type}(C)$ and σ^{2}_{type} is a type map.

The condition (4.12) for type maps can be shown as follows. Let $C \in \Sigma_{type}(\mathcal{B})$ with $rank_{\mathcal{B}}(C) = n$. Then, if $C \in \Sigma_{type}(\mathcal{B}) \setminus \Sigma_{type}(\mathcal{P})$ holds, we can conclude

$$\begin{array}{ll}
(\sigma^{\sharp}_{type} \circ \sigma'_{type})(C) & = \\
\sigma'_{type}(C) * \sigma^{\sharp}_{type} & = \\
 & \text{where } C' \in \Sigma_{type}(\mathcal{T}_{inst}) \setminus \Sigma_{type}(\mathcal{I}) \\
((`1, \ldots, `n) \cdot C') * \sigma^{\sharp}_{type} & = \\
\sigma^{\sharp}_{type}(C') \rhd \langle `1, \ldots, `n \rangle & = \\
\sigma^{1}_{type}(C) \rhd \langle `1, \ldots, `n \rangle & = \\
\sigma^{1}_{type}(C) &
\end{array}$$

Next, if $C \in \Sigma_{type}(\mathcal{P})$ then $\sigma'_{type}(C) = \sigma_{type}(C) \in \widehat{Types}(\Sigma_{type}(\mathcal{P}))$ holds. We need then to show $\sigma_{type}(C) * \sigma^{\sharp}_{type} = \sigma^{1}_{type}(C)$, which is done by proving the following more general claim.

Claim. *For any $\varsigma \in \widehat{Types}(\Sigma_{type}(\mathcal{P}))$ with $Bound(\varsigma) \leq n$*

$$\varsigma * \sigma^{\sharp}_{type} = \varsigma * \sigma^{2}_{type}$$

holds.

Proof. (by structural induction on ς)
(i) If $\varsigma = `i$ with $1 \leq i \leq n$ then both *lhs* and *rhs* reduce to $`i$.
(ii) Let $\varsigma = (\varsigma_1, \ldots, \varsigma_m) \cdot C_1$ with $C_1 \in \Sigma_{type}(\mathcal{P})$. Then

$$\begin{array}{ll}
\varsigma * \sigma^{\sharp}_{type} & = \\
\sigma^{\sharp}_{type}(C_1) \rhd \langle \varsigma_1 * \sigma^{\sharp}_{type}, \ldots, \varsigma_m * \sigma^{\sharp}_{type} \rangle & = \\
 & \text{by the definition of } \sigma^{\sharp}_{type} \\
\sigma^{2}_{type}(C_1) \rhd \langle \varsigma_1 * \sigma^{\sharp}_{type}, \ldots, \varsigma_m * \sigma^{\sharp}_{type} \rangle & = \\
 & \text{by ind. hyp.} \\
\sigma^{2}_{type}(C_1) \rhd \langle \varsigma_1 * \sigma^{2}_{type}, \ldots, \varsigma_m * \sigma^{2}_{type} \rangle & = \\
\varsigma * \sigma^{2}_{type} &
\end{array}$$

□

In particular, for the fixed $C \in \Sigma_{type}(\mathcal{P})$ we obtain

$$\begin{array}{ll}
\sigma_{type}(C) * \sigma^{\sharp}_{type} & = \\
\sigma_{type}(C) * \sigma^{2}_{type} & = \\
(\sigma^{2}_{type} \circ \sigma_{type})(C) & = \\
 & \text{since } \sigma^{1}_{type}, \sigma^{2}_{type} \text{ make the diagram (4.8)} \\
 & \text{for type maps commute} \\
\sigma^{1}_{type}(C). &
\end{array}$$

Finally, the condition (4.13) is obtained directly from the definition of $\sigma^{\sharp}_{type}$.

Next, for the uniqueness of $\sigma^{\sharp}_{type}$, let $\sigma^{\flat}_{type}$ be another type map satisfying (4.12) and (4.13). Further, let $C \in \Sigma_{type}(\mathcal{T}_{inst})$ with $rank_{\mathcal{T}_{inst}}(C) = n$. If additionally $C \in \Sigma_{type}(\mathcal{T}_{inst}) \setminus \Sigma_{type}(\mathcal{I})$ holds then there exists unique $C' \in \Sigma_{type}(\mathcal{B}) \setminus \Sigma_{type}(\mathcal{P})$ with $\sigma'_{type}(C') = (\text{`}1, \ldots, \text{`}n) \cdot C$. So,

$$\begin{array}{ll}
\sigma^{\flat}_{type}(C) & = \\
\sigma^{\flat}_{type}(C) \rhd \langle \text{`}1, \ldots, \text{`}n \rangle & = \\
\sigma^{\flat}_{type}(C) \rhd \langle \text{`}1 * \sigma^{\flat}_{type}, \ldots, \text{`}n * \sigma^{\flat}_{type} \rangle & = \\
((\text{`}1, \ldots, \text{`}n) \cdot C) * \sigma^{\flat}_{type} & = \\
\sigma'_{type}(C') * \sigma^{\flat}_{type} & = \\
(\sigma^{\flat}_{type} \circ \sigma'_{type})(C') & = \\
 & \text{by (4.12) for } \sigma^{\flat}_{type} \\
\sigma^{1}_{type}(C') & = \\
 & \text{by the definition of } \sigma^{\sharp}_{type} \\
\sigma^{\sharp}_{type}(C). &
\end{array}$$

On the other hand, if $C \in \Sigma_{type}(\mathcal{I})$ then

$$\begin{array}{ll}
\sigma^{\flat}_{type}(C) & = \\
\sigma^{\flat}_{type}(C) \rhd \langle \text{`}1, \ldots, \text{`}n \rangle & = \\
\sigma^{\flat}_{type}(C) \rhd \langle \text{`}1 * \sigma^{\flat}_{type}, \ldots, \text{`}n * \sigma^{\flat}_{type} \rangle & = \\
((\text{`}1, \ldots, \text{`}n) \cdot C) * \sigma^{\flat}_{type} & = \\
\iota'_{type}(C) * \sigma^{\flat}_{type} & = \\
(\sigma^{\flat}_{type} \circ \iota'_{type})(C) & = \\
 & \text{by (4.13) for } \sigma^{\flat}_{type} \\
\sigma^{2}_{type}(C) & = \\
 & \text{by the definition of } \sigma^{\sharp}_{type} \\
\sigma^{\sharp}_{type}(C). &
\end{array}$$

Operation map of $\tau^\sharp$. Let $f \in \Sigma_{op}(\mathcal{T}_{inst})$. There are the following cases:

1. $f \in \Sigma_{op}(\mathcal{T}_{inst}) \setminus \Sigma_{op}(\mathcal{I})$; independently of whether f is logical or derived in $\mathcal{T}_{inst}$, by the construction of instantiation there exists unique $f' \in \Sigma_{op}(\mathcal{B}) \setminus \Sigma_{op}(\mathcal{P})$ with $\sigma'_{op}(f') = f$, such that in this case we define

$$\sigma^\sharp_{op}(f) \stackrel{def}{=} \sigma^1_{op}(f')$$

2. otherwise, if $f \in \Sigma_{op}(\mathcal{I})$ then

$$\sigma^\sharp_{op}(f) \stackrel{def}{=} \sigma^2_{op}(f).$$

First of all we show $\Sigma_{op}(\mathcal{T}_{inst}) \overset{\sigma^\sharp_{op}}{\longmapsto}_{\sigma^\sharp_{class},\sigma^\sharp_{type}} \Sigma_{op}(\mathcal{B}^\tau)$, i.e. that $\sigma^\sharp_{op}$ is an operation map. So, the operation map condition (3.9) need to be verified. To this end, let $f \in \Sigma_{op}(\mathcal{T}_{inst})$ and $\kappa \stackrel{def}{=} \textit{type-of}_{\mathcal{T}_{inst}}(f)$. We have to show

$$\textit{type-of}_{\mathcal{B}^\tau}(\sigma^\sharp_{op}(f)) \stackrel{\alpha}{=} \overline{\sigma}^{\sigma^\sharp_{class}}_{type}(\kappa).$$

1. If $f \in \Sigma_{op}(\mathcal{T}_{inst}) \setminus \Sigma_{op}(\mathcal{I})$ then $\sigma^\sharp_{op}(f) = \sigma^1_{op}(f')$ holds by the definition, where $f' \in \Sigma_{op}(\mathcal{B}) \setminus \Sigma_{op}(\mathcal{P})$ with $\sigma'_{op}(f') = f$. Moreover, let $\kappa' \stackrel{def}{=} \textit{type-of}_{\mathcal{B}}(f')$. Since σ'_{op} is an operation map we further have

$$\kappa = \textit{type-of}_{\mathcal{T}_{inst}}(f) = \textit{type-of}_{\mathcal{T}_{inst}}(\sigma'_{op}(f')) \stackrel{\alpha}{=} \overline{\sigma}^{\sigma'_{class}}_{type}(\kappa') \tag{4.14}$$

Using this, we can reason

$$\begin{array}{ll} \textit{type-of}_{\mathcal{B}^\tau}(\sigma^\sharp_{op}(f)) & = \\ \textit{type-of}_{\mathcal{B}^\tau}(\sigma^1_{op}(f')) & \stackrel{\alpha}{=} \\ & \text{since } \sigma^1_{op} \text{ is an operation map} \\ \underline{\overline{\sigma}^{\sigma^1_{class}}_{type}(\kappa')} & = \\ (\sigma^\sharp_{type} \circ \sigma'_{type})(\kappa') & = \\ (\overline{\sigma}^{\sigma^\sharp_{class}}_{type} \circ \overline{\sigma}^{\sigma'_{class}}_{type})(\kappa') & \stackrel{\alpha}{=} \\ & \text{by (4.14) and Lemma 3.5} \\ \overline{\sigma}^{\sigma^\sharp_{class}}_{type}(\kappa) & \end{array}$$

2. If $f \in \Sigma_{op}(\mathcal{I})$ then

$$\kappa = \textit{type-of}_{\mathcal{T}_{inst}}(f) = \textit{type-of}_{\mathcal{I}}(f) \in \textit{Types}(\Sigma_{type}(\mathcal{I}))$$

holds. Using this, we can reason:

$$
\begin{array}{ll}
\textit{type-of}_{\mathcal{B}^{\tau}}(\sigma^{\sharp}_{op}(f)) & = \\
\textit{type-of}_{\mathcal{B}^{\tau}}(\sigma^{2}_{op}(f)) & \stackrel{\alpha}{=} \\
& \text{since } \sigma^{2}_{op} \text{ is an operation map} \\
\overline{\sigma}^{\sigma^{2}_{class}}_{type}(\kappa) & = \\
\hline
(\sigma^{\sharp}_{type} \circ \iota'_{type})(\kappa) & = \\
\overline{\sigma}^{\sigma^{\sharp}_{class}}_{type}(\kappa). &
\end{array}
$$

In order to show (4.12) for operation maps, let $f \in \Sigma_{op}(\mathcal{B})$. Further, if $\sigma'_{op}(f) \in \Sigma_{op}(\mathcal{T}_{inst}) \setminus \Sigma_{op}(\mathcal{I})$ then $(\sigma^{\sharp}_{op} \circ \sigma'_{op})(f) = \sigma^{1}_{op}(f)$ holds immediately by the definition of $\sigma^{\sharp}_{op}$. Otherwise, if $\sigma'_{op}(f) \in \Sigma_{op}(\mathcal{I})$ then $\sigma'_{op}(f) = \sigma_{op}(f)$ holds by construction of σ'_{op}, and thus

$$
\begin{array}{ll}
(\sigma^{\sharp}_{op} \circ \sigma'_{op})(f) & = \\
& \text{by the definition of } \sigma^{\sharp}_{op} \\
(\sigma^{2}_{op} \circ \sigma_{op})(f) & = \\
(\sigma^{1}_{op} \circ \iota_{op})(f) & = \\
\sigma^{1}_{op}(f). &
\end{array}
$$

holds. For the proof of (4.13), let $f \in \Sigma_{op}(\mathcal{I})$. Then $\sigma^{\sharp}_{op}(\iota'_{op}(f)) = \sigma^{\sharp}_{op}(f) = \sigma^{2}_{op}(f)$ follows directly by the definition of $\sigma^{\sharp}_{op}$.

Finally, for the uniqueness of $\sigma^{\sharp}_{op}$, let $\sigma^{\flat}_{op}$ be another operation map satisfying (4.12) and (4.13). Further, let $f \in \Sigma_{op}(\mathcal{T}_{inst})$ be a constant. If $f \in \Sigma_{op}(\mathcal{T}_{inst}) \setminus \Sigma_{op}(\mathcal{I})$ then there is unique $f' \in \Sigma_{op}(\mathcal{B}) \setminus \Sigma_{op}(\mathcal{P})$ with $\sigma'_{op}(f') = f$. So,

$$\sigma^{\flat}_{op}(f) = \sigma^{\flat}_{op}(\sigma'_{op}(f')) = \sigma^{1}_{op}(f') = \sigma^{\sharp}_{op}(f)$$

holds. Otherwise, if $f \in \Sigma_{op}(\mathcal{I})$ then we have

$$\sigma^{\flat}_{op}(f) = \sigma^{\flat}_{op}(\iota'_{op}(f)) = \sigma^{2}_{op}(f) = \sigma^{\sharp}_{op}(f).$$

The signature morphism $\sigma^{\sharp}$. Altogether, we obtain the operation map $\Sigma_{op}(\mathcal{T}_{inst}) \xmapsto[\sigma^{\sharp}_{class}, \sigma^{\sharp}_{type}]{\sigma^{\sharp}_{op}} \Sigma_{op}(\mathcal{B}^{\tau})$, uniquely satisfying (4.12) and (4.13).

However, it still has to be shown, that

$$\sigma^{\sharp} \stackrel{def}{=} \langle \mathcal{T}_{inst}, \mathcal{B}^{\tau}, \sigma^{\sharp}_{class}, \sigma^{\sharp}_{type}, \Sigma_{op}(\mathcal{T}_{inst}), \sigma^{\sharp}_{op} \rangle$$

is indeed a complete signature morphism. That is, we additionally have to verify for any $f \in \Sigma^{der}_{op}(\mathcal{T}_{inst})$ the definitional conditions:

1. $\sigma^{\sharp}_{op}(f) \in \Sigma^{der}_{op}(\mathcal{B}^{\mathcal{T}})$, i.e. is a derived constant, and

2. $\textit{def-of}_{\mathcal{B}^{\mathcal{T}}}(\sigma^{\sharp}_{op}(f)) \stackrel{\alpha}{=} \overline{\sigma}_{op}^{\sigma^{\sharp}_{class},\sigma^{\sharp}_{type}}(\textit{def-of}_{\mathcal{T}_{inst}}(f))$ where $\overline{\sigma}_{op}^{\sigma^{\sharp}_{class},\sigma^{\sharp}_{type}}$ denotes the homomorphic extension of $\sigma^{\sharp}_{op}$.

To this end, let $f \in \Sigma^{der}_{op}(\mathcal{T}_{inst})$ be a derived constant. We consider the following cases.

1. If $f \in \Sigma_{op}(\mathcal{T}_{inst}) \setminus \Sigma_{op}(\mathcal{I})$ then $\sigma^{\sharp}_{op}(f) = \sigma^{1}_{op}(f')$ with unique $f' \in \Sigma_{op}(\mathcal{B}) \setminus \Sigma_{op}(\mathcal{P})$ such that $\sigma'_{op}(f') = f$. Moreover, by the construction of σ'_{op} we have $f' \in \Sigma^{der}_{op}(\mathcal{B}) \setminus \Sigma_{op}(\mathcal{P})$, i.e. f' is derived as well, since f is derived and σ'_{op} by definition cannot map a logical constant in $\Sigma_{op}(\mathcal{B}) \setminus \Sigma_{op}(\mathcal{P})$ to a derived constant. Furthermore, since σ^1 is a complete signature morphism, $\sigma^{1}_{op}(f') \in \Sigma^{der}_{op}(\mathcal{B}^{\mathcal{T}})$ holds.

   The construction of σ'_{op}, or more exactly the translation of definitions by instantiation, ensures the property:

$$\overline{\sigma}_{op}^{\sigma'_{class},\sigma'_{type}}(\textit{def-of}_{\mathcal{B}}(f')) = \textit{def-of}_{\mathcal{T}_{inst}}(f) \tag{4.15}$$

   Using this, we can reason:

$$\begin{array}{ll}
\textit{def-of}_{\mathcal{B}^{\mathcal{T}}}(\sigma^{\sharp}_{op}(f)) & = \\
\textit{def-of}_{\mathcal{B}^{\mathcal{T}}}(\sigma^{1}_{op}(f')) & \stackrel{\alpha}{=} \\
 & \text{since } \sigma^1 \text{ is a complete} \\
 & \text{signature morphism} \\
\overline{\sigma}_{op}^{\sigma^{1}_{class},\sigma^{1}_{type}}(\textit{def-of}_{\mathcal{B}}(f')) & = \\
 & \text{by (4.12) for operation maps} \\
\overline{(\sigma^{\sharp}_{op} \circ \sigma'_{op})}^{\sigma^{\sharp}_{class} \circ\, \sigma'_{class},\sigma^{\sharp}_{type} \circ\, \sigma'_{type}}(\textit{def-of}_{\mathcal{B}}(f')) & = \\
\overline{\sigma}_{op}^{\sigma^{\sharp}_{class},\sigma^{\sharp}_{type}}(\overline{\sigma}_{op}^{\sigma'_{class},\sigma'_{type}}(\textit{def-of}_{\mathcal{B}}(f'))) & = \\
 & \text{by (4.15)} \\
\overline{\sigma}_{op}^{\sigma^{\sharp}_{class},\sigma^{\sharp}_{type}}(\textit{def-of}_{\mathcal{T}_{inst}}(f)) &
\end{array}$$

2. Otherwise, if $f \in \Sigma_{op}(\mathcal{I})$ holds then we have $\sigma^{\sharp}_{op}(f) = \sigma^{2}_{op}(f)$. It can be either the case that $f \in \Sigma^{der}_{op}(\mathcal{I})$ holds, i.e. f has been defined already in $\mathcal{I}$, or $f \in \Sigma^{log}_{op}(\mathcal{I})$, i.e. f has been defined via translation of a definition by the construction of instantiation. So, we consider these two cases separately.

(a) If $f \in \Sigma_{op}^{log}(\mathcal{I})$ then by the construction of instantiation there exist (not necessarily unique) $f' \in \Sigma_{op}^{log}(\mathcal{P})$ and $d' \in \mathit{Defs}(\mathcal{B}) \backslash \mathit{Defs}(\mathcal{P})$ such that $f' \in \Sigma_{op}^{der}(\mathcal{B})$ with $\mathit{prop\text{-}of}_{\mathcal{B}}(d') = f' \equiv \mathit{def\text{-}of}_{\mathcal{B}}(f')$, $\sigma'_{op}(f') = \sigma_{op}(f') = f$, and

$$\overline{\sigma_{op}}^{\sigma'_{class}, \sigma'_{type}}(\mathit{def\text{-}of}_{\mathcal{B}}(f')) \stackrel{\alpha}{=} \mathit{def\text{-}of}_{\mathcal{T}_{inst}}(f) \tag{4.16}$$

Now, from $\sigma^1_{op} \circ \iota_{op} = \sigma^2_{op} \circ \sigma_{op}$ we can firstly conclude $\sigma^1_{op}(f') = \sigma^2_{op}(f)$. But since $f' \in \Sigma_{op}^{der}(\mathcal{B})$ and σ^1 is a complete signature morphism, we can also conclude $\sigma^1_{op}(f') \in \Sigma_{op}^{der}(\mathcal{B}^{\tau})$, i.e. that $\sigma^{\sharp}_{op}(f) \in \Sigma_{op}^{der}(\mathcal{B}^{\tau})$ holds. Further, we can reason

$\mathit{def\text{-}of}_{\mathcal{B}^{\tau}}(\sigma^{\sharp}_{op}(f))$	$=$
$\mathit{def\text{-}of}_{\mathcal{B}^{\tau}}(\sigma^{2}_{op}(\sigma_{op}(f'))$	$=$
$\mathit{def\text{-}of}_{\mathcal{B}^{\tau}}(\sigma^{1}_{op}(f'))$	$\stackrel{\alpha}{=}$ since σ^1 is a complete signature morphism
$\overline{\sigma_{op}}^{\sigma^1_{class}, \sigma^1_{type}}(\mathit{def\text{-}of}_{\mathcal{B}}(f'))$	$=$ by (4.12) for operation maps
$\overline{(\sigma^{\sharp}_{op} \circ \sigma'_{op})}^{\sigma^{\sharp}_{class} \circ \sigma'_{class}, \sigma^{\sharp}_{type} \circ \sigma'_{type}}(\mathit{def\text{-}of}_{\mathcal{B}}(f'))$	$=$
$\overline{\sigma_{op}}^{\sigma^{\sharp}_{class}, \sigma^{\sharp}_{type}}(\overline{\sigma_{op}}^{\sigma'_{class}, \sigma'_{type}}(\mathit{def\text{-}of}_{\mathcal{B}}(f')))$	$\stackrel{\alpha}{=}$ by (4.16) and Lemma 3.12
$\overline{\sigma_{op}}^{\sigma^{\sharp}_{class}, \sigma^{\sharp}_{type}}(\mathit{def\text{-}of}_{\mathcal{T}_{inst}}(f))$.	

(b) If $f \in \Sigma_{op}^{der}(\mathcal{I})$, i.e. f is derived already in $\mathcal{I}$, then $\mathit{def\text{-}of}_{\mathcal{T}_{inst}}(f) = \mathit{def\text{-}of}_{\mathcal{I}}(f) \in \mathit{Terms}(\Sigma_{op}(\mathcal{I}))$ holds. Since σ^2 is a complete signature morphism, $\sigma^{\sharp}_{op}(f) = \sigma^2_{op}(f) \in \Sigma_{op}^{der}(\mathcal{B}^{\tau})$ holds. So, we can reason as follows, where ι'_{class} and ι'_{type} denote the class and type map of the inclusion ι', respectively:

$\mathit{def\text{-}of}_{\mathcal{B}^{\tau}}(\sigma^{\sharp}_{op}(f))$	$=$ by the definition of $\sigma^{\sharp}_{op}$
$\mathit{def\text{-}of}_{\mathcal{B}^{\tau}}(\sigma^{2}_{op}(f))$	$\stackrel{\alpha}{=}$ since σ^2 is a complete signature morphism

$$\overline{\sigma}_{op}^{\sigma^2_{class},\sigma^2_{type}}(\textit{def-of}_{\mathcal{I}}(f)) =$$

by (4.13) for operation maps

$$\overline{(\sigma^{\sharp}_{op} \circ \iota'_{op})}^{\sigma^{\sharp}_{class} \circ \iota'_{class}, \sigma^{\sharp}_{type} \circ \iota'_{type}}(\textit{def-of}_{\mathcal{I}}(f)) =$$

$$\overline{\sigma}_{op}^{\sigma^{\sharp}_{class},\sigma^{\sharp}_{type}}(\textit{def-of}_{\mathcal{T}_{inst}}(f)).$$

Altogether, these considerations establish, that $\sigma^{\sharp}$ defined above is the complete signature morphism $\sigma^{\sharp} : \mathcal{T}_{inst} \longrightarrow \mathcal{B}^{\tau}$ uniquely satisfying the equations $\sigma_1 = \sigma^{\sharp} \circ \sigma'$ and $\sigma_2 = \sigma^{\sharp} \circ \sigma(\iota')$, where $\sigma(\iota')$ denotes the underlying signature morphism of the injection ι'.

Theorem map of $\tau^{\sharp}$. Now, the signature morphism $\sigma^{\sharp}$ will be extended to a theory morphism by construction of the theorem map $\xi^{\sharp}$. Let

1. $\mathcal{G}(\mathcal{P}) = \langle D_{\mathcal{P}}, \textit{thm-of}_{\mathcal{P}} \rangle$;
2. $\mathcal{G}(\mathcal{B}) = \langle D_{\mathcal{B}}, \textit{thm-of}_{\mathcal{B}} \rangle$;
3. $\mathcal{G}(\mathcal{I}) = \langle D_{\mathcal{I}}, \textit{thm-of}_{\mathcal{I}} \rangle$;
4. $\mathcal{G}(\mathcal{T}_{inst}) = \langle D_{inst}, \textit{thm-of}_{\mathcal{T}_{inst}} \rangle$;
5. $\mathcal{G}(\mathcal{B}^{\tau}) = \langle D_{\mathcal{B}^{\tau}}, \textit{thm-of}_{\mathcal{B}^{\tau}} \rangle$;

and $\xi' : D_{\mathcal{B}} \to D_{inst}$ the theorem map of τ'. Further, let $d \in D_{inst}$.

1. If $d \in D_{inst} \setminus D_{\mathcal{I}}$ then there is unique $d' \in D_{\mathcal{B}} \setminus D_{\mathcal{P}}$ such that $\xi'(d') = d$ holds. Moreover, by construction of ξ', the theorem map condition (4.1) is satisfied as follows:

$$\textit{thm-of}_{\mathcal{T}_{inst}}(d) = \overline{\sigma}'(\textit{thm-of}_{\mathcal{B}}(d')). \tag{4.17}$$

   So, we define $\xi^{\sharp}(d) \stackrel{def}{=} \xi^1(d')$;

2. If $d \in D_{\mathcal{I}}$ then $\xi^{\sharp}(d) \stackrel{def}{=} \xi^2(d)$.

In order to justify, that $\xi^{\sharp}$ is a theorem map, the theorem map conditions (4.1) and (4.2) have to be verified. So for (4.1), let $d \in D_{inst}$ be arbitrary. If additionally $d \in D_{inst} \setminus D_{\mathcal{I}}$ holds then

$$
\begin{array}{ll}
\textit{thm-of}_{\mathcal{B}_\tau}(\xi^\sharp(d)) & = \\
 & \text{using the unique } d' \in D_{\mathcal{B}} \setminus D_{\mathcal{P}} \\
\textit{thm-of}_{\mathcal{B}_\tau}(\xi^1(d')) & \stackrel{\alpha}{=} \\
 & \text{since } \xi^1 \text{ is a theorem map} \\
\overline{\sigma}^1(\textit{thm-of}_{\mathcal{B}}(d')) & = \\
\overline{(\sigma^\sharp \circ \sigma')}(\textit{thm-of}_{\mathcal{B}}(d')) & = \\
 & \text{by (4.17)} \\
\overline{\sigma}^\sharp(\textit{thm-of}_{\mathcal{T}_{inst}}(d)) &
\end{array}
$$

Otherwise, $d \in D_{\mathcal{I}}$ holds, such that

$$
\begin{array}{ll}
\textit{thm-of}_{\mathcal{B}_\tau}(\xi^\sharp(d)) & = \\
\textit{thm-of}_{\mathcal{B}_\tau}(\xi^2(d)) & \stackrel{\alpha}{=} \\
 & \text{since } \xi^2 \text{ is a theorem map} \\
\overline{\sigma}^2(\textit{thm-of}_{\mathcal{T}}(d)) & = \\
\overline{(\sigma^\sharp \circ \sigma(\iota'))}(\textit{thm-of}_{\mathcal{I}}(d)) & = \\
\overline{\sigma}^\sharp(\textit{thm-of}_{\mathcal{T}_{inst}}(d)) &
\end{array}
$$

The condition (4.2) holds as well, since axioms in $D_{inst} \setminus D_{\mathcal{I}}$ are mapped to axioms, and theorems to theorems, provided by the construction of $\mathcal{T}_{inst}$. For $d \in D_{\mathcal{I}}$, $\xi^\sharp(d)$ satisfies (4.2), because ξ^2 is a theorem map.

For the proof of (4.12), let $d \in D_{\mathcal{B}}$. Then if $\xi'(d) \in D_{inst} \setminus D_{\mathcal{I}}$ then $\xi^\sharp(\xi'(d)) = \xi^1(d)$ by the definition of $\xi^\sharp$. Otherwise, if $\xi'(d) \in D_{\mathcal{I}}$ then $\xi'(d) = \xi(d)$ holds as well, and, hence $\xi^\sharp(\xi'(d)) = \xi^2(\xi(d)) = \xi^1(d)$ holds, by the definition of $\xi^\sharp$ and since $\tau^2 \circ \tau = \tau^1 \circ \iota$ is assumed. The proof of (4.13) is obtained immediately from the definition of $\xi^\sharp$.

For the proof of the uniqueness of $\xi^\sharp$, let $\xi^\flat$ be some theorem map satisfying (4.12) and (4.13). Further, let $d \in D_{inst}$. If $d \in D_{inst} \setminus D_{\mathcal{I}}$ then there is unique $d' \in D_{\mathcal{B}} \setminus D_{\mathcal{P}}$ with $\xi'(d') = d$. So,

$$\xi^\flat(d) = \xi^\flat(\xi'(d')) = \xi^1(d') = \xi^\sharp(d)$$

holds. Otherwise, if $d \in D_{\mathcal{I}}$ then

$$\xi^\flat(d) = \xi^2(d) = \xi^\sharp(d)$$

holds as well.

The theory morphism $\tau^\sharp$. Finally, we can conclude that $\tau^\sharp = \langle \sigma^\sharp, D_{inst}, \xi^\sharp \rangle$ is the theory morphism satisfying (4.12) and (4.13).

4.5.2 Instantiation of parameterised theories in Isabelle

Returning again to the connection between inclusions and the theory hierarchy, in Isabelle any inclusion $\mathcal{T}_1 \hookrightarrow \mathcal{T}_2$ implies $\mathcal{T}_1 \in Anc(\mathcal{T}_2)$. This holds, because the elements, e.g. type constructors, from two different theories are always distinct in the context of Isabelle. This is provided by the fact that any theory in Isabelle has its unique identifier, i.e. its name.

The instantiation procedure in Isabelle proceeds almost exactly as the procedure described in the previous section. Firstly, an important technical difference is, that the instantiating theory $\mathcal{I}$ is assumed to be a *state* of the currently developed theory with the name, e.g. *Thy*, which is not fixed such that the extension steps

$$\mathcal{I} \hookrightarrow \mathcal{T}_0 \hookrightarrow \ldots \hookrightarrow \mathcal{T}_{inst}$$

are interpreted as a sequence of transition steps between the states of *Thy*. So, e.g. $\mathcal{T}_{inst}$ is the state of *Thy* which extends $\mathcal{I}$ by all the elements introduced via previous transitions.

Secondly, axioms and theorems the treatment of axioms and theorems is somewhat refined, in comparison to the last step of the instantiation procedure. In particular, the procedure, integrated with Isabelle, distinguishes between *proper* and *axiomatic* instantiations, which is sketched below.

To this end, we note that the instantiation procedure constructs in the fifth step the theory $\mathcal{T}_4$ and the signature morphism $\sigma' : \mathcal{B} \longrightarrow \mathcal{T}_4$. Now, let $AX(\mathcal{B}) \setminus (AX(\mathcal{P}) \cup Defs(\mathcal{B})) = \{a_1, \ldots, a_n\}$ be the set of 'proper' axioms introduced in the ancestors of the body theory $AX(\mathcal{B})$ which are not ancestors of the parameter $AX(\mathcal{P})$. So, we can consider the following set

$$G \stackrel{def}{=} \{\overline{\sigma}'(prop\text{-}of_{\mathcal{B}}(a_1)), \ldots, \overline{\sigma}'(prop\text{-}of_{\mathcal{B}}(a_n))\}$$

It comprises exactly those propositions which have to be present in the context of the target theory in order to extend σ' to a theory morphism. Consequently, these propositions are called *theory morphism goals* of σ'.

In this situation we can get benefit of the fact that the instantiation procedure is integrated with a powerful LCF-style prover. This means that different generic automated proof tactics of Isabelle can be applied directly to the emerging theory morphism goals. Furthermore, these basic tactics can be combined to more special tactics solving special kinds of goals, leaving space for improvements. However, such tactics will by far not always succeed, but the set G can be possibly reduced this way to a set $G' \subseteq G$.

Now, proper and axiomatic differently treat the set G':

- A proper instantiation delegates the decision, which of the goals in G' can be proved in $\mathcal{T}_4$ and which can be again asserted as axioms, to the user. As soon as the theory morphism goals in G' are solved in some way, the signature morphism σ' can be extended to a theory morphism τ by e.g.
thymorph τ **by_sigmorph** σ'

- In contrast, an axiomatic instantiation asserts all theory morphism goals in G' as axioms, i.e. constructs the theory $\mathcal{T}_5$ and the theory morphism $\tau' : \mathcal{B} \longrightarrow \mathcal{T}_5$, as defined by the instantiation procedure. So, axiomatic instantiations should be used carefully, since this way inconsistencies could be introduced. On the other hand, they could be necessary as, e.g. the stack-example in the introduction shows: the theory *QuotientType* introduces new type and thus a type definition axiom, such that the instantiation in *StackImpl* has to be axiomatic. But in this case it is reasonable, since this single axiom forms a conservative extension.

Finally, in both these cases the translation of theorems in $Thms(\mathcal{B}) \setminus Thms(\mathcal{P})$ to $\mathcal{T}_5$ is skipped for practical reasons, in contrast to the last step in the definition of the instantiation procedure. As already mentioned, it is more practical to consider the theorems as 'implicitly' translated, i.e. translate these 'on-demand'.

At the top-level of Isabelle a proper instantiation of a parametrised theory $\langle \mathcal{P}, \mathcal{B} \rangle$ by a theory morphism $\tau : \mathcal{B} \longrightarrow \mathcal{I}$ is represented by the command
instantiate_theory $\mathcal{B}$ **by_thymorph** τ
renames*:* $\mathcal{D}_{renames}$
In contrast, an axiomatic instantiation requires the additional keyword **axiomatic**, i.e.
instantiate_theory $\mathcal{B}$ **by_thymorph** τ
renames*:* $\mathcal{D}_{renames}$
axiomatic
Also similar to the theorem translation is the **renames** clause: it provides the possibility for the user to give explicit new names to type classes, type constructors, constants, and axioms, which are translated from $Anc(\mathcal{B}) \setminus Anc(\mathcal{P})$ by the instantiation procedure. Although, this renaming facility is not essential for instantiation from the theoretical point of view, it turns out to be very useful in the practice.

Theory morphism goals arising by proper instantiations, as the set G' mentioned above, can be also considered more generally for any signature morphism σ as the answer to the question 'what is to prove in the target theory of σ in order to extend it to a theory morphism?'. At the top-level of Isabelle the answer to this question can be obtained via the special command **thymorph_goals** σ.

4.6 Conclusion

So far all notions and concepts required for the extension of a logical framework, specified in Chapter 2, by morphisms and instantiation of theories have been introduced. The case study in the next part of this thesis shows how these can be employed in formal developments in the most developed object logic of the theorem prover Isabelle, namely in the higher order logic.

Finally, we conclude this part with some considerations about some problems, which arise for generalisation of the instantiation of theories to computation of pushouts. Let

$$\mathcal{T}_1 \overset{\tau^1}{\longleftarrow} \mathcal{T} \overset{\tau^2}{\longrightarrow} \mathcal{T}_2$$

be a span without the restriction that τ_1 is an inclusion, as made in Definition 22. Then we are looking for the theory $\mathcal{T}_{pushout}$ together with theory morphisms $\tau^L : \mathcal{T}_1 \longrightarrow \mathcal{T}_{pushout}$, $\tau^R : \mathcal{T}_2 \longrightarrow \mathcal{T}_{pushout}$. The construction of the type maps σ^L_{type}, σ^R_{type} for τ^L and τ^R, respectively, is the main problem which will be sketched in the sequel.

Let $C \in \Sigma_{type}(\mathcal{T})$ with $rank_{\mathcal{T}}(C) = n$. Further, let $\varsigma_1 \in \widehat{Types}(\Sigma_{type}(\mathcal{T}_1))$, $\varsigma_2 \in \widehat{Types}(\Sigma_{type}(\mathcal{T}_2))$ with $\sigma^1_{type}(C) = \varsigma_1$ and $\sigma^2_{type}(C) = \varsigma_2$.

We are interested in type maps $\sigma^L_{type} : \Sigma_{type}(\mathcal{T}_1) \to \widehat{Types}(\Sigma_{type}(\mathcal{T}_{pushout}))$ and $\sigma^R_{type} : \Sigma_{type}(\mathcal{T}_2) \to \widehat{Types}(\Sigma_{type}(\mathcal{T}_{pushout}))$ such that

$$\sigma^L_{type} \circ \sigma^1_{type} = \sigma^R_{type} \circ \sigma^2_{type} \tag{4.18}$$

holds, and any further $\sigma^{L'}_{type}$ and $\sigma^{R'}_{type}$, satisfying (4.18), factor through σ^L_{type} and σ^R_{type}, respectively. Using the definition of the type map composition, the equation (4.18) can be reformulated:

$$\sigma^1_{type}(C) * \sigma^L_{type} = \sigma^2_{type}(C) * \sigma^R_{type}$$

has to be satisfied for any type constructor $C \in \Sigma_{type}(\mathcal{T})$. In turn, this is essentially the same as to solve

$$\varsigma_1 * \sigma^L_{type} = \varsigma_2 * \sigma^R_{type} \tag{4.19}$$

for some fixed arbitrary type schemes ς_1, ς_2 over the particular type signatures.

In the case when σ^1_{type} and σ^2_{type} are just rank-preserving mappings of type constructors from $\Sigma_{type}(\mathcal{T})$ to $\Sigma_{type}(\mathcal{T}_1)$ and $\Sigma_{type}(\mathcal{T}_2)$, respectively, we have $\varsigma_1 = (`1,\ldots,`n)\cdot C_1$ and $\varsigma_2 = (`1,\ldots,`n)\cdot C_2$. So, we can build the closure identifying C_1 and C_2. Further, for any equivalence class we introduce a type constructor $\overline{C}$ with the rank n in $\mathcal{T}_{pushout}$, such that we can assign $C_1 \mapsto (`1,\ldots,`n)\cdot\overline{C}$ and $C_2 \mapsto (`1,\ldots,`n)\cdot\overline{C}$.

Generally, (4.19) is not always solvable since, e.g. for $\varsigma_1 = `i$ and $\varsigma_2 = `j$ with $i \neq j$ no such σ^L_{type} and σ^R_{type} exist.

Next, let us consider a simple example when $\varsigma_1 = (`1,`2)\cdot C_1$ and $\varsigma_2 = (`1\cdot C_2)\cdot C_2$ in order to sketch some problems with the general treatment of pushouts. So, we enumerate the possible solutions to (4.19), for a start without taking the universal property into account.

1. $\sigma^L_{type}(C_1) \overset{def}{=} `1$, $\sigma^R_{type}(C_2) \overset{def}{=} `1$ is a solution, since

$$\begin{array}{ll} ((`1,`2)\cdot C_1) * \sigma^L_{type} & = \\ `1 \triangleright \langle `1 * \sigma^L_{type}, `2 * \sigma^L_{type}\rangle & = \\ `1 * \sigma^L_{type} & = \\ `1 & \end{array}$$

and

$$\begin{array}{ll} ((`1\cdot C_2)\cdot C_2) * \sigma^R_{type} & = \\ `1 \triangleright \langle (`1\cdot C_2) * \sigma^R_{type}\rangle & = \\ (`1\cdot C_2) * \sigma^R_{type} & = \\ `1 \triangleright \langle `1\rangle & = \\ `1 & \end{array}$$

2. Further, let $\overline{C} \in \Sigma_{type}(\mathcal{T}_{pushout})$ be a type constructor with the rank 0. Then $\sigma^L_{type}(C_1) \overset{def}{=} \overline{C}$, $\sigma^R_{type}(C_2) \overset{def}{=} \overline{C}$ is also a trivial solution, but in this case both *lhs* and *rhs* reduce to $\overline{C}$.

3. Next, let $\overline{\overline{C}} \in \Sigma_{type}(\mathcal{T}_{pushout})$ be a type constructor with the rank 1. Then $\sigma^L_{type}(C_1) \overset{def}{=} (`1\cdot\overline{\overline{C}})\cdot\overline{\overline{C}}$, $\sigma^R_{type}(C_2) \overset{def}{=} `1\cdot\overline{\overline{C}}$ is also a solution, since

$$\begin{array}{ll} ((`1,`2)\cdot C_1) * \sigma^L_{type} & = \\ ((`1\cdot\overline{\overline{C}})\cdot\overline{\overline{C}}) \triangleright \langle `1 * \sigma^L_{type}, `2 * \sigma^L_{type}\rangle & = \\ ((`1\cdot\overline{\overline{C}})\cdot\overline{\overline{C}}) \triangleright \langle `1, `2\rangle & = \\ (`1\cdot\overline{\overline{C}})\cdot\overline{\overline{C}}) & \end{array}$$

and

$$
\begin{array}{ll}
((`1 \cdot C_2) \cdot C_2) * \sigma^R_{type} & = \\
(`1 \cdot \overline{\overline{C}}) \rhd \langle (`1 \cdot C_2) * \sigma^R_{type} \rangle & = \\
(`1 \rhd \langle (`1 \cdot C_2) * \sigma^R_{type} \rangle) \cdot \overline{\overline{C}} & = \\
((`1 \cdot C_2) * \sigma^R_{type}) \cdot \overline{\overline{C}} & = \\
((`1 \cdot \overline{\overline{C}}) \rhd \langle `1 \rangle) \cdot \overline{\overline{C}} & = \\
(`1 \cdot \overline{\overline{C}}) \cdot \overline{\overline{C}} &
\end{array}
$$

This leads to the following observations.

1. The third solution is more general than the both previous, since it uses a type constructor $\overline{\overline{C}}$ with the rank 1. Therefore, the first solution factors through the last, provided by the assignment $\overline{\overline{C}} \mapsto$ '1, whereas the second by the assignment $\overline{\overline{C}} \mapsto \overline{C}$.

2. On the other hand, this solution cannot be further generalised by taking another type constructor with the rank > 1 in $\Sigma_{type}(\mathcal{T}_{pushout})$. That is, the rank of the 'unifying' type constructor in $\Sigma_{type}(\mathcal{T}_{pushout})$ is bounded by the minimum of the ranks of type constructors occurring in ς_1 and ς_2.

3. If we try to solve

$$((`1, `2) \cdot C_1) * \sigma^L_{type} = ((`1 \cdot C_2) \cdot C_2) * \sigma^R_{type}$$

systematically, using the hierarchy given by the structure of type schemes, we would firstly solve the problem for the corresponding sub-terms, i.e.

$$`1 * \sigma^L_{type} = (`1 \cdot C_2) * \sigma^R_{type}$$

This gives the solution $\sigma^R_{type}(C_2) \stackrel{def}{=} `1$. Though, extending this solution to the original problem leads to $\sigma^L_{type}(C_1) \stackrel{def}{=} `1$, and thus to the first solution, which is not the most general.

4. Thus, in order to find the most general solution, any sub-term of ς_1 which contain at least one type constructor has to be considered in connection with all sub-terms of ς_2, and then vice versa. That is, in this example we firstly try the assignment $C_1 \mapsto ((`1 \cdot C_2) \cdot C_2)$ and evaluate the *lhs*, which succeeds, i.e. yields the *rhs*. As the next step we would take $C_1 \mapsto (`1 \cdot C_2)$, but this assignment fails.

In general, not only one but all possible solutions could be required, because several solutions could be excluded if we proceed with further type constructors in $\Sigma_{type}(\mathcal{T})$.

At this point, it should be noted, why this efficiently works in the case of the computation of instantiation. There we act on the assumption that τ^1 is an inclusion. Thus, σ^1_{type} is an inclusion type map, i.e. we have the situation where $\sigma^1_{type}(C) = (`1, \ldots, `n) \cdot C$ for any $C \in \Sigma_{type}(\mathcal{T})$. This immediately allows us to define $\sigma^L_{type}(C) \stackrel{def}{=} \sigma^2_{type}(C)$, whereas σ^R_{type} is then an inclusion type map. This is also what the instantiation procedure implicitly does.

5. Altogether, we can conclude that the decision whether a pushout exists is in principle possible. Even though this problem is similar to the second order unification, it is in fact simpler, because of the bounded rank of type constructors introduced in $\mathcal{T}_{pushout}$. On the other hand, translating this problem of type constructor assignments to a second order unification problem, and then applying some of the already known algorithms to it, might be a reasonable approach.

Part II

Structured Development in Isabelle

Chapter 5

Allegories in the Higher Order Logic

5.1 Introduction

At this point the main structuring concepts have been formally founded and introduced in Isabelle such that we can start to employ these in concrete developments.

The presented case study is mainly inspired by [Bird and de Moor 1997], and therefore, not surprisingly, based on the notion of allegory, which is a specialisation of the notion of category, enriched with some specific properties of the category of relations **Rel**. It is a well-known fact that the set-theoretical relational calculus is a very suitable formalism for the development of correct programs, mainly because of the uniform treatment of programs (or more precisely of their relational semantics) and specifications as relations, providing a powerful framework for reasoning and transformation. From the relational point of view, any deterministic (or single-valued) relation can be seen as a (deterministic, possibly partial) program, and any relation as a specification. Thus, one can say that a program meets a specification if the latter relation includes the former.

In Isabelle/HOL a relation between the terms of type α and the terms of type β is usually represented as a function with the type $\alpha \times \beta \Rightarrow bool$, i.e. as a polymorphic predicate. Since sets and predicates are basically the same in HOL, this type is also abbreviated by $(\alpha \times \beta)set$. The representation $\alpha \Rightarrow$

β *set* can be taken as well. The advantage of the former is that it allows to use the common set operations directly, while the latter representation is more 'functional style' and stronger reflects non-determinism in specifications. So, with the former representation we can define the range of a relation by e.g. $ran\, r \equiv \{b \mid \exists a.\ (a, b) \in r\}$, whereas using the latter one by $ran\, r \equiv \{b \mid \exists a.\ b \in r\ a\}$.

Now, allegories lift the relational calculus to an abstract, categorical level. One of the effects is that no point-wise specifications, like the definitions of the range above, and reasoning are possible. That is, categorical point-free specifications and proofs are required. An interesting thing is, that such an abstraction of notions and properties to the categorical level is largely possible, which firstly brings more general structural insights, and secondly allows to re-use these in many different contexts.

5.2 Allegories

We start with the general definition of an allegory.

Definition 1 (allegory)
Let **C** be a category. Further, let denote the composition of two arrows $f : A \to B$ and $g : B \to C$ by $(f\ ;\ g) : A \to C$. Then **C** forms an *allegory* if

1. there is a *partial order* operator $\sqsubseteq$, comparing any two arrows $f : A \to B$ and $g : A \to B$ such that the composition of arrows is monotone w.r.t. $\sqsubseteq$,

2. there is a *converse* operator, which assigns to any arrow $f : A \to B$ its *converse arrow* $f^\circ : B \to A$, and which is a monotone (w.r.t. $\sqsubseteq$) and contra-variant involution (i.e. $(f^\circ)^\circ = f$ and $(f\ ;\ g)^\circ = g^\circ\ ;\ f^\circ$ hold),

3. there is a *meet* operator assigning to any two arrows $f : A \to B$ and $g : A \to B$ the arrow $(f \sqcap g) : A \to B$ which is the greatest lower bound of f and g w.r.t. $\sqsubseteq$, and, moreover satisfying the following so-called *modular law*:

$$(f\ ;\ g) \sqcap h \sqsubseteq ((h\ ;\ g^\circ) \sqcap f)\ ;\ g.$$

The basic problem is to find an appropriate way to formalise allegories in the higher order logic of Isabelle. Since arrows form the basic concept of categories, we first concentrate the efforts to find an appropriate type classifying arrows. Let again take the type $\alpha \Rightarrow \beta \Rightarrow bool$ of relations. This we can now consider from the point that there is an identity class map σ_{class}, some type map σ_{type}, and a type κ_0, such that

$$\overline{\sigma}_{type}^{\sigma_{class}}(\kappa_0) = \alpha \Rightarrow \beta \Rightarrow bool \tag{5.1}$$

holds. The most basic observation about κ_0 is that it has to contain at least the type variables α and β. On the other hand, this two variables are sufficient: assuming $\kappa_0 = (\alpha, \beta)\ C$ with some fixed type constructor C with the rank 2 and the type map σ_{type} constructed by
type_map: $[(\alpha, \beta)\ C \ \mapsto\ \alpha \Rightarrow \beta \Rightarrow bool]$
then the condition (5.1) is satisfied. Alternatively, we could also set e.g. $\kappa_0 \stackrel{def}{=} \alpha \Rightarrow \beta\, C'$ taking a type constructor C' with the rank 1 and the type assignment
type_map: $[\beta\ C' \ \mapsto\ \beta \Rightarrow bool]$
but the former setting is more general, since the set of types, generated from $\alpha \Rightarrow \beta\, C'$ by all possible type assignments for C', is a subset of types, generated from $(\alpha, \beta)\, C$. In other words, the former setting allows more type instantiations.

As the result of these considerations, we start the theory development with the theory *ArrowType*, importing the theory *Fun* (a short description can be found in Section 2.12.1) and just declaring the single type constructor $\rightarrow$ with the rank 2 (used as an infix operator) by
typedecl $\alpha \rightarrow \beta$
As mentioned in Section 2.12.1, this declaration also introduces the standard HOL-arity $(\{type\}, \{type\}, \{type\})$ for the type constructor $\rightarrow$, since *HOL* is an ancestor of *Fun*.

Further, we start the theory *Allegory* importing *ArrowType*. By this setting we obtain the parametrisation $\langle$*ArrowType*, *Allegory*$\rangle$, emphasising that further development in the context of *Allegory* is parameterised over the type constructor $\rightarrow$.

Then, we declare the logical constants for the identity arrows, as well as for composition, order, converse and meet operations in *Allegory* by:
consts

Id	::	$\alpha \rightarrow \alpha$
$\_\ ;\ \_$	::	$(\alpha \rightarrow \beta) \Rightarrow (\beta \rightarrow \gamma) \Rightarrow (\alpha \rightarrow \gamma)$
$\_ \sqsubseteq \_$	::	$(\alpha \rightarrow \beta) \Rightarrow (\alpha \rightarrow \beta) \Rightarrow bool$
$\_^{\circ}$	::	$(\alpha \rightarrow \beta) \Rightarrow (\beta \rightarrow \alpha)$
$\_ \sqcap \_$	::	$(\alpha \rightarrow \beta) \Rightarrow (\alpha \rightarrow \beta) \Rightarrow (\alpha \rightarrow \beta)$

Next, we formalise that identities and composition form a category by:
axioms

`Left_Id`	:	$Id\ ;\ f = f$
`Right_Id`	:	$f\ ;\ Id = f$
`comp_assoc`	:	$f\ ;\ g\ ;\ h = f\ ;\ (g\ ;\ h)$

and also axiomatise the additional properties of allegories, corresponding exactly to Definition 1.

Now let us consider in more detail, what has been axiomatised here. Any term $t \in \mathit{Terms}(\Sigma_{op}(\mathsf{Allegory}))$, having the type $\kappa_1 \to \kappa_2$ with arbitrary $\kappa_1, \kappa_2 \in \mathit{Types}(\Sigma_{type}(\mathsf{Allegory}))$, forms an arrow such that, e.g. $t \,;\, \mathit{Id} = t$ holds. Here, Id actually denotes the instance of Id for the type κ_2, i.e. the term

$$\mathit{Id} ::_{type} \mathit{type\text{-}of}_{\mathsf{Allegory}}(\mathit{Id})[\alpha := \kappa_2] = \mathit{Id} ::_{type} \kappa_2 \to \kappa_2$$

giving the identity arrow for κ_2. This justifies the following interpretation: the theory $\mathsf{Allegory}$ specifies an 'abstract' allegory $\mathbf{A}$, where

1. $\mathit{Obj}(\mathbf{A}) = \mathit{Types}(\Sigma_{type}(\mathsf{Allegory}))$,

2. $\mathit{hom}_{\mathbf{A}}(\kappa_1, \kappa_2) = \{t \mid t \ \mathit{of\text{-}type}_{\mathsf{Allegory}} \ \kappa_1 \to \kappa_2\}$ for any $\kappa_1, \kappa_2 \in \mathit{Obj}(\mathbf{A})$, i.e. $\mathit{hom}_{\mathbf{A}}(\kappa_1, \kappa_2) \subset \mathit{Terms}(\Sigma_{op}(\mathsf{Allegory}))$ holds.

Notice that no arrows apart of the identities are required to exist in $\mathbf{A}$, since $\mathit{Id} \,;\, \mathit{Id} = \mathit{Id}$, $\mathit{Id}^{\circ} = \mathit{Id}$ and $\mathit{Id} \sqcap \mathit{Id} = \mathit{Id}$ hold for all identities, because these propositions can be easily proved for $\mathit{Id} :: \alpha \to \alpha$ with a fixed α, i.e. independently of the choice of a particular type instance. This also justifies that any identity arrow forms a trivial sub-allegory of $\mathbf{A}$. On the other hand, $\mathbf{A}$ *could* have arrows beside the identities, which will be addressed via the set $\mathcal{X}_{term}$, i.e. our fixed infinite supply of term variables.

5.3 Predicates on arrows, range of an arrow

Many properties of relations can be reformulated for arrows in the theory $\mathsf{Allegory}$ in the point-free manner. So, an arrow $f :: \alpha \to \alpha$ is

1. *reflexive* if $\mathit{Id} \sqsubseteq f$

2. *symmetric* if $f \sqsubseteq f^{\circ}$

3. *transitive* if $f \,;\, f \sqsubseteq f$

4. *coreflexive* if $f \sqsubseteq \mathit{Id}$

Moreover, an arrow $f :: \alpha \to \beta$ is

1. *simple* if $f^{\circ} \,;\, f$ is coreflexive

2. *entire* if $f\ ;\ f^\circ$ is reflexive

3. a *mapping* if it is simple and entire

In the following, the most prominent rôle will be played by coreflexive arrows and mappings. The reason is that coreflexive arrows correspond to injections under the standard relational interpretation, and hence capture predicates. That is, any coreflexive arrow $f :: \alpha \rightarrow \alpha$ corresponds to an injection $r :: \alpha \Rightarrow \alpha\, set$ under the type assignment $\alpha \rightarrow \beta \mapsto \alpha \Rightarrow \beta\ set$, and hence to a subset of elements of the type α, i.e. to $\alpha\, set$. Therefore, coreflexive arrows are also called *monotypes*. In order to emphasise this, coreflexive arrows will be also denoted by capitals $A, B, \ldots$. Furthermore, any simple arrow corresponds to a single-valued relation (in other words to a partial map), and any entire arrow to a total relation, such that any mapping corresponds to a relation, which is indeed a mapping in the set-theoretical sense. Many important properties of these special sorts of arrows can be summarised, emphasising substructures constituted by them.

1. Coreflexive arrows form a sub-allegory denoted by *Corefl*($\mathbf{A}$), where the composition coincides with the greatest lower bound, i.e. $A\ ;\ B = A \sqcap B$. Thus, any coreflexive arrow is also symmetric and transitive. Moreover, coreflexive arrows are closed under the order 'downwards', i.e. if f is coreflexive and $g \sqsubseteq f$ holds then g is coreflexive as well.

2. Reflexive arrows also form a sub-allegory and are closed under the order 'upwards': if f is reflexive and $f \sqsubseteq g$ holds then g is reflexive as well.

3. Simple arrows form a sub-category and are closed under $\_ \sqcap \_$. Actually, even the stronger proposition holds: $f \sqcap g$ is simple if f does. Further, simple arrows distribute over $\_ \sqcap \_$ operator as follows:

   (a) $f\ ;\ (g \sqcap h) = (f\ ;\ g) \sqcap (f\ ;\ h)$ holds, provided f is simple,

   (b) $(g \sqcap h)\ ;\ f = (g\ ;\ f) \sqcap (h\ ;\ f)$ holds, provided f° is simple,

   such that the modular law becomes an equality

   $$(f\ ;\ g) \sqcap h = ((h\ ;\ g^\circ) \sqcap f)\ ;\ g$$

   provided g is simple. Similarly to coreflexive arrows, simple arrows are closed under the order 'downwards'.

4. Entire arrows form a sub-category and are closed under the order 'upwards'.

5. Two last points yield that mappings form a sub-category, denoted by $\mathit{Mappings}(\mathbf{A})$.

Moreover, transitive arrows are closed under $\_ \sqcap \_$ and converse, while symmetric arrows are basically the fixed points of the converse operator and are closed under $\_ \sqcap \_$.

Arrow transformers. Let us call terms, having the type of the form $(\_ \to \_) \Rightarrow (\_ \to \_)$, *arrow transformers*. Due to the higher order environment, we can use higher order predicates in order to qualify arrow transformers. So, we can define the predicate $\mathit{Mono} :: ((\alpha \to \beta) \Rightarrow (\gamma \to \delta)) \Rightarrow \mathit{bool}$ for *monotone* arrow transformers by $\mathit{Mono}\, F \equiv (\forall f\ g.\, f \sqsubseteq g \longrightarrow F f \sqsubseteq F g)$. This approach allows us to derive properties for a complete class of arrow transformers, and further obtain these properties immediately for any instance of the class. Actually, this treatment is closely related to the local proof context approach, sketched in the introduction.

For example, the greatest lower bound property $F(f \sqcap g) \sqsubseteq F f \sqcap F g$ can be shown for any monotone F. Further, for a fixed arrow $f :: \alpha \to \beta$, the term $(\lambda(x :: (\gamma \to \alpha)).\, x \ ;\ f)$ is of type $(\gamma \to \alpha) \Rightarrow (\gamma \to \beta)$, i.e. an arrow transformer. That it is also monotone, follows immediately from the definition of an allegory. Thus, we can directly conclude

$$(f \sqcap g) \ ;\ h \sqsubseteq f \ ;\ h \sqcap g \ ;\ h$$

for any f, g, h. Similarly, for a fixed arrow $f :: \alpha \to \beta$, the arrow transformer $(\lambda(x :: (\alpha \to \beta)).\, f \sqcap x)$ is monotone as well.

Moreover, the partial order $\sqsubseteq$ on arrows induces the partial order on arrow transformers, defined point-wise by $F \mathrel{\dot{\sqsubseteq}} G \equiv (\forall f.\, F f \sqsubseteq G f)$.

Adjoints. We can also consider the pairs of *adjoint* arrow transformers: two transformers $F :: (\alpha \to \beta) \Rightarrow (\gamma \to \delta)$ and $G :: (\gamma \to \delta) \Rightarrow (\alpha \to \beta)$ form an adjoint pair if

$$(F f \sqsubseteq g) = (f \sqsubseteq G g)$$

holds for all arrows $f :: \alpha \to \beta$ and $g :: \gamma \to \delta$, and F, G are called *lower* and *upper* adjoints, respectively. Thus, adjoint pairs are special case of Galois connections (e.g. [Backhouse 2002]), which are generally defined between two, possibly different, partial orders.

Range and domain. One of the basic properties of arrows, which relies on the modular law, is the following: $f \sqsubseteq f \,;\, (f^{\circ} \,;\, f)$. That is, $f^{\circ} \,;\, f$ is an arrow satisfying $f \sqsubseteq f \,;\, x$. Inspired by the set-theoretical notion of the range of a relation, the *range* of an arrow $f :: \alpha \to \beta$ is the arrow $\mathit{ran}\, f :: \beta \to \beta$, defined by $\mathit{ran} f \equiv (f^{\circ} \,;\, f) \sqcap \mathit{Id}$. Obviously, $\mathit{ran} f$ is always coreflexive by the definition. Furthermore, the crucial property of $\mathit{ran} f$ is that it is the least coreflexive arrow satisfying $f \sqsubseteq f \,;\, x$, i.e. the following universal property can be proved:

$$(\mathit{ran}\, f \sqsubseteq x) \;=\; (f \sqsubseteq f \,;\, x) \quad \bullet \text{ provided } x \text{ is coreflexive} \tag{5.2}$$

More precisely, the direction $\mathit{ran} f \sqsubseteq x \Longrightarrow f \sqsubseteq f \,;\, x$ holds for any $x :: \beta \to \beta$, whereas the direction $f \sqsubseteq f \,;\, x \Longrightarrow \mathit{ran} f \sqsubseteq x$ requires the assumption that x is coreflexive.

Some of the further derived properties of the range are the following:

1. $f \,;\, \mathit{ran}\, f = f$ holds for any $f :: \alpha \to \beta$,
2. $\mathit{ran}\, f = f$ holds for any coreflexive f,
3. ran is a monotone arrow transformer, which in turn implies that $\mathit{ran}\,(f \sqcap g) \sqsubseteq \mathit{ran}\, f \sqcap \mathit{ran}\, g$ holds,
4. regarding the range of the composition of arrows, we firstly have the inequality $\mathit{ran}\,(f \,;\, g) \sqsubseteq \mathit{ran}\, g$, and secondly the equality $\mathit{ran}\,(f \,;\, g) = \mathit{ran}\,(\mathit{ran} f \,;\, g)$,
5. regarding the range of the greatest lower bound of arrows, we have the equality $\mathit{ran}\,(f \sqcap g) = (f^{\circ} \,;\, g) \sqcap \mathit{Id}$.

The *domain* $\mathit{dom} f$ of an arrow $f :: \alpha \to \beta$ is then defined to be the range of the converse arrow of f, i.e. $\mathit{dom} f \equiv \mathit{ran} f^{\circ}$. This notion is especially interesting in connection with entire arrows. So, the following property of entire arrows is proved: an arrow $f :: \alpha \to \beta$ is entire iff $\mathit{dom}\, f = \mathit{Id}$ holds. In particular, this can be used to show that whenever an arrow $f \,;\, g$ is entire, so is the arrow f.

Image. In the relational calculus the image of a relation r under a set A is defined to be the set $\{b \mid \exists a \in A.\, b \in r\, a\}$ and hence can be seen as a generalisation of the notion of the range. In the context of allegories we do not have an explicit notion of a set, but, as already mentioned, can use coreflexive arrows to mimic sets. Furthermore, a composition $A \,;\, f$, where A is coreflexive, can be seen as a restriction of f to the subset, represented by A.

Let $f :: \alpha \to \beta$ and $g :: \alpha \to \alpha$ be two arrows. Then the *image* of f *under* g is denoted by $Im\, f\ g$ and defined by $ran\ (g\ ;\ f)$. So, $Im\, f\ A$ corresponds to the relational image if A is a coreflexive arrow. Remarkable here is that image is defined as a special case of range. Therefore, several properties of the range are inherited by the image, e.g. monotonicity (in the first and the second argument, i.e. $Mono(Im\, f)$ and $Mono(\lambda x.\ Im\ x\ g)$), as well as coreflexivity of $Im\, f\ g$ for any f and g. Furthermore, the following properties of the image can be proved in the context of allegories.

1. $Im\, f\ g \sqsubseteq ran\, f$,
2. $Im\, f\ (A \sqcap B) = Im\, f\ A \sqcap Im\, f\ B$, provided f° is simple and A, B coreflexive,
3. $Im\ (f\ ;\ g)\ A = Im\ g\ (Im\, f\ A)$,
4. $Im\, f\ (dom\, f) = ran\, f$ and also $Im\, f\ Id = ran\, f$.

This notion of the image will also play an important rôle in the next chapter regarding locally complete allegories.

5.4 Two example instances

The next two sections give firstly a quite concrete model of an allegory and secondly a special class of allegories, namely bounded semi-lattices. The aim here is also to show in more detail how theory morphisms and instantiation of theories work in practice.

5.4.1 A 4-element allegory

Let us start the theory *Mod4* importing the theory *Datatype*, which provides the derived concept for declaration of algebraic data types in Isabelle/HOL. Moreover, *Datatype* has the theory *Fun* as an ancestor such that the common ancestors of *Mod4* and *ArrowType* are the ancestors of *Fun*. This means, especially, that any morphism from *ArrowType* to *Mod4* has only *ArrowType* in its domain.

So, using the datatype concept, we specify the type *arrow*, representing a set containing 4 elements, as follows:
datatype $arrow = A \mid B \mid C \mid D$
In the first step we declare a theory morphism from *ArrowType* to *Mod4* by
thymorph *mod4* : *ArrowType* $\longrightarrow$ *Mod4*
type_map: $[(\alpha \to \beta) \mapsto arrow]$

That is, we interpret here the formal type parameter $\rightarrow$ of allegories by *arrow* in *Mod4*. Further, let us consider the span

$$\mathsf{Allegory} \hookleftarrow \mathsf{ArrowType} \stackrel{\mathsf{mod4}}{\longrightarrow} \mathsf{Mod4}$$

and the following proper instantiation
instantiate_theory *Allegory* **by_thymorph** *mod4*
which translates the logical constants from *Allegory* to *Mod4*, i.e.

$\_ \sqcap \_$	::	$arrow \Rightarrow arrow \Rightarrow arrow$
$\_ \sqsubseteq \_$	::	$arrow \Rightarrow arrow \Rightarrow bool$
Id	::	$arrow$
$\_^{\circ}$	::	$arrow \Rightarrow arrow$
$\_ ; \_$	::	$arrow \Rightarrow arrow \Rightarrow arrow$

are added to *Mod4*. Since *Allegory* contains an axiomatic specification and this instantiation is proper, it returns merely the signature morphism *mod4'* : *Allegory* $\longrightarrow$ *Mod4*, generating additionally the theory morphism goals, comprising allegory laws for the operations above. These propositions have to be provided by *Mod4* in order to extend *mod4'* to a theory morphism. To this end, we make the logical constants above derived, defining these as follows:

1. the definitions of the order and the meet operation are represented by the Hasse-diagram shown in Figure 5.1
2. the identity is defined by $Id \equiv B$
3. the converse is defined by $x^{\circ} \equiv x$
4. the composition is defined by means of the following table:

;	A	B	C	D
A	A	A	A	A
B	A	B	C	D
C	A	C	C	D
D	A	D	D	C

Provided by these definitions, allegory laws can be proved, such that the signature morphism *mod4'* can be extended to a theory morphism by
thymorph *mod4_as_allegory* **by_sigmorph** *mod4'*
This theory morphism imposes the allegory **4** where

1. $Obj(\mathbf{4}) = Types(\Sigma_{type}(\mathsf{Mod4}))$;

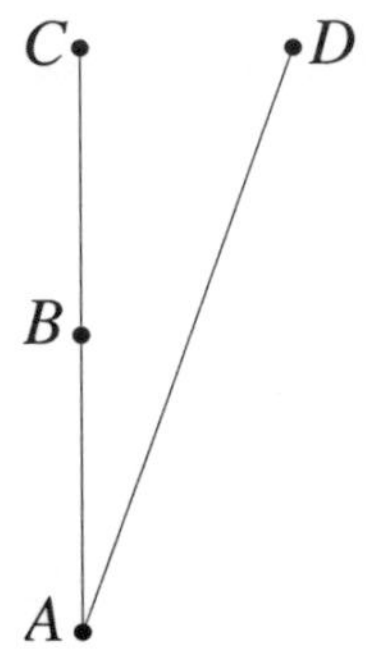

Figure 5.1: The order in *arrow*

2. $hom_{\mathbf{4}}(\kappa_1, \kappa_2) = \{t \mid t \ \textit{of-type}_{\mathsf{Mod4}} \ \textit{arrow}\}$ for any $\kappa_1, \kappa_2 \in \mathit{Obj}(\mathbf{4})$, which reduces to $hom_{\mathbf{4}}(\kappa_1, \kappa_2) = \{A, B, C, D\}$. That is, there is always the same set of arrows between any two objects in **4**, which means that **4** is basically an allegory with one object and four arrows.

Finally, interpreting the abstractly defined predicates on arrows in this special case we obtain, e.g. that A is coreflexive and therefore simple, C is reflexive and therefore entire, while D is just entire.

5.4.2 Bounded semi-lattices

This example instance should emphasise some aspects in the connection between allegories and lattices. To this end, we start the theory *BoundedSemilattice* having only *Fun* as its parent theory. This way, any morphism from *Allegory* to *BoundedSemilattice* will have only *ArrowType* and *Allegory* in its domain.

Firstly we declare the type constructor with the rank 0 by
typedecl L
together with the logical constants

$$\_ \sqcap \_ \ :: \ L \Rightarrow L \Rightarrow L$$
$$\top \ :: \ L$$

and the well-known axioms for these, specifying $a \sqcap b$ as the greatest lower bound of a and b, and $\top$ as the greatest element:

`Idem`	:	$a \sqcap a = a$
`Assoc`	:	$(a \sqcap b) \sqcap c = a \sqcap (b \sqcap c)$
`Comm`	:	$a \sqcap b = b \sqcap a$
`Top`	:	$\top \sqcap a = a$

Then, the order on L is defined in the standard way by $a \sqsubseteq b \equiv (a \sqcap b = a)$, while the converse operator is simply defined to be the identity on L.

So, we can declare the following signature morphism

sigmorph *semilattice* : *Allegory* $\longrightarrow$ *BoundedSemilattice*

type_map: $[(\alpha \to \beta) \mapsto L]$

op_map: $[\_ \sqcap \_ \mapsto \_ \sqcap \_,$
$\_ ; \_ \mapsto \_ \sqcap \_,$
$Id \mapsto \top,$
$\ldots]$

Firstly, the type assignment imposes the interpretation of any bounded semilattice with elements of a monomorphic type as an allegory with a single object and possibly infinitely many arrows. Secondly, both operations $\_ \sqcap \_$ and $\_ ; \_$ from *Allegory* are mapped to $\_ \sqcap \_$ in *BoundedSemilattice*, which also explains the assignment of $\top$ to Id. The remaining assignments for the order and the converse are clear and hence omitted. In fact, in this case the assignment $Id \mapsto \top$ can be omitted from the declaration above as well, because $\top$ is the unique constant in the operation signature of *BoundedSemilattice* satisfying the operation map condition for Id w.r.t. the type assignment $(\alpha \to \beta) \mapsto L$, such that $Id \mapsto \top$ is inferred.

In order to extend this signature morphism to a theory morphism, the allegory laws for this interpretation have to be shown. So, for example, the modular law obtains the following form here:

$$(a \sqcap b) \sqcap c \sqsubseteq (a \sqcap (c \sqcap b)) \sqcap b$$

and follows from the axioms for the greatest lower bound. This also emphasises that the modular law should not be confused with the modular equality, characterising modular lattices (where $a \sqcup b$ denotes the least upper bound of a and b):

$$(a \sqcap c) \sqcup (b \sqcap c) = ((a \sqcap c) \sqcup b) \sqcap c$$

which does not hold here, primarily because least upper bounds not necessarily exist.

Now we are able to extend the constructed signature morphism to a theory morphism by, e.g.

thymorph *semilattice_as_allegory* **by sigmorph** *semilattice*

Finally, interpreting the abstractly defined predicates on arrows in this particular context, we obtain that all arrows are coreflexive, and therefore transitive and simple, while only $\top$ is additionally entire, and hence the unique mapping.

5.5 Conclusion

In this chapter the notion of allegory has been introduced and axiomatically formalised in the higher order logic of Isabelle abstracting over the particular type of arrows. Further, the central predicates on arrows, e.g. mappings, as well as arrow constructions, e.g. image of an arrow, have been considered. Moreover, two particular structures have been formalised in Isabelle, and the fact that these structures model allegories has been established via construction of theory morphisms.

Chapter 6

Locally Complete Allegories

In the following we want to consider a special class of allegories, namely so-called *locally complete*. The axiomatisation of allegories in the previous chapter has not required neither the existence of the least upper bound of two arrows nor the existence of the greatest lower bound of an arbitrary set of arrows of the same type. So, introducing an additional operation we obtain both of them.

Definition 2 (locally complete allegory)
An allegory **A** is said to be *locally complete* if for any set $\mathcal{S}$ of arrow from A to B there exists an arrow $\bigsqcup \mathcal{S} : A \to B$, called the *join* of $\mathcal{S}$, satisfying the following three rules:

1. $(\bigsqcup \mathcal{S} \sqsubseteq f) = (\forall g \in \mathcal{S}.\, g \sqsubseteq f)$, i.e. $\bigsqcup \mathcal{S}$ is the least upper bound of $\mathcal{S}$,
2. $f \sqcap \bigsqcup \mathcal{S} \sqsubseteq \bigsqcup\{f \sqcap g \mid g \in \mathcal{S}\}$,
3. $f \,;\, \bigsqcup \mathcal{S} \sqsubseteq \bigsqcup\{f \,;\, g \mid g \in \mathcal{S}\}$

In order to extend allegories in Isabelle by the join operator, the theory *LocallyComplete*, importing only the theory *Allegory*, is started. Here, we firstly introduce the logical constant for join
consts
$\bigsqcup \_ :: (\alpha \to \beta)set \Rightarrow (\alpha \to \beta)$
and, secondly, formulate the three axioms, as stated in the definition above. The join operator strongly increases the expressivity, reducing the set of models of allegories. As a simple example we can take the 4-element allegory from

Section 5.4.1, which is not locally complete, since e.g. the set $\{C, D\}$ does not have any upper bound.

So, already by the first axiom from Definition 2 we obtain

1. the polymorphic *bottom arrow* $\bot :: \alpha \to \beta$, defined by $\bot \equiv \bigsqcup \emptyset$, and the polymorphic *top arrow* $\top :: \alpha \to \beta$, defined by $\top \equiv \bigsqcup \mathit{UNIV}$, ($\mathit{UNIV}$ is basically the polymorphic predicate $(\lambda x.\ \mathit{True})$, and so denotes in this case the set of all arrows $\alpha \to \beta$),

2. the least upper bound of two arrows (*binary join*) $f \sqcup g \equiv \bigsqcup \{f, g\}$,

3. the greatest lower bound of a set S of arrows can be now defined by
$$\bigsqcup \{x \mid \forall f \in S.\, x \sqsubseteq f\}$$

4. the property $(\bigsqcup S)^\circ = \bigsqcup \{f^\circ \mid f \in S\}$, since the converse operator is a monotone involution,

5. If all arrows $f \in S$ with $S :: (\alpha \to \alpha)\mathit{set}$ are coreflexive then the arrow $\bigsqcup S$ is coreflexive as well, which essentially establishes that the sub-allegory of coreflexive arrows is locally complete. The corresponding proposition does not hold for the sub-category of reflexive arrows exactly because $\bot$ is (by far) not always reflexive.

Furthermore, in connection with monotone arrow transformers, from the universal property of the join operator we can also deduce

$$F\left(\bigsqcup S\right) \sqsupseteq \bigsqcup \{F\ x \mid x \in S\} \quad \bullet \text{ provided } F \text{ is monotone}$$

Then,

1. $f \sqcap \bigsqcup S \sqsupseteq \bigsqcup \{f \sqcap g \mid g \in S\}$ holds, since $(\lambda x. f \sqcap x)$ is a monotone arrow transformer, and hence together with the second axiom the equality
$$f \sqcap \bigsqcup S = \bigsqcup \{f \sqcap g \mid g \in S\}$$
holds,

2. $f\,;\, \bigsqcup S \sqsupseteq \bigsqcup \{f\,;\, g \mid g \in S\}$ holds, because $(\lambda x. f\,;\, x)$ is a monotone arrow transformer, and hence together with the third axiom the equality
$$f\,;\bigsqcup S = \bigsqcup \{f\,;\, g \mid g \in S\} \tag{6.1}$$
holds as well,

3. $(\bigsqcup S) ; f \sqsupseteq \bigsqcup\{g ; f \mid g \in S\}$ holds, also because $(\lambda x . x ; f)$ is a monotone arrow transformer, but we cannot say anything yet about the $\sqsubseteq$-direction.

Division. The definition of the greatest lower bound of a set of arrows above has already pointed at the possibility to define the greatest arrow (w.r.t. $\sqsubseteq$) satisfying a given condition, by means of the join operator. Now we will use this property of join to define the (right) division on arrows.

Let $f :: \alpha \rightarrow \beta$ and $g :: \alpha \rightarrow \gamma$ be arbitrary arrows. Then the *right division* arrow $f \swarrow g$ is defined to be the greatest arrow $x :: \gamma \rightarrow \beta$ satisfying $g ; x \sqsubseteq f$, i.e. $f \swarrow g \equiv (\bigsqcup \{x \mid g ; x \sqsubseteq f\})$. By this definition the following universal property is satisfied:

$$(g ; x \sqsubseteq f) = (x \sqsubseteq f \swarrow g) \tag{6.2}$$

In other words, the arrow transformer $(\lambda f .\ f \swarrow g)$ is the upper adjoint for $(\lambda x .\ g ; x)$. The proof of (6.2) mainly relies on the universal property of the join and the equation (6.1).

Consequently, for any $f :: \alpha \rightarrow \beta$ and $g :: \gamma \rightarrow \beta$ the *left division* arrow $f \nearrow g$ is the greatest $x :: \gamma \rightarrow \alpha$ satisfying $x ; f \sqsubseteq g$, but can be now defined in terms of the right division and converse: $f \nearrow g \equiv (g^{\circ} \swarrow f^{\circ})^{\circ}$. Then the proof of the universal property

$$(x ; f \sqsubseteq g) = (x \sqsubseteq f \nearrow g) \tag{6.3}$$

uses the universal property of the right division and that the converse operator is a monotone involution. Now, a proof of

$$\left(\bigsqcup S\right) ; f \sqsubseteq \bigsqcup\{g ; f \mid g \in S\}$$

is possible, because the proposition is equivalent to $\bigsqcup S \sqsubseteq f \nearrow (\bigsqcup\{g ; f \mid g \in S\})$ by (6.3), such that the universal property of the join operator can be applied. Altogether, we obtain also the equality

$$\left(\bigsqcup S\right) ; f = \bigsqcup\{g ; f \mid g \in S\} \tag{6.4}$$

The monotype factor. An operation, similar to the division, will be now defined for coreflexive arrows. The motivation behind this operation is to lift the notion of the inverse image of a relation under a set, familiar from the relational calculus, to the level of locally complete allegories. So, if $r :: \alpha \Rightarrow \beta\ set$ is a relation and $B :: \beta\ set$ a set of elements of type β, then the set-theoretical inverse image of r under B is the set $\{a \mid \forall b \in r\ a.\ b \in B\} :: \alpha\ set$. Furthermore, one of the basic properties of the inverse image, proved for relations, is that it forms the upper adjoint of the image operator. This fact and the already in Section 5.3 defined notion of the image of an arrow provide the abstraction of the notion of the inverse image to the notion of the monotype factor.

Let $f :: \alpha \to \beta$ and $g :: \beta \to \beta$ be two arrows. Then the *monotype factor* $f \setminus g$ of f under g is the greatest *coreflexive* arrow $X :: \alpha \to \alpha$ satisfying the condition $Im\ f\ X \sqsubseteq g$, i.e. $f \setminus g \equiv \bigsqcup\{X \mid corefl\ X \wedge Im\ f\ X \sqsubseteq g\}$. By this definition and the properties of join, the universal property of the monotype factor, namely

$$(Im\ f\ A \sqsubseteq g) = (A \sqsubseteq f \setminus g) \quad \bullet \text{ provided } A \text{ is coreflexive} \tag{6.5}$$

can be deduced. Some of the further properties of the monotype factor are:

1. $(f \setminus g)\ ;\ f \sqsubseteq f\ ;\ g$ holds for any f and g,
2. $f \setminus (g \setminus B) \sqsubseteq (f\ ;\ g) \setminus B$ holds for any f, g and coreflexive B,
3. $(f \setminus g)\ ;\ dom\ f \sqsubseteq dom\,(f\ ;\ g)$ holds for any f and g,
4. $f \setminus (g \sqcap h) = (f \setminus g) \sqcap (f \setminus h)$ holds for any f, g, h.

6.1 Fixed points and Kleene algebras

Now we can turn towards the construction of fixed points in locally complete allegories. To this end, we start the theory *Fixed_Points* importing *LocallyComplete* as well as *Nat*. As already mentioned in Section 2.12.1, the theory *Nat* formalises the natural numbers (type *nat*) in Isabelle/HOL and provides recursive definitions and inductive proofs for *nat*. The reason for the use of the natural numbers is that least fixed points will be constructed by ascending chains of arrows. So, a *sequence* of arrows of type $\alpha \to \beta$ is represented by a function $nat \Rightarrow (\alpha \to \beta)$, while an (ascending) *chain* of arrows of type $\alpha \to \beta$ is a sequence s, satisfying $s_n \sqsubseteq s_{n+1}$.

One of the basic properties of chains in connection with the join operator is that the composition of the least upper bounds of two chains is the same as the least upper bound of their point-wise composition:

$$\left(\bigsqcup_n s_n\right)\ ;\ \left(\bigsqcup_n s'_n\right) = \bigsqcup_n (s_n\ ;\ s'_n) \quad \bullet \text{ provided } s, s' \text{ are chains.} \tag{6.6}$$

This equality provides a proof of the following important lemma.

Lemma 6.1 *For any chain s of simple arrows, the least upper bound $\bigsqcup_n s_n$ is a simple arrow as well.*

Proof. By the definition of simple arrows, we need to show that the arrow $(\bigsqcup_n s_n)^\circ$; $(\bigsqcup_n s_n)$ is coreflexive. By the properties of the join we have:

$$\begin{array}{ll} (\bigsqcup_n s_n)^\circ \,;\, (\bigsqcup_n s_n) & = \\ (\bigsqcup_n s_n{}^\circ) \,;\, (\bigsqcup_n s_n) & = \\ \bigsqcup_n (s_n{}^\circ \,;\, s_n) & \end{array}$$

As already shown, $\bigsqcup_n (s_n{}^\circ \,;\, s_n)$ is coreflexive if $s_n{}^\circ \,;\, s_n$ is coreflexive for any n, or in other words if s_n is simple for any n, which has been assumed. □

Continuity. Similarly to the monotonicity predicate on arrow transformers, we can now also define the continuity predicate. So, an arrow transformer $F :: (\alpha \to \beta) \Rightarrow (\gamma \to \delta)$ is *continuous*, if for any chain s

$$F\left(\bigsqcup_n s_n\right) \sqsubseteq \bigsqcup_n F(s_n)$$

holds. This formulation differs from the usual, where the equality $F(\bigsqcup_n s_n) = \bigsqcup_n F(s_n)$ is required, and is somewhat weaker. The reason for this is that in this context basically only monotone arrow transformers are considered, such that the direction $F(\bigsqcup_n s_n) \sqsupseteq \bigsqcup_n F(s_n)$ can be derived.

Least fixed points. Let $F :: (\alpha \to \beta) \Rightarrow (\alpha \to \beta)$ be an arrow transformer. First of all, it induces the sequence *Iter* F, which is recursively defined by

$$\begin{array}{lll} (\mathit{Iter}\ F)_0 & = & \bot \\ (\mathit{Iter}\ F)_{n+1} & = & F((\mathit{Iter}\ F)_n) \end{array}$$

So, *Iter* F is a chain for any monotone F. The *least fixed point* of F is the arrow $\mu F :: \alpha \to \beta$ defined by

$$\mu F \equiv \bigsqcup_n (\mathit{Iter}\ F)_n$$

By this definition, we can conclude

1. $F(\mu F) \sqsubseteq \mu F$, provided F is monotone and continuous,
2. for any f with $F\,f \sqsubseteq f$, $\mu F \sqsubseteq f$ holds, provided F is monotone,

from which follows that μF is indeed the least fixed point of a monotone and continuous arrow transformer F.

Furthermore, the construction of μF by chains provides a proof by induction on *nat* of the following general fixed point induction rule.

Proposition 6.1 *Let F be a monotone arrow transformer and P a predicate on arrows satisfying the following properties:*

1. *$P\bot$ holds,*
2. *$P(\bigsqcup_n s_n)$ holds for any chain s with $P(s_n)$ for any n,*
3. *$P\,f$ implies $P(F\,f)$ for any arrow f.*

Then $P(\mu F)$ holds.

In particular, taking P to be the predicate for simple arrows, we obtain that μF is simple if F is monotone and preserves simple arrows.

μ-fusion. Another general derived proposition about least fixed points is the following so-called μ-fusion rule (see e.g. [Backhouse 2002]).

Proposition 6.2 *Let $H :: (\alpha \to \beta) \Rightarrow (\alpha \to \beta)$ and $G :: (\gamma \to \delta) \Rightarrow (\gamma \to \delta)$ be monotone arrow transformers, where G is additionally also assumed to be continuous. Moreover, let $F :: (\alpha \to \beta) \Rightarrow (\gamma \to \delta)$ be an arrow transformer for which an upper adjoint $U :: (\gamma \to \delta) \Rightarrow (\alpha \to \beta)$ exists. Then $F \circ H \mathrel{\dot{\sqsubseteq}} G \circ F$ implies $F(\mu H) \sqsubseteq \mu G$.*

In particular, as already mentioned, for the arrow transformer $(\lambda x.\, x \,;\, f)$ the left division $(\lambda y.\, f \nearrow y)$ forms the upper adjoint, such that taking $(\lambda x.\, x \,;\, f)$ for F, we obtain:

Corollary 6.2.1 *$\mu H \,;\, f \sqsubseteq \mu G$ holds if $(H\,x) \,;\, f \sqsubseteq G(x \,;\, f)$ holds for any x, where H is monotone and G is monotone and continuous.*

This special case of the μ-fusion will be used later on in the context of hylomorphisms.

Greatest fixed points. Finally, the *greatest fixed point* of $F :: (\alpha \to \beta) \Rightarrow (\alpha \to \beta)$ can be defined by $\nu F \equiv \bigsqcup \{f \mid f \sqsubseteq F\,f\}$, such that $\nu F = F(\nu F)$ can be shown for any monotone F.

6.1.1 Kleene algebra

At this point the connection of the locally complete allegories to typed Kleene algebras can be elaborated. More precisely, using theory morphisms we can show that any locally complete allegory forms a typed Kleene algebra, already mentioned in the introduction (Section 1.4). In the following, typed Kleene algebras will be just called Kleene algebras, and are formalised in a similar manner as in [Kozen 1998] by means of the axiomatic theory *KleeneAlgebra* importing only *HOL*.

As the type of elements in the domain we take the fixed type constructor K with the rank 2:
typedecl $(\alpha, \beta)\ K$
This declaration provides the intended interpretation of domain elements in a Kleene algebra as arrows in locally complete allegories, i.e. the intended type assignment is $(\alpha, \beta)\ K \mapsto (\alpha \rightarrow \beta)$. Further, the operations on $(\alpha, \beta)\ K$ are specified by the following polymorphic logical constants:
consts

0	::	$(\alpha, \beta)\ K$
1	::	$(\alpha, \alpha)\ K$
$\_ + \_$	::	$(\alpha, \beta)\ K \Rightarrow (\alpha, \beta)\ K \Rightarrow (\alpha, \beta)\ K$
$\_ \cdot \_$	::	$(\alpha, \beta)\ K \Rightarrow (\beta, \gamma)\ K \Rightarrow (\alpha, \gamma)\ K$
$\_^\star$	::	$(\alpha, \alpha)\ K \Rightarrow (\alpha, \alpha)\ K$

such that the allegorical interpretation is almost complete:
sigmorph *lcomplete_as_Kleene_alg* : *KleeneAlgebra* $\longrightarrow$ *Fixed_Points*
type_map: $[(\alpha, \beta)K \mapsto (\alpha \rightarrow \beta)]$
op_map:

$$[\_ + \_ \mapsto \_ \sqcup \_,\quad \_ \cdot \_ \mapsto \_ \,;\, \_,\quad 0 \mapsto \bot,\quad 1 \mapsto Id,\quad \_^\star \mapsto ?]$$

So, it remains to give a definition for the Kleene-star in locally complete allegories. This is, of course, the reflexive transitive closure of an arrow: let $f :: \alpha \rightarrow \alpha$ be an arrow then the *reflexive transitive* closure of f is the arrow $f^\star$, defined by $f^\star \equiv \mu(\lambda x.\ Id \sqcup (f\ ;\ x))$.

In order to extend the signature morphism above to a theory morphism, the axioms of the Kleene algebra have to be proved in the context of locally complete allegories. The essential part form the proofs of the four Kleene-star axioms for the definition of the reflexive transitive closure of an arrow, which comprise the following propositions:

1. $f\ ;\ x \sqsubseteq x \Longrightarrow f^\star\ ;\ x \sqsubseteq x$

2. $x ; f \sqsubseteq x \Longrightarrow x \sqsubseteq f^\star ; x$
3. $Id \sqcup (f ; f^\star) \sqsubseteq f^\star$
4. $Id \sqcup (f^\star ; f) \sqsubseteq f^\star$

Finally, we obtain the theory morphism
thymorph *lcomplete_as_Kleene_alg* **by_sigmorph** *lcomplete_as_Kleene_alg*
from *KleeneAlgebra* to *Fixed_Points*.

Languages. In order to justify the formalisation of Kleene algebras above, we firstly explore the connection to formal languages, and secondly to the special case of regular languages. To this end, we start the theory *RegularLang* importing the main HOL-theory *Main*. Here we declare
typedecl Σ
which should represent the type of elements in a fixed alphabet. So, words are then of the type Σ *list*, i.e. finite lists (or strings) over the alphabet Σ. Consequently, any set of strings forms a language, i.e. a language is a term having the type (Σ *list*) *set*. Now, the Kleene algebra operations can be defined in the standard way:

0	$\equiv$	$\emptyset$
1	$\equiv$	$\{[\,]\}$ i.e. the set containing only the empty list
$L_1 + L_2$	$\equiv$	$L_1 \cup L_2$
$L_1 \cdot L_2$	$\equiv$	$\{u@v \mid u \in L_1, v \in L_2\}$ where @ denotes the concatenation
$L^\star$	$\equiv$	$\bigcup_n L^n$ where $L^n \equiv \underbrace{L \cdot \ldots \cdot L}_{n-times}$

This establishes the theory morphism
thymorph *lang_as_Kleene_algebra* : *KleeneAlgebra* $\longrightarrow$ *RegularLang*
type_map: $[(\alpha, \beta)\ K \mapsto (\Sigma\ list)\ set]$
op_map: ...
There are two important notes on this language algebra:

1. The type Σ basically does not say anything about the finiteness of our alphabets. So, e.g. by the type assignment $\Sigma \mapsto nat$ we can have natural numbers as the alphabet. However, neither the definitions above nor the proofs of the Kleene algebra axioms rely on the infinity of Σ.

2. The type (Σ *list*) *set* of languages covers all possible languages over Σ, i.e. in particular also non-regular. So, for instance, $L \equiv \{u@v \mid length\ u = length\ v\}$ defines the language $u^n v^n$, which is well-known to be non-regular.

Based on the well-known Kleene theorem, we can use the regular expressions in order to pick out the set of the regular languages. To this end, we firstly specify the regular expressions over Σ by the algebraic data type *RegExpr*

$$\begin{array}{lcl} RegExpr & = & Empty \\ & | & Eps \\ & | & Inj\ \Sigma \\ & | & Alt\ RegExpr\ RegExpr \\ & | & Con\ RegExpr\ RegExpr \\ & | & Star\ RegExpr \end{array}$$

and secondly the evaluation $\|\_\| :: RegExpr \Rightarrow (\Sigma\ list)\ set$, using the language operations defined above, by:

$$\begin{array}{lcl} \|\ Empty\ \| & = & 0 \\ \|\ Eps\ \| & = & 1 \\ \|\ Inj\ a\ \| & = & \{\ [a]\ \} \\ \|\ Alt\ r_1\ r_2\ \| & = & \|\ r_1\ \| + \|\ r_2\ \| \\ \|\ Con\ r_1\ r_2\ \| & = & \|\ r_1\ \| \cdot \|\ r_2\ \| \\ \|\ Star\ r\ \| & = & \|\ r\ \|^{\star} \end{array}$$

Thus, $\|\_\|$ is the standard denotational semantics of regular expressions, and its range defines the type of regular languages for finite alphabets Σ:
typedef $RegLang = \|(UNIV :: RegExpr\ set)\|$
These form a proper sub-algebra, i.e. the language operations defined above can be restricted to *RegLang* such that another theory morphism
thymorph *reg_lang_as_Kleene_algebra* : *KleeneAlgebra* $\longrightarrow$ *RegularLang*
type_map: $[(\alpha, \beta)\ K \mapsto RegLang]$
op_map: $[\ 0 \mapsto 0_{RegLang},\ 1 \mapsto 1_{RegLang},\ \_+\_ \mapsto \_+_{RegLang}\_,\ \_\cdot\_ \mapsto \_\cdot_{RegLang}\_,\ \_^{\star} \mapsto \_^{\star_{RegLang}}\ ]$

can be constructed (here, $0_{RegLang}$ and $1_{RegLang}$ are essentially the same as 0 and 1, respectively).

6.1.2 Church-Rosser property and confluence

In the next step, the connection to abstract reduction systems will be considered. This should be another example, emphasising how we can proceed from concrete to more abstract notions and propositions.

In the context of higher order logic, an abstract reduction system on terms of type α can be modelled by a relation $\_ \rightsquigarrow \_ :: (\alpha \times \alpha)set$. Such a reduction system is then called

1. *Church-Rosser* (*CR*), if for all u, v with $u \overset{\star}{\leftrightsquigarrow} v$, there exists z with $u \overset{\star}{\rightsquigarrow} z$ and $v \overset{\star}{\rightsquigarrow} z$;

2. *confluent* (*Conf*), if for all u, v such that there exists z with $z \overset{\star}{\rightsquigarrow} u$ and $z \overset{\star}{\rightsquigarrow} v$, there exists z' with $u \overset{\star}{\rightsquigarrow} z'$ and $v \overset{\star}{\rightsquigarrow} z'$;

3. *semi-confluent* (*SConf*) [Baader and Nipkow 1998], if for all u, v such that there exists z with $z \rightsquigarrow u$ and $z \overset{\star}{\rightsquigarrow} v$, there exists z' with $u \overset{\star}{\rightsquigarrow} z'$ and $v \overset{\star}{\rightsquigarrow} z'$.

These three properties have been shown to be equivalent (e.g. also [Baader and Nipkow 1998]).

Now, since the types $(\alpha \times \alpha)set$ and $\alpha \Rightarrow \alpha\ set$ are isomorphic, the interpretation of an arrow $f :: \alpha \to \alpha$ in the allegory **A** as an abstract reduction system on terms of type α is allowed. Then, it is also possible to adapt the Church-Rosser property as the predicate $CR :: (\alpha \to \alpha) \Rightarrow bool$ on arrows, namely by

$$CR\, f \equiv f^{\overset{\star}{\leftrightarrow}} \sqsubseteq f^\star \,;\, (f^\star)^\circ \tag{6.7}$$

where $f^{\overset{\star}{\leftrightarrow}} \equiv (f \sqcup f^\circ)^\star$ is the reflexive transitive symmetric closure of f.

At this point, one can also take into account the theory morphism *lcomplete_as_Kleene_alg : KleeneAlgebra* $\longrightarrow$ *Fixed_Points* and try firstly to abstract the Church-Rosser property to the Kleene algebras. To this end, we need to define the reflexive transitive symmetric closure for any element $a :: (\alpha, \alpha)K$, which, in turn, relies on the converse operator in *Fixed_Points*, and which is not present in *KleeneAlgebra*. To repair this, we extend *KleeneAlgebra* by the theory *KleeneAlgebraConv*, where the logical constant $\_^\circ :: (\alpha, \beta)\ K \Rightarrow (\beta, \alpha)\ K$ together with the three axioms

$$\begin{aligned}(a^\circ)^\circ &= a\\(a+b)^\circ &= a^\circ + b^\circ\\(a \cdot b)^\circ &= b^\circ \cdot a^\circ\end{aligned}$$

are introduced. The consistency (at least relative) of this new axiomatic theory *KleeneAlgebraConv* is justified in two ways:

1. The theory morphism *lcomplete_as_Kleene_alg* can be obviously extended to *KleeneAlgebraConv* by the operation mapping, assigning to the converse in *KleeneAlgebraConv* the converse in *Allegory*.

2. More interesting is that in the language model in the theory *RegularLang*, presented above, the converse operator can be naturally defined by

$$L^\circ \equiv \{reverse\ u \mid u \in L\}$$

where *reverse* refers to the standard reverse function on lists. This provides another theory morphism from *KleeneAlgebraConv*, but now with the target in *RegularLang*.

Some of the properties of $\_^\circ$, derived in *KleeneAlgebraConv* are

1. $0^\circ = 0$,
2. $1^\circ = 1$,
3. $(a^\star)^\circ = (a^\circ)^\star$.

Turning back to the original problem of the specification of the reflexive transitive symmetric closures in the context of Kleene algebras, we can now define $a^{\overset{\star}{\leftrightarrow}} \equiv (a + a^\circ)^\star$ in the theory *KleeneAlgebraConv*. Based on the identity $(a^\star)^\circ = (a^\circ)^\star$ the next two (somehow expected) properties of $\_^{\overset{\star}{\leftrightarrow}}$ can be proved:

1. $\left(a^{\overset{\star}{\leftrightarrow}}\right)^\circ = a^{\overset{\star}{\leftrightarrow}}$,
2. $(a^\circ)^{\overset{\star}{\leftrightarrow}} = a^{\overset{\star}{\leftrightarrow}}$.

Further, the Church-Rosser property (6.7) in locally complete allegories can be abstracted to $CR :: (\alpha, \alpha)\ K \Rightarrow bool$ for the Kleene algebras with converse by

$$CR\ a \equiv a^{\overset{\star}{\leftrightarrow}} \leq a^\star \cdot (a^\star)^\circ$$

Similarly, also the confluence and semi-confluence properties:

$$\begin{aligned} Conf\ a &\equiv (a^\star)^\circ \cdot a^\star \leq a^\star \cdot (a^\star)^\circ \\ SConf\ a &\equiv a^\circ \cdot a^\star \leq a^\star \cdot (a^\star)^\circ \end{aligned}$$

What ultimately motivates this abstraction step is that the equivalence of these three properties can be shown in *KleeneAlgebraConv*. The proof proceeds, as in [Baader and Nipkow 1998], in three steps.

1. $CR\ a \Longrightarrow Conf\ a$. Unfolding the definitions of CR, $Conf$ and $\_^{\overset{\star}{\leftrightarrow}}$, and using the transitivity of $\leq$ and the properties of the converse, this goal reduces to

$$(a^{\circ})^{\star} \cdot a^{\star} \leq (a + a^{\circ})^{\star}$$

By the star decomposition rule $(a + b)^{\star} = a^{\star} \cdot (b \cdot a^{\star})^{\star}$ we have to show

$$(a^{\circ})^{\star} \cdot a^{\star} \leq a^{\star} \cdot (a^{\circ} \cdot a^{\star})^{\star}$$

Provided by the rule $u \cdot v \leq v \cdot w \Longrightarrow u^{\star} \cdot v \leq v \cdot w^{\star}$, this reduces to

$$a^{\circ} \cdot a^{\star} \leq a^{\star} \cdot (a^{\circ} \cdot a^{\star})$$

which is true, since $1 \cdot u = u$ and $1 \leq u^{\star}$ hold for any u.

2. $Conf\ a \Longrightarrow SConf\ a$. Unfolding the definitions of $Conf$ and $SConf$, and using the transitivity of $\leq$ and the properties of the converse, this goal reduces to

$$a^{\circ} \cdot a^{\star} \leq (a^{\circ})^{\star} \cdot a^{\star}$$

which is true, since the multiplication is monotone and $u \leq u^{\star}$ holds for any u.

3. $SConf\ a \Longrightarrow CR\ a$. Unfolding the definitions of CR, $SConf$ and $\_^{\overset{\star}{\leftrightarrow}}$, and applying the properties of the converse, we have to show

$$(a + a^{\circ})^{\star} \leq a^{\star} \cdot (a^{\circ})^{\star}$$

under the assumption $a^{\circ} \cdot a^{\star} \leq a^{\star} \cdot (a^{\circ})^{\star}$. By the star decomposition rule $(a + b)^{\star} = (a^{\star} \cdot b)^{\star} \cdot a^{\star}$ we have then to show

$$(a^{\star} \cdot a^{\circ})^{\star} \cdot a^{\star} \leq a^{\star} \cdot (a^{\circ})^{\star}$$

Provided by the rule $u \cdot v \leq v \cdot w^{\star} \Longrightarrow u^{\star} \cdot v \leq v \cdot w^{\star}$, this reduces to

$$a^{\star} \cdot (a^{\circ} \cdot a^{\star}) \leq a^{\star} \cdot (a^{\circ})^{\star}$$

Using the assumption $a^{\circ} \cdot a^{\star} \leq a^{\star} \cdot (a^{\circ})^{\star}$ and monotonicity, we can show

$$a^{\star} \cdot a^{\star} \cdot (a^{\circ})^{\star} \leq a^{\star} \cdot (a^{\circ})^{\star}$$

which is true, since $u^{\star} \cdot u^{\star} = u^{\star}$ holds for any u.

Finally, it should be noted that these abstract results can be transported to any Kleene algebra with a converse operator, and therefore to any locally complete allegory, via theorem translation. In particular, in the language model we obtain that a language L is 'confluent', i.e. *Conf* L holds, iff

$$\{\mathit{reverse}\ w\ @\ w' \mid w, w' \in L^\star\} \subseteq \{w\ @\ \mathit{reverse}\ w' \mid w, w' \in L^\star\}$$

holds. So, for instance all palindrome languages are confluent in this sense. Figure 6.1 sketches the development so far.

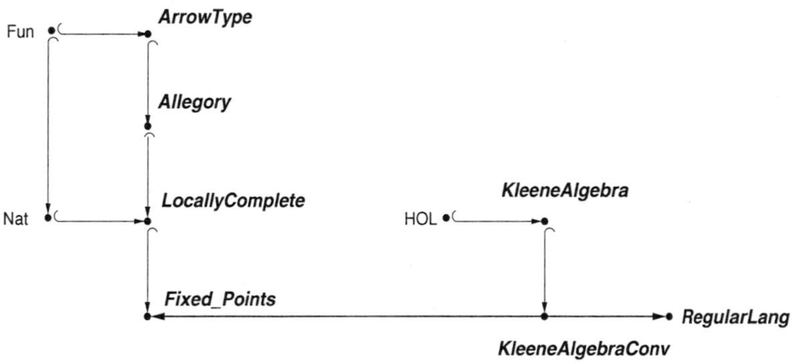

Figure 6.1: The current state of the theory development

6.2 Relators and hylomorphisms

Since allegories are special kind of categories, the notion of functors between categories will be now specialised to the notion of relators between allegories.

Definition 3 (relators)
Let $\mathbf{A}$ and $\mathbf{B}$ be locally complete allegories. Then a *relator* $\mathsf{F} : \mathbf{A} \Rrightarrow \mathbf{B}$ is a functor, additionally satisfying

1. $f \sqsubseteq_{\mathbf{A}} g$ implies $\mathsf{F} f \sqsubseteq_{\mathbf{B}} \mathsf{F} g$, for any two arrows f, g in $\mathbf{A}$, i.e. is monotone,
2. $\mathsf{F}\,(f^{\circ_{\mathbf{A}}}) = (\mathsf{F} f)^{\circ_{\mathbf{B}}}$ for any arrow f in $\mathbf{A}$, i.e. preserves converse,
3. $\mathsf{F}\,(\bigsqcup_n^{\mathbf{A}} s_n) \sqsubseteq_{\mathbf{B}} \bigsqcup_n^{\mathbf{B}} (\mathsf{F}\, s_n)$ for any chain s of arrows in $\mathbf{A}$, i.e. is continuous.

These definition slightly differs from the definition of relators in [Bird and de Moor 1997]. The reason is that in [Bird and de Moor 1997] relators between so-called tabular allegories are considered. Then, a relator is just a monotone functor, and the authors show that for tabular allegories this already implies that it also preserves converse. Here tabular allegories are not considered, and we focus on locally complete allegories instead. This explains the additionally required continuity property for relators.

Now, in the presented Isabelle/HOL development, we concentrate on properties of a single locally complete allegory **A**, such that only endo-relators will be axiomatised and considered. But for simplicity reasons, from now on endo-relators in **A** will be just called relators.

The theory *Relator* importing *Fixed_Points* formalises relators from **A** to **A** by means of an axiomatisation of the type constructor class R as follows. Since objects in **A** are types, the object part of a relator can be represented by a declaration of a type constructor with the rank 3:
typedecl $(\gamma_1, \gamma_2, \alpha)\,\mathsf{R}$
such that the arrow part can be then specified by the logical constant:
consts $\mathsf{R} :: (\alpha \to \beta) \Rightarrow ((\gamma_1, \gamma_2, \alpha)\,\mathsf{R} \to (\gamma_1, \gamma_2, \beta)\,\mathsf{R})$
The reasons for taking a type constructor with rank 3 instead of 1 and for the carrying of the auxiliary type variables γ_1, γ_2 will be given later on in the formalisation of the Divide-and-Conquer scheme.

Furthermore, the properties required by Definition 3 are then formulated for R in form of axioms. So, we can consider the arrow part R as a fixed, monotone and continuous arrow transformer. Based on this, the following properties of R are derived:

1. R preserves coreflexive arrows,
2. then R preserves also simple arrows,
3. R preserves reflexive arrows,
4. then R preserves entire arrows, and hence also preserves mappings;
5. using the fact that the composition and meet coincides for coreflexive arrows, we also obtain

$$\mathsf{R}(A \sqcap B) = \mathsf{R}A \sqcap \mathsf{R}B$$

for any coreflexive A, B.

R-algebras. Inspired by the corresponding notion, well-known from the category theory, an arrow $a :: (\gamma_1, \gamma_2, \alpha)\ \mathsf{R} \to \alpha$ will be called an R-algebra. Consequently, an arrow $c :: \alpha \to (\gamma_1, \gamma_2, \alpha)\,\mathsf{R}$ will be called an R-coalgebra. In contrast to the general categories, in the context of allegories, the converse arrow of an algebra gives a coalgebra and vice versa, making these notions interchangeable.

Hylomorphisms. Now we can formalise the well-known notion of hylomorphisms (e.g. [Doornbos and Backhouse 1995]) for the relator R. Hylomorphisms play an important rôle in structured program development, because they capture the idea of involving an intermediate data structure (represented by R in this case) and give a recursion pattern that occurs frequently in practice. So, in the sequel hylomorphisms will be used for derivation of functions in correct-by-construction manner, based on the divide-and-conquer principle.

Firstly, let $c :: \alpha \to (\gamma_1, \gamma_2, \alpha)\ \mathsf{R}$ and $a :: (\gamma_1, \gamma_2, \beta)\ \mathsf{R} \to \beta$ be an R-coalgebra and an R-algebra, respectively. They define the arrow transformer $\mathcal{HT}^{\mathsf{R}}(c, a) :: (\alpha \to \beta) \Rightarrow (\alpha \to \beta)$ by

$$\mathcal{HT}^{\mathsf{R}}(c, a) \equiv (\lambda x.\ c\ ;\ \mathsf{R}\, x\ ;\ a) \tag{6.8}$$

Provided by the monotonicity and continuity of the relator R, $\mathcal{HT}^{\mathsf{R}}(c, a)$ is monotone and continuous. Moreover, if both c, a are simple then $\mathcal{HT}^{\mathsf{R}}(c, a)$ preserves simple arrows, and similarly $\mathcal{HT}^{\mathsf{R}}(c, a)$ preserves entire arrows for entire c, a. Thus, if both c, a are mappings then $\mathcal{HT}^{\mathsf{R}}(c, a)$ also preserves mappings.

Next, the least fixed point of $\mathcal{HT}^{\mathsf{R}}(c, a)$, i.e. the arrow $\mathcal{H}^{\mathsf{R}}(c, a) :: \alpha \to \beta$ defined by $\mathcal{H}^{\mathsf{R}}(c, a) \equiv \mu(\mathcal{HT}^{\mathsf{R}}(c, a))$ is called the R-*hylomorphism* from the coalgebra c to the algebra a. Based on the properties of least fixed points, we can derive the following central properties of $\mathcal{H}^{\mathsf{R}}$.

1. If $c' \sqsubseteq c$ and $a' \sqsubseteq a$ then $\mathcal{H}^{\mathsf{R}}(c', a') \sqsubseteq \mathcal{H}^{\mathsf{R}}(c, a)$. In particular, this property gives the possibility to reason abstractly about hylomorphisms. So, e.g. thinking of c' and a' as of implementations satisfying c, a, respectively, in order to prove that $\mathcal{H}^{\mathsf{R}}(c', a')$ satisfies some specification f we can reason that $\mathcal{H}^{\mathsf{R}}(c, a) \sqsubseteq f$ holds, using the (possibly) more abstract specifications c, a.

2. $\mathcal{HT}^{\mathsf{R}}(c, a)\, f \sqsubseteq f$ implies $\mathcal{H}^{\mathsf{R}}(c, a) \sqsubseteq f$ for any f, which provides a way to establish that $\mathcal{H}^{\mathsf{R}}(c, a)$ satisfies some specification f by a proof, that f is a prefixed point of $\mathcal{HT}^{\mathsf{R}}(c, a)$.

3. $\mathcal{H}^{\mathsf{R}}(c, a)$ is a simple arrow if both c, a are simple, since in this case $\mathcal{HT}^{\mathsf{R}}(c, a)$ preserves simple arrows.

The condition that both c, a are entire arrows is necessary, but not sufficient in order to conclude that $\mathcal{H}^{\mathsf{R}}(c, a)$ is entire. More precisely, whether $\mathcal{H}^{\mathsf{R}}(c, a)$ is entire also depends on an additional property of the coalgebra c, which is the subject of the next section.

6.3 Reductivity

Thinking of the coalgebra $c :: \alpha \to (\gamma_1, \gamma_2, \alpha)\ \mathsf{R}$ in c ; $\mathsf{R}\,x$; a as of a function, which decomposes an input and supplies the result to $\mathsf{R}x$, the sequence of arrows c, c ; $\mathsf{R}\,c$, ... has to reach some fixed point c ; $\mathsf{R}\,c$; ... ; $\mathsf{R}^n\,c$ with $n \in \mathbb{N}$, because otherwise $\mathcal{H}^{\mathsf{R}}(c, a)$ cannot be computed by a finite iteration of $\mathcal{HT}^{\mathsf{R}}(c, a)$, and hence cannot be an entire arrow. To this end, three characterisations of R-*reductive coalgebras* have been proposed in [Doornbos and Backhouse 1995]. Two of these nice and abstract conditions are considered below in the context of locally complete allegories..

Reductivity predicates. Let $c :: \alpha \to (\gamma_1, \gamma_2, \alpha)\ \mathsf{R}$ be an R-coalgebra. Further, let the arrow transformer $c \setminus^{\mathsf{R}} :: (\alpha \to \alpha) \Rightarrow (\alpha \to \alpha)$ be defined by

$$c\setminus^{\mathsf{R}} \equiv (\lambda x.\ c \setminus\ x) \circ \mathsf{R}$$

Then c is called R-*reductive* if is satisfies the equation

$$\mu(c\setminus^{\mathsf{R}}) = \mathit{Id} \tag{6.9}$$

Alternatively, the following characterisation is also given in [Doornbos and Backhouse 1995]

$$c \setminus^{\mathsf{R}} A \sqsubseteq A \text{ implies } \mathit{Id} \sqsubseteq A \text{ for any coreflexive } A \tag{6.10}$$

and shown to be equivalent to (6.9). This result cannot be obtained in this form in this context, because in contrast to [Doornbos and Backhouse 1995] the construction of least fixed points here requires the continuity argument, whereas the arrow transformer $c\setminus^{\mathsf{R}}$ is in general not continuous. The reason for this is that the monotype factor is in general not continuous. This can be seen as follows: in set-theoretical context the inverse image of the set $\bigcup_n s_n$, for a chain s of sets under an arbitrary relation r, is by far not always a subset of the union of the inverse images of s_n under r for any n. On the other hand, this property holds if r is a mapping, and indeed one can also show that $c\backslash^{\mathsf{R}}$ is continuous for any arrow c which is a mapping. So, taking continuity into account, the condition (6.9) is stronger than (6.10), but the both are equivalent if c is a mapping. Therefore, we will use (6.9) as the definition of R-reductive coalgebras, because it is compact and well-suited for abstract reasoning, while the condition (6.10) will be then used later for the Divide-and-Conquer design tactic in relations. There, it will give a condition which is more suitable than (6.9) if we want to establish that a function is reductive.

Furthermore, we can adapt the proof in [Doornbos and Backhouse 1995] that $x = \mathcal{H}^{\mathsf{R}}(c,a)$ holds for any x satisfying $\mathcal{HT}^{\mathsf{R}}(c,a)\ x = x$ provided c is R-reductive. That is, in this case the hylomorphism $\mathcal{H}^{\mathsf{R}}(c,a)$ is the unique fixed point of $\mathcal{HT}^{\mathsf{R}}(c,a)$. This proof is based on the special case of the μ-fusion, formulated in Corollary 6.2.1.

Finally, the following lemma gives sufficient conditions for entire hylomorphisms.

Lemma 6.2 *Let c be an* R*-reductive coalgebra, and a an* R*-algebra. Then $\mathcal{H}^{\mathsf{R}}(c,a)$ is an entire arrow if both c, a are entire.*

Proof. We have to show $Id \sqsubseteq \mathcal{H}^{\mathsf{R}}(c,a)\ ;\ (\mathcal{H}^{\mathsf{R}}(c,a))^{\circ}$. Using that c is R-reductive and the definition of hylomorphisms, this is equivalent to

$$\mu(c\backslash^{\mathsf{R}}) \sqsubseteq \mu(\mathcal{HT}^{\mathsf{R}}(c,a))\ ;\ (\mathcal{H}^{\mathsf{R}}(c,a))^{\circ}$$

In turn, this can be resolved by the following derived property of least fixed points, which can be seen as dual to the μ-fusion:

- *$(H \circ F) \stackrel{.}{\sqsubseteq} (F \circ G)$ implies $\mu H \sqsubseteq F(\mu G)$ for any monotone F, H, and monotone and continuous G*

instantiating H by $c\backslash^{\mathsf{R}}$, G by $\mathcal{HT}^{\mathsf{R}}(c,a)$ and F by $(\lambda x.\ x\ ;\ (\mathcal{H}^{\mathsf{R}}(c,a))^{\circ})$. Then we have to show

$$c\backslash^{\mathsf{R}}\ (x\ ;\ (\mathcal{H}^{\mathsf{R}}(c,a))^{\circ}) \sqsubseteq \mathcal{HT}^{\mathsf{R}}(c,a)\ x\ ;\ (\mathcal{H}^{\mathsf{R}}(c,a))^{\circ}$$

for any x. This can be done as follows:

$$
\begin{array}{ll}
c\backslash^{\mathsf{R}}\ (x\ ;\ (\mathcal{H}^{\mathsf{R}}(c,a))^{\circ}) & = \\
 & \text{by the definition of } c\backslash^{\mathsf{R}} \\
c\backslash\ \mathsf{R}(x\ ;\ (\mathcal{H}^{\mathsf{R}}(c,a))^{\circ}) & \sqsubseteq \\
 & \text{since } c \text{ is entire} \\
c\backslash\ \mathsf{R}(x\ ;\ (\mathcal{H}^{\mathsf{R}}(c,a))^{\circ})\ ;\ (c\ ;\ c^{\circ}) & \sqsubseteq \\
 & \text{monotype factor rule}
\end{array}
$$

$$
\begin{array}{ll}
c \,;\, \mathsf{R}(x \,;\, (\mathcal{H}^{\mathsf{R}}(c,a))^{\circ}) \,;\, c^{\circ} & = \\
 & \text{since R is a functor} \\
 & \text{and preserves converse} \\
c \,;\, \mathsf{R}\,x \,;\, (\mathsf{R}\,(\mathcal{H}^{\mathsf{R}}(c,a)))^{\circ} \,;\, c^{\circ} & \sqsubseteq \\
 & \text{since } a \text{ is entire} \\
c \,;\, \mathsf{R}\,x \,;\, (a \,;\, a^{\circ}) \,;\, (\mathsf{R}\,(\mathcal{H}^{\mathsf{R}}(c,a)))^{\circ} \,;\, c^{\circ} & = \\
 & \text{since converse is contra-variant} \\
c \,;\, \mathsf{R}\,x \,;\, a \,;\, (c \,;\, \mathsf{R}\,(\mathcal{H}^{\mathsf{R}}(c,a)) \,;\, a)^{\circ} & = \\
 & \text{by the definition of } \mathcal{HT}^{\mathsf{R}} \text{ and since} \\
 & \mathcal{H}^{\mathsf{R}} \text{ is a fixed point} \\
\mathcal{HT}^{\mathsf{R}}(c,a)\,x \,;\, (\mathcal{H}^{\mathsf{R}}(c,a))^{\circ} &
\end{array}
$$

□

Using the previous results, an immediate consequence of this lemma is that $\mathcal{H}^{\mathsf{R}}(c, a)$ is a mapping if both c, a are, and if c is additionally R-reductive.

6.4 The Divide-and-Conquer design tactic

At this point all concepts have been formalised and propositions proved in order to formulate the Divide-and-Conquer design tactic in the context of locally complete allegories, as a transformation *in-the-large.* In comparison to [Bortin, Johnsen and Lüth 2006], the design tactic presented here differs in the following points:

- the previous tactic has been formulated only for the particular case of relations, in contrast to locally complete allegories here,
- consequently, here we get rid of the pre/post-condition style of the specification of the divide-and-conquer function,
- here we take a fixed arbitrary relator R to represent the intermediate data structure used by decomposition and composition, whereas the previous tactic was constructed only for the special relator $D\_G \times \_ \times \_ + \mathbf{1}$; more precisely, the relator $D\_G \times \_ \times \_$ has been considered, and the $\mathbf{1}$-case has been treated in the special way using the predicate *Primitive*, recognising 'primitive' inputs to which the function *Dir_Solve* could be applied,
- the existence of a measure function for decomposition is captured by R-reductivity in this context,

- finally, here the synthesised divide-and-conquer function will be polymorphic, in contrast to the function $F :: D\_F \Rightarrow R\_F$, with the fixed types D_F, R_F obtained before.

The first part of formal parameters: specifications. Since the main idea of the design tactic is to construct functions in the correct-by-construction style, the particular specifications will play the central rôle here. So, we start the theory *Divide_and_Conquer*, extending *Hylomorphisms*, which will contain the formal parameters regarding the specifications of the three essential parts: decomposition, composition, and of the divide-and-conquer function, synthesised from these.

Firstly, in *Divide_and_Conquer* we specify two fixed type constructors *Domain* and *Range* with the rank 2:
typedecl (γ_1, γ_2) *Domain*
typedecl (γ_1, γ_2) *Range*
representing the types of the elements in the domain and the range of the target mapping. Consequently, the target specification *dac_spec* has to be an arrow connecting *Domain* and *Range*, which might be declared as follows:
consts
$dac\_spec :: (\gamma_1, \gamma_2)\ Domain \rightarrow (\gamma_1, \gamma_2)\ Range$
By this declaration, the default HOL-sort $\{type\}$ has been assigned to the type variables γ_1, γ_2. Now, taking into account that we can also introduce formal type class parameters, we declare
classes
$class_1 \subseteq type$
$class_2 \subseteq type$
and can then generalise the declaration of *dac_spec* by
consts
$dac\_spec :: (\gamma_1 ::_{sort} \{class_1\}, \gamma_2 ::_{sort} \{class_2\})\ Domain \rightarrow (\gamma_1, \gamma_2)\ Range$
which is more general than the former, since it can be derived from the latter by a class map $(class_1 \mapsto type), (class_2 \mapsto type)$. On the other hand, now we can also assign any of the subclasses of *type* to the formal parameters $class_1$ and $class_2$.

Further, fixing the formal parameters for the specifications of decomposition and composition steps, the relator R comes into play:
consts
$dcmp\_spec :: (\gamma_1 ::_{sort} \{class_1\}, \gamma_2 ::_{sort} \{class_2\})\ Domain \rightarrow (\gamma_1, \gamma_2, (\gamma_1, \gamma_2)\ Domain)\ \mathsf{R}$
$cmps\_spec :: (\gamma_1, \gamma_2, (\gamma_1, \gamma_2)\ Range)\ \mathsf{R} \rightarrow (\gamma_1 ::_{sort} \{class_1\}, \gamma_2 ::_{sort} \{class_2\})\ Range$
That is, the specification of decomposition *dcmp_spec* has to be an R-coalgebra

and the specification of composition *cmps_spec* an R-algebra. More precisely, *dcmp_spec* is considered to specify, how a decomposition step can decompose the input values using the intermediate data structure R over the domain. Consequently, *cmps_spec* is considered to specify, how a composition step can compose the output values in the intermediate data structure R over the range to a single output value.

Moreover, these declarations emphasise, why the type constructor R has been introduced with the rank 3 instead of 1: the intermediate data structure, represented by R, can this way use both of the polymorphic type parameters of *Domain* and *Range* as constant factors. So, it will be possible to interpret R by, e.g. $\gamma_1 \times \_ \times \_ + \mathbf{1}$, where γ_1 will refer exactly to the first type parameter of *Domain* and *Range*.

Finally, the single axiom combining *dcmp_spec*, *cmps_spec* and *dac_spec* is formulated by

axioms

SPRP : $\mathcal{HT}^{\mathsf{R}}(dcmp\_spec, cmps\_spec)\ dac\_spec \sqsubseteq dac\_spec$

That is, it essentially requires the specification *dac_spec* to be a prefixed point of $\mathcal{HT}^{\mathsf{R}}(dcmp\_spec, cmps\_spec)$. In the context of divide-and-conquer algorithms, this condition is sometimes also called *strong problem reduction principle (SPRP)*. However, substituting $\perp$ for e.g. *dcmp_spec*, will trivially satisfy this axiom for any *cmps_spec* and *dac_spec*. In order to ensure that both *dcmp_spec* and *cmps_spec* can be implemented by functions they have to be bounded below, which is done in the next step by the corresponding decomposition and composition mappings.

The second part of formal parameters: implementations. For the implementation of *dcmp_spec* we introduce the logical constant *decompose*, which has clearly the same type as *dcmp_spec*, and, hence is an R-coalgebra as well. Similarly, the logical constant *compose* represents the implementation of *cmps_spec*. The required properties of *decompose* and *compose* are the following:

1. *decompose* is a mapping, additionally satisfying $decompose \sqsubseteq dcmp\_spec$,
2. *compose* is a mapping, additionally satisfying $compose \sqsubseteq cmps\_spec$,
3. *decompose* is an R-reductive coalgebra.

These properties are also specified via axioms in Isabelle.

Deriving a correct mapping. Now, the synthesised arrow, representing the divide-and-conquer function which satisfies the given specification *dac_spec*, is the derived constant

consts $dac :: (\gamma_1 ::_{sort} \{class_1\}, \gamma_2 ::_{sort} \{class_2\})\ Domain \rightarrow (\gamma_1, \gamma_2)\ Range$

It is defined to be the hylomorphism $\mathcal{H}^{\mathsf{R}}(decompose, compose)$. Indeed, the properties of hylomorphisms and assumptions on the formal Divide-and-Conquer parameters, specified via axioms above, provide the following derived properties of *dac*:

1. *dac* is a mapping,
2. $dac = decompose\ ;\ \mathsf{R}\ dac\ ;\ compose$,
3. $dac \sqsubseteq dac\_spec$.

Figure 6.2 sketches the development so far (omitting the Kleene algebra branch).

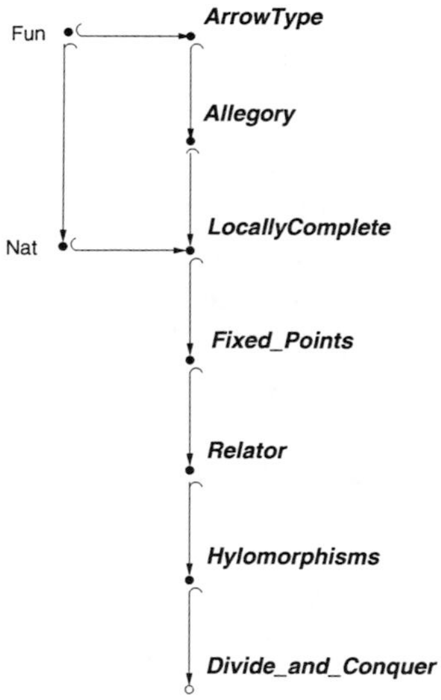

Figure 6.2: The current state of the theory development

6.5 Construction of relators

The relator R forms one of the essential parts of the presented Divide-and-Conquer design tactic, because it provides the intermediate structure used by the decomposition and composition steps. Thus, in order to apply the tactic in concrete situations, concrete relator instances are required as well. Among all the instances, the polynomial relators, i.e. built by products and coproducts, play the most important rôle in construction of programs. Therefore, similarly to [Bird and de Moor 1997], we will concentrate on the construction of polynomial relators, which is already not straightforward. One of the main problems is mentioned in [Bird and de Moor 1997]: since any allegory is identical to its opposite, the dual categorical universal constructions of product and coproduct coincide, which is not desirable. To avoid this problem, [Bird and de Moor 1997] propose to construct products and coproducts in the sub-category *Mappings*($\mathbf{A}$) and then extend these to $\mathbf{A}$ in some canonical way. As already mentioned, in [Bird and de Moor 1997] relators and, hence their construction are considered in the context of tabular allegories which are not considered here. However, we will pursue a similar strategy, but also take into account that in the context of Isabelle/HOL the products and coproducts in the sub-category *Mappings*($\mathbf{Rel}$) of functions are already available: these are exactly the product type constructor $\_ \times \_$, introduced in the theory *Product_Type*, and the coproduct type constructor $\_ + \_$, introduced in the theory *Sum_Type*. This is not the most general treatment, which would require a formalisation of (co)products for *Mappings*($\mathbf{A}$), but already sufficient, because we focus on the allegory of relations.

As a motivation let us explicitly consider the coproducts in $\mathbf{Set}$. For any two functions $f : A \to B$ and $g : C \to D$ the function $f+g : A+C \to B+D$ is defined by $(f+g)(\iota_A(a)) \stackrel{def}{=} \iota_B(f(a))$ and $(f+g)(\iota_C(c)) \stackrel{def}{=} \iota_D(g(c))$, or categorically as $[\iota_B \circ f, \iota_D \circ g]$. The question is now how to extend this operation to relations. That is, for any two relations $r : A \to \mathcal{P}B$ and $s : C \to \mathcal{P}D$ we are looking for a relation $r \mathbin{+\!\!+} s : A + C \to \mathcal{P}(B + D)$, such that in particular $r \mathbin{+\!\!+} s = r + s$ holds if r and s are functions. This we can do as follows

$$\begin{aligned} (r \mathbin{+\!\!+} s)(\iota_A(a)) &\stackrel{def}{=} \{\iota_B(b) \mid b \in r(a)\} \\ (r \mathbin{+\!\!+} s)(\iota_C(c)) &\stackrel{def}{=} \{\iota_D(d) \mid d \in s(c)\} \end{aligned}$$

Obviously, this definition provides the identity $r \mathbin{+\!\!+} s = r + s$ for functions r and s, because $r(a)$ and $s(c)$ are singleton sets in this case. Furthermore, since $f + g$ can be considered as the action on functions of a bi-functor from **Set** $\times$ **Set** to **Set**, $r +\!\!+ s$ extends this to the action of a bi-relator from **Rel** $\times$ **Rel** to **Rel**. Important is that the object part of this bi-relator is the same as the object part of the bi-functor: it sends any two sets to their disjoint sum, i.e. the coproduct in **Set**. Altogether, we can say that the allegory of relations allows the extension of coproducts from its sub-category of mappings.

In a similar way we can extend the product $f \times g : A \times C \to B \times D$ of functions $f : A \to B$ and $g : C \to D$ to product of relations $r ** s$. Now we can consider the so-called *polynomial* functors on **Set**, i.e. those functors which send any set X to $C_0 + C_1 \times X + \ldots + C_n \times X^n$ and act on functions using $+$ and $\times$ operators, where $n \in \mathbb{N}$ and $C_0, \ldots, C_n$ are fixed objects. Consequently, a polynomial relator acts in the same way on sets, but uses $+\!\!+$ and $*$ to combine relations. So, for instance, the *square* relator Sq sends any X to $X \times X$ and any $r : A \to \mathcal{P}B$ to $r ** r : A \times A \to \mathcal{P}(B \times B)$. An important observation about this is, that the square relator can be seen as a combination of the identity relator $\mathsf{I} : \mathbf{Rel} \longrightarrow \mathbf{Rel}$ with itself, because $r ** r = \mathsf{I}(r) ** \mathsf{I}(r)$ obviously holds for any r. By construction, the restriction of Sq to the sub-category **Set** yields the corresponding square functor. Furthermore, beside the identity relator we can also consider the collection of constant relators R_C, sending any X to C and any r to $id_C : C \to \mathcal{P}C$. Thus, we can use these and the square relator to build, e.g. the relator Sq_C, sending any X to $C + X \times X$ and $r : A \to \mathcal{P}B$ to $\mathsf{R}_C(r) +\!\!+ \mathsf{Sq}(r) : C + A \times A \to \mathcal{P}(C + B \times B)$, for some fixed C.

Altogether, this shows how polynomial relators can be built of the identity and constant relators, and next paragraphs show how this construction can be utilised in Isabelle using successive instantiations of parametrisations by theory morphisms. This will be used later for derivations of concrete relators, representing intermediate data types in the applications of the Divide-and-Conquer tactic.

The theory *ProdArrow*. To keep the things simple, we firstly take the theory *ProdArrow* which extends *Product_Type* and *Fixed_Points* and declares the logical constant

consts

$\_ ** \_ :: (\alpha \to \beta) \Rightarrow (\gamma \to \delta) \Rightarrow ((\alpha \times \gamma) \to (\beta \times \delta))$

Basically, this operation should represent nothing else as the arrow part of the bi relator, the object part of which sends any two types κ_1, κ_2 to their product type $\kappa_1 \times \kappa_2$. These also explains the following axioms for $\_ ** \_$:

1. $Id ** Id = Id$,
2. $(f\ ;\ g) ** (h\ ;\ k) = (f ** h)\ ;\ (g ** k)$,
3. $f \sqsubseteq h$ and $g \sqsubseteq k$ implies $(f ** g) \sqsubseteq (h ** k)$,
4. $f^\circ ** g^\circ \sqsubseteq (f ** g)^\circ$,
5. $(\bigsqcup_n s_n) ** (\bigsqcup_n s'_n) \sqsubseteq \bigsqcup_n (s_n ** s'_n)$ for any two chains s and s',

i.e. the corresponding bi-relator axioms.

The essential observation about this axiomatic theory *ProdArrow* is that any theory morphism τ : *ProdArrow* $\longrightarrow \mathcal{T}$ with some target theory $\mathcal{T}$ establishes that the theory $\mathcal{T}$ provides a locally complete allegory **B**, which moreover allows an extension of the categorical products in its sub-category *Mappings*(**B**). This however does not say that $\mathcal{T}$ provides this in a conservative way, but it will be shown that this is quite possible, for instance in **Rel**.

The theory *SumArrow*. Almost the same we do for the coproducts: we take the theory *SumArrow*, which extends *Sum_Type* and *Fixed_Points*, and declares the logical constant
consts
$\_ ++ \_ :: (\alpha \to \beta) \Rightarrow (\gamma \to \delta) \Rightarrow ((\alpha + \gamma) \to (\beta + \delta))$
and the corresponding bi-relator axioms for it. In order to put products and coproducts together we just take the empty theory *Polynomial*, extending both *ProdArrow* and *SumArrow*, such that we obtain the theory development, sketched in Figure 6.3.

So, the idea is that, e.g. in a theory having *ProdArrow* as an ancestor we can derive new relators building products of existing relators. In order to provide a basis for such derivations, two most important basic relators are described below.

The identity relator. In the theory *Id_Relator* importing only *Fixed_Points* we construct the following theory morphism
thymorph *Id_Rel* : *Relator* $\longrightarrow$ *Id_Relator*
type_map: $[\,(\gamma_1, \gamma_2, \alpha)\ \mathsf{R} \mapsto \alpha\,]$
op_map: $[\,\mathsf{R} \mapsto \mathsf{IdR}\,]$
where the constant $\mathsf{IdR} :: (\alpha \to \beta) \Rightarrow (\alpha \to \beta)$, representing the arrow part of the identity relator, is consequently defined by $\mathsf{IdR}\, f \equiv f$.

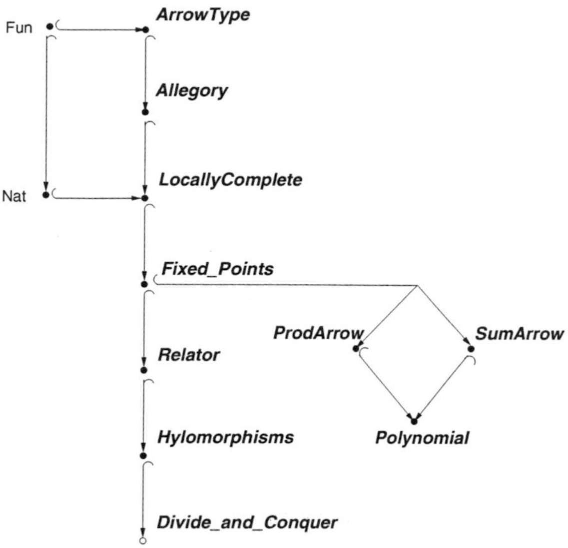

Figure 6.3: The current state of the theory development

The constant relator. In the theory *Const_Relator* importing only *Fixed_Points* we firstly declare the following type constructor with the rank 2:
typedecl (γ_1, γ_2) ConstR
Then we construct the following theory morphism
thymorph *Const_Rel* : *Relator* $\longrightarrow$ *Const_Relator*
type_map: $[\,(\gamma_1, \gamma_2, \alpha)\,\mathsf{R} \mapsto (\gamma_1, \gamma_2)\,\mathsf{ConstR}\,]$
op_map: $[\,\mathsf{R} \mapsto \mathsf{ConstR}\,]$
where the constant ConstR :: $(\alpha \to \beta) \Rightarrow ((\gamma_1, \gamma_2)\,\mathsf{ConstR} \to (\gamma_1, \gamma_2)\,\mathsf{ConstR})$, representing the arrow part of the constant relator, is consequently defined by $\mathsf{ConstR}\, f \equiv \mathit{Id}$.

6.5.1 Deriving new polynomial relators

In order to derive new polynomial relators from existing ones, two parameterised theories *ProductRule* and *SumRule* are considered. Both of them are parameterised over two axiomatic relator instances, which are then combined to their product in the first case and to their sum in the second. Consequently, *ProductRule* imports *ProdArrow*, and *SumRule* imports *SumArrow*. Here we will consider only *ProductRule* in more detail, because *SumRule* proceeds in a similar way.

The parameters of *ProductRule* comprise the theories *Parameter_Relator1* and *Parameter_Relator2*, which are constructed by axiomatic instantiations of *Relator*. In order to distinguish between this two axiomatic relator instances, we employ renaming, such that we can refer to the instance in *Parameter_Relator1* by R_1, and, similarly, by R_2 in *Parameter_Relator2*.

ProductRule. The combination of the parameters is explained in terms of the following theory morphism
thymorph *product_rule* : *Relator* $\longrightarrow$ *ProductRule*
type_map: $[\,(\gamma_1,\gamma_2,\alpha)\,\mathsf{R} \mapsto (\gamma_1,\gamma_2,\alpha)\,\mathsf{R}_1 \times (\gamma_1,\gamma_2,\alpha)\,\mathsf{R}_2\,]$
op_map: $[\,\mathsf{R} \mapsto \mathsf{PR}\,]$
where the constant

$$\mathsf{PR} :: (\alpha \to \beta) \Rightarrow ((\gamma_1,\gamma_2,\alpha)\mathsf{R}_1 \times (\gamma_1,\gamma_2,\alpha)\mathsf{R}_2) \to (\gamma_1,\gamma_2,\beta)\mathsf{R}_1 \times (\gamma_1,\gamma_2,\beta)\mathsf{R}_2)$$

representing the arrow part of the combined relator PR, is simply defined via the product operation _ ** _ from *ProdArrow*:

$$\mathsf{PR}\,f \equiv \mathsf{R}_1\,f \;**\; \mathsf{R}_2\,f$$

Applying ProductRule. Figure 6.4 sketches, how the *ProductRule* tactic can be applied in order to derive the square relator. In the theory *Square_Relator*, the identity relator IdR, imported from *Id_Relator*, is used as the actual parameter for the both formal parameters R_1 and R_2. More precisely, this is done by the type mappings $((\gamma_1,\gamma_2,\alpha)\ \mathsf{R}_1 \mapsto \alpha)$, $((\gamma_1,\gamma_2,\alpha)\ \mathsf{R}_2 \mapsto \alpha)$, and operation mappings $(\mathsf{R}_1 \mapsto \mathsf{IdR})$, $(\mathsf{R}_2 \mapsto \mathsf{IdR})$.

So, computing the pushout of the span *ProductRule* $\hookleftarrow$ *Parameter_Relator2* $\longrightarrow$ *Square_Relator*, we obtain the extended theory *Square_Relator'*, containing the relator SqR, which object part maps any type κ to $\kappa \times \kappa$, and arrow part any arrow $f :: (\alpha \to \beta)$ to the arrow $f \;**\; f :: (\alpha \times \alpha) \to (\beta \times \beta)$. The fact, that SqR indeed gives a relator, is finally established by the theory morphism from *Relator* to *Square_Relator'*, obtained by composition. So, we can use SqR for construction of further relators, like the relator $\kappa \mapsto (\kappa_1,\kappa_2)\,\mathsf{ConstR} \times \kappa \times \kappa + \mathbf{1}$, where κ_1,κ_2 are some fixed types and $\mathbf{1}$ is a special instance of ConstR, namely the singleton relator sending any type to $\mathbf{1}$, which contains a single element. In particular, the relator $(\gamma_1,\gamma_2)\,\mathsf{ConstR} \times \_ \times \_ + \mathbf{1}$ will be used later on, as the intermediate data type for the derivation of a quicksort function.

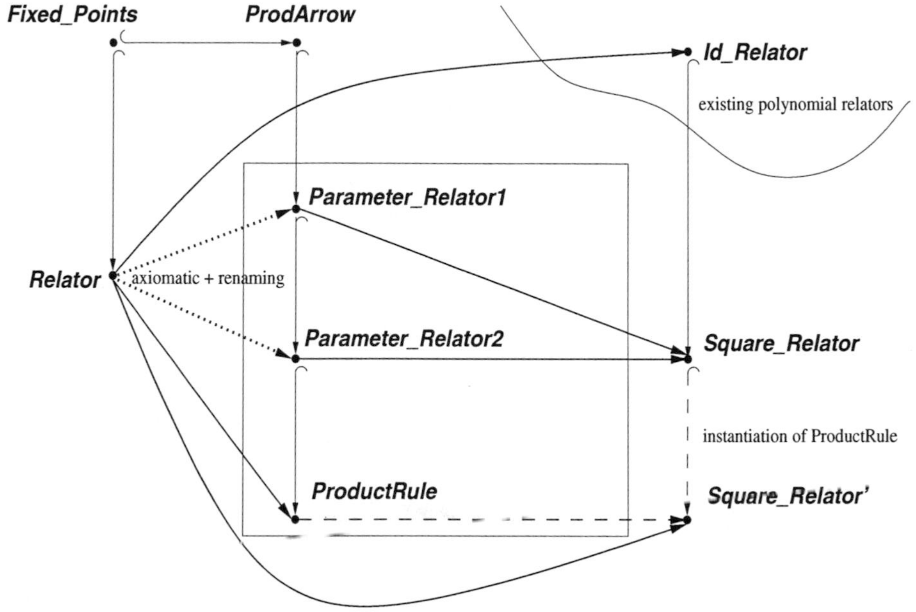

Figure 6.4: Deriving the square relator

6.6 Conclusion

In this chapter locally complete allegories and endo-relators have been treated. It has been shown how the join operator provides the construction of upper adjoints and fixed points for monotone and continuous arrow transformers. Based on this, hylomorphism arrows have been defined by means of least fixed points. Moreover, the behaviour of hylomorphisms (especially regarding the mapping property) depending on the coalgebra and algebra parameters has been studied. Using the fact that the Divide-and-Conquer principle can be formalised via hylomorphisms, it has been represented in form of a design tactic on the level of locally complete allegories.

On the other hand, typed Kleene algebras, additionally equipped with an appropriate converse operator, have been considered as a generalisation of locally complete allegories. It has been shown, how for such Kleene algebras we can specify and reason with an abstract counterpart of the well-known confluence property.

Chapter 7

Relations

Now we can turn towards relations, establishing that the interpretation $(\alpha \to \beta) \mapsto (\alpha \Rightarrow \beta \Rightarrow bool)$ of the arrow type gives the locally complete allegory of relations. But it is also worth to make the following, somewhat smaller step $(\alpha \to \beta) \mapsto (\alpha \Rightarrow \beta \Rightarrow V)$ with a fixed type V to the locally complete allegory of so-called V-valued relations.

7.1 V-valued relations

We start with the formal definition of a locale, which is almost the same as in [Bird and de Moor 1997]:

Definition 4 (locale)
A *locale* is a partial order $(\sqsubseteq_V, V)$, where for any $X \subseteq V$ there exists the least upper bound $\bigsqcup_V X$ and for any two $a, b \in V$ there is the greatest lower bound $a \sqcap_V b$, such that

$$\left(\bigsqcup_V X\right) \sqcap_V a \sqsubseteq_V \bigsqcup_V \{x \sqcap_V a \mid x \in X\}$$

holds for any $a \in V$.

This structure is formalised in Isabelle/HOL in the theory *Locale*, importing *Nat*, *Product_Type* and *Sum_Type*, as follows. Firstly, the type constructor V with the rank 0 is declared by
typedecl V

Then, the three logical constants
consts
$\_ \sqsubseteq_V \_$ :: $V \Rightarrow V \Rightarrow bool$
$\_ \sqcap_V \_$ :: $V \Rightarrow V \Rightarrow V$
$\bigsqcup_V$ :: $V\ set \Rightarrow V$
are specified via axioms, as stated in Definition 4.

Next, the theory *V_valued* of the V-valued relations imports *Locale*, such that *V_valued* can be seen as parameterised over some locale V, i.e. we have the parametrisation $\langle$*Locale*, *V_valued*$\rangle$.

Further, we can construct the following theory morphism
thymorph *v_val_arrow* : *ArrowType* $\longrightarrow$ *V_valued*
type_map: $[(\alpha \to \beta) \mapsto (\alpha \Rightarrow \beta \Rightarrow V)]$

A proper instantiation of the parametrisation $\langle$*ArrowType*, *Allegory*$\rangle$ by *v_val_arrow* introduces the following logical constants in *V_valued*:
Id :: $\alpha \Rightarrow \alpha \Rightarrow V$
$\_ ; \_$ :: $(\alpha \Rightarrow \beta \Rightarrow V) \Rightarrow (\beta \Rightarrow \gamma \Rightarrow V) \Rightarrow (\alpha \Rightarrow \gamma \Rightarrow V)$
$\_ \sqcap \_$ :: $(\alpha \Rightarrow \beta \Rightarrow V) \Rightarrow (\alpha \Rightarrow \beta \Rightarrow V) \Rightarrow (\alpha \Rightarrow \beta \Rightarrow V)$
$\_^\circ$ :: $(\alpha \Rightarrow \beta \Rightarrow V) \Rightarrow (\beta \Rightarrow \alpha \Rightarrow V)$
$\_ \sqsubseteq \_$ :: $(\alpha \Rightarrow \beta \Rightarrow V) \Rightarrow (\alpha \Rightarrow \beta \Rightarrow V) \Rightarrow bool$
and generates the theory morphism goals, i.e. the proof obligations comprising translated axioms from *Allegory*. In order to prove these, the allegory operations above can be defined in *V_valued* as follows:

1. $\_ \sqsubseteq \_$ and $\_ \sqcap \_$ are just point-wise extension of $\_ \sqsubseteq_V \_$ and $\_ \sqcap_V \_$, respectively, and the definition of the converse $\_^\circ$ is obvious.

2. the identity can be defined by

   $$Id\ a\ b\ \equiv\ if\ a = b\ then\ \top_V\ else\ \bot_V$$

   where $\top_V$ and $\bot_V$ are the top and the bottom of the locale V.

3. the definition of the composition is more interesting:

   $$(f\ ;\ g)\ a\ c \equiv \bigsqcup\nolimits_V \{f\ a\ b \sqcap_V\ g\ b\ c \mid b \in UNIV_V\}$$

Provided by these definitions, the theory morphism goals can be proved in *V_valued*, which establishes the theory morphism *v_val_as_allegory* : *Allegory* $\longrightarrow$ *V_valued*.

In order to show that the allegory of V-valued relations is also locally complete, we proceed with a proper instantiation of the theory *LocallyComplete* by the theory morphism *v_val_as_allegory*. This step introduces the logical constant $\bigsqcup :: (\alpha \Rightarrow \beta \Rightarrow V)\ set \Rightarrow (\alpha \Rightarrow \beta \Rightarrow V)$ and generates the theory morphism goals comprising the translated axioms from *LocallyComplete*. In *V_valued* the least upper bound is defined by $(\bigsqcup S)\ a\ b \equiv \bigsqcup_V \{f\ a\ b \mid f \in S\}$. This definition provides the proofs of the three theory morphism goals. So, we can also construct the theory morphism *v_val_as_lc_allegory* : *LocallyComplete* $\longrightarrow$ *V_valued*, which establishes that V-valued relations form a locally complete allegory. Finally, a proper instantiation of *Fixed_Points* by *v_val_as_lc_allegory* immediately yields the theory morphism *v_val_fixpoints* : *Fixed_Points* $\longrightarrow$ *V_valued*, since the theory *Fixed_Points* does not contain neither logical constants nor axioms (apart of definitions, of course).

Products and coproducts for V-valued relations. One of the main benefits of the consideration of V-valued relations here is, that the extensions of categorical products and coproducts in *ProdArrow* and *SumArrow*, introduced axiomatically in the previous chapter, can be now defined in *V_valued*.

The instantiation of *ProdArrow* by the theory morphism *v_val_fixpoints* introduces the logical constant

$$\_ \ast\ast \_ :: (\alpha \Rightarrow \beta \Rightarrow V) \Rightarrow (\gamma \Rightarrow \delta \Rightarrow V) \Rightarrow ((\alpha \times \gamma) \Rightarrow (\beta \times \delta) \Rightarrow V)$$

which can be defined by

$$(f \ast\ast g)\ (a, c)\ (b, d) \equiv f\ a\ b \sqcap_V\ g\ c\ d$$

such that the theory morphism goals, comprising the translated axioms in *ProdArrow* (Section 6.5), can be shown. The most complicated case here forms the continuity property of $\_ \ast\ast \_$. So, we obtain the theory morphism *v_val_products* : *ProdArrow* $\longrightarrow$ *V_valued*.

We proceed further in the same manner, instantiating *SumArrow* by the theory morphism *v_val_fixpoints*. This gives the logical constant

$$\_ +\!\!+ \_ :: (\alpha \Rightarrow \beta \Rightarrow V) \Rightarrow (\gamma \Rightarrow \delta \Rightarrow V) \Rightarrow ((\alpha + \gamma) \Rightarrow (\beta + \delta) \Rightarrow V)$$

which is basically defined by the following four equations:

1. $(f +\!\!+ g)\ (\mathit{Inl}\ a)\ (\mathit{Inl}\ b) = f\ a\ b$,
2. $(f +\!\!+ g)\ (\mathit{Inr}\ c)\ (\mathit{Inr}\ d) = g\ c\ d$,
3. $(f +\!\!+ g)\ (\mathit{Inl}\ a)\ (\mathit{Inr}\ d) = \bot_V$,

4. $(f \;+\!\!+\; g)\,(Inr\ c)\,(Inl\ b) = \bot_V$,

where $Inl :: \alpha \Rightarrow \alpha + \beta$ and $Inr :: \beta \Rightarrow \alpha + \beta$ are the both constructors of the sum type $\alpha + \beta$. Using these identities, the axioms translated from *SumArrow* can be proved, and thus the theory morphism *v_val_coproducts* : *SumArrow* $\longrightarrow$ *V_valued* can be constructed.

Altogether, this means that in the allegory of V-valued relations, the extensions of categorical products and coproducts to arrows are definable. Furthermore, we obtain that V-valued relations allow the construction of polynomial relators, which can be now manifested by the theory morphism *v_val_polynomial* : *Polynomial* $\longrightarrow$ *V_valued*. The current state of the development in allegories is sketched in Figure 7.1.

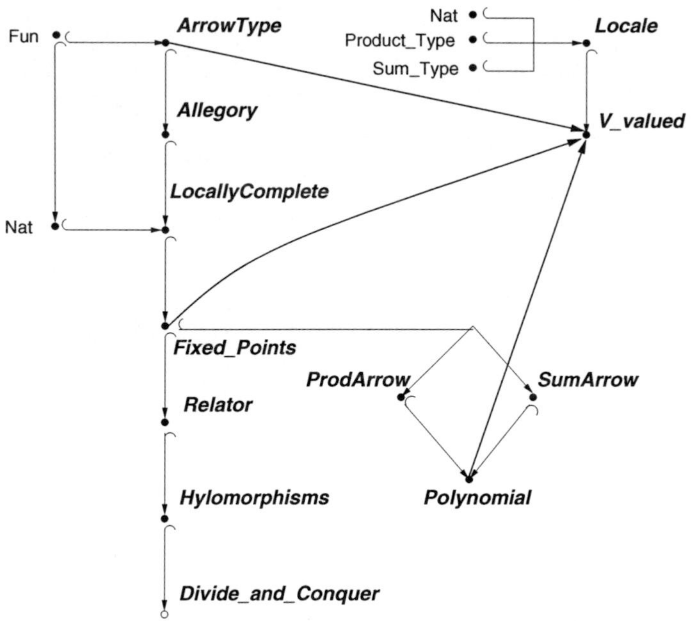

Figure 7.1: The current state of the theory development

7.2 The allegory of relations

Now we have arrived at the point where we will reuse all the presented axiomatic developments in the conservative development branch of Isabelle/HOL (in the sense of Figure 1.1).

The development in the allegory of relations **Rel** is started in the theory *Rel*, importing the theory *Datatype*, which basically summarises the results from *Product_Type* and *Sum_Type*.

As described above, the allegory of V-valued relations is parameterised over a locale structure on V. It is not difficult to prove that the type *bool* of Boolean values forms a locale, which can be now expressed by means of the following theory morphism:

thymorph *bool_as_locale* : *Locale* $\longrightarrow$ *Rel*
type_map: $[V \mapsto bool]$
op_map: $[\ \_ \sqsubseteq_V \_ \ \mapsto\ \_ \longrightarrow \_,$
$\qquad\quad \_ \sqcap_V \_ \ \mapsto\ \_ \wedge \_,$
$\qquad\quad \bigsqcup_V \ \mapsto\ \exists\ ]$

A proper instantiation of the parametrisation $\langle$*Locale*, *V_valued*$\rangle$ by the theory morphism *bool_as_locale* immediately yields the theory morphism *rel_as_v_val* : *V_valued* $\longrightarrow$ *Rel*, since *V_valued* is a definitional extension of *Locale*. Furthermore, now we can compose *rel_as_v_val* with any theory morphism, having *V_valued* as its target (as shown in Figure 7.1), in order to get 'access' to the abstract developments, done for allegories. So, in particular, we have *rel_fixpoints* : *Fixed_Points* $\longrightarrow$ *Rel* and *rel_polynomial* : *Polynomial* $\longrightarrow$ *Rel*.

However, the abstract allegorical results, obtained via instantiation for relations, have to be slightly adapted for this context, which is described in the sequel.

Conversion for sets. First of all, the already mentioned correspondence between coreflexive relations and sets need to be made explicit. This can be done by the following two derived constants:

$rep\_set :: \alpha\ set \Rightarrow (\alpha \Rightarrow \alpha\ set)$
$abs\_set :: (\alpha \Rightarrow \alpha\ set) \Rightarrow \alpha\ set$

where $rep\_set\ A \equiv (\lambda a.\{a\})$ and $abs\_set\ r \equiv \{a \mid a \in r\ a\}$. In other words, rep_set embeds any set as the corresponding coreflexive relation, while abs_set projects any coreflexive relation to the corresponding set. Indeed, $abs\_set(rep\_set\ A) = A$, $rep\_set(abs\_set\ r) \sqsubseteq r$ hold for any $A :: \alpha\ set$ and

$r :: (\alpha \Rightarrow \alpha\ set)$. Moreover, $rep\_set(abs\_set\ r) = r$ holds for any coreflexive relation r.

The constants *rep_set* and *abs_set* provide a conversion between sets and coreflexive relations, such that we will identify these in this context, even though we cannot do this in Isabelle. Let us consider, for instance the image operator. By the instantiation of *Allegory*, the image operator is translated to *Rel* as the derived constant
$Im\ :: (\alpha \Rightarrow \beta\ set) \Rightarrow (\alpha \Rightarrow \alpha\ set) \Rightarrow (\beta \Rightarrow \beta\ set)$
Based on the set conversions, the application $Im\,r\,A$ to $A :: \alpha\,set$ will be allowed here and yields the set $Im\ r\ A :: \beta\ set$, whereas in Isabelle we first need to lift Im explicitly to sets as follows. We define the constant
$Im'\ :: (\alpha \Rightarrow \beta\ set) \Rightarrow (\alpha\ set) \Rightarrow (\beta\ set)$
via the conversions *rep_set* and *abs_set* by
$Im'\,r\ \equiv abs\_set \circ Im\ r\ \circ rep\_set$
such that Im' is the explicit restriction of Im to the coreflexive relations.

Conversion for mappings. We proceed similarly with the mappings. The conversions
$rep\_map :: (\alpha \Rightarrow \beta) \Rightarrow (\alpha \Rightarrow \beta\ set)$
$abs\_map :: (\alpha \Rightarrow \beta\ set) \Rightarrow (\alpha \Rightarrow \beta)$
can be defined by $rep\_map\ f \equiv Id \circ f$ and $abs\_map\ r \equiv (\lambda a.\ THE\,b.\ r\,a = \{b\})$, such that the properties

1. $rep\_map\,f$ is a mapping for any f,

2. $abs\_map(rep\_map\,f) = f$ for any f,

3. $rep\_map(abs\_map\ r) = r$ for any mapping r

hold. As with coreflexive relations and sets, relations, which are mappings, will be now identified with functions, such that e.g. the compositions $f\ ;\ g$ and $g \circ f$ will be considered as essentially the same for any two functions $f :: \alpha \Rightarrow \beta$, $g :: \beta \Rightarrow \gamma$, as well as for any two relations $f :: \alpha \Rightarrow \beta\ set$, $g :: \beta \Rightarrow \gamma\ set$, satisfying the mapping condition.

7.2.1 The Divide-and-Conquer tactic for relations

We can now adapt the Divide-and-Conquer tactic, formulated generally for locally complete allegories, to the special case of relations.

The first parameter part: relator. An *axiomatic* instantiation of the theory *Relator* by the theory morphism *rel_fixpoints* : *Fixed_Points* $\longrightarrow$ *Rel* translates the axiomatisation of the relator R to relations. The object part of R stays the same, while the arrow part is now the logical constant $\mathsf{R} :: (\alpha \Rightarrow \beta\ set) \Rightarrow ((\gamma_1, \gamma_2, \alpha)\ \mathsf{R} \Rightarrow ((\gamma_1, \gamma_2, \beta)\ \mathsf{R})\ set)$.

Based on this, we can firstly extract the endo-functor RF on mappings from R as follows. We define the constant
$\mathsf{RF} :: (\alpha \Rightarrow \beta) \Rightarrow ((\gamma_1, \gamma_2, \alpha)\ \mathsf{R} \Rightarrow (\gamma_1, \gamma_2, \beta)\ \mathsf{R})$
by $\mathsf{RF} \equiv abs\_map \circ \mathsf{R} \circ rep\_map$, which is well-defined, provided by the fact that R preserves mappings. This property has been already shown generally for any relator. Based on this definition of RF, the functor property $\mathsf{RF}(f \circ g) = \mathsf{RF}(f) \circ \mathsf{RF}(g)$ can be derived.

Secondly, we also extract the endo-relator $\mathsf{RS} : \mathit{Corefl}(\mathbf{Rel}) \longrightarrow \mathit{Corefl}(\mathbf{Rel})$ on the sub-allegory of coreflexive relations, i.e. we define the constant
$\mathsf{RS} :: \alpha\ set \Rightarrow ((\gamma_1, \gamma_2, \alpha)\ \mathsf{R})\ set$
by $\mathsf{RS} \equiv abs\_set \circ \mathsf{R} \circ rep\_set$, and reuse the general property that R preserves coreflexive arrows. So, e.g. the property $\mathsf{RS}(A \cap B) = \mathsf{RS}(A) \cap \mathsf{RS}(B)$ can be derived.

The second parameter part. The second part of the parameters of the Divide-and-Conquer tactic is obtained by the next instantiation of the general Divide-and-Conquer formalisation for locally complete allegories. So, by translation we obtain the type classes $class_1$, $class_2$, the type constructors *Domain*, *Range*, as well as the logical constants

$$
\begin{array}{lll}
dac\_spec & :: & (\gamma_1 ::_{sort} \{class_1\}, \gamma_2 ::_{sort} \{class_2\})\ Domain \Rightarrow \\
 & & ((\gamma_1, \gamma_2)\ Range)\ set \\[1ex]
dcmp\_spec & :: & (\gamma_1 ::_{sort} \{class_1\}, \gamma_2 ::_{sort} \{class_2\})\ Domain \Rightarrow \\
 & & ((\gamma_1, \gamma_2, (\gamma_1, \gamma_2)\ Domain)\ \mathsf{R})\ set \\[1ex]
cmps\_spec & :: & (\gamma_1, \gamma_2, (\gamma_1, \gamma_2)\ Range)\ \mathsf{R} \Rightarrow \\
 & & ((\gamma_1 ::_{sort} \{class_1\}, \gamma_2 ::_{sort} \{class_2\})\ Range)\ set \\[1ex]
decompose & :: & (\gamma_1 ::_{sort} \{class_1\}, \gamma_2 ::_{sort} \{class_2\})\ Domain \Rightarrow \\
 & & (\gamma_1, \gamma_2, (\gamma_1, \gamma_2)\ Domain)\ \mathsf{R} \\[1ex]
compose & :: & (\gamma_1, \gamma_2, (\gamma_1, \gamma_2)\ Range)\ \mathsf{R} \Rightarrow \\
 & & (\gamma_1 ::_{sort} \{class_1\}, \gamma_2 ::_{sort} \{class_2\})\ Range
\end{array}
$$

The axioms formulated here are almost the same as in the abstract version.

These firstly comprise:

1. $\mathcal{HT}^{\mathsf{R}}(dcmp\_spec, cmps\_spec)\ dac\_spec \sqsubseteq dac\_spec$, corresponding to the *SPRP*-axiom,

2. $decompose \sqsubseteq dcmp\_spec$, and

3. $compose \sqsubseteq cmps\_spec$.

It remains to resolve the R-reductivity axiom for *decompose* in some way, i.e. we need to provide the proposition $\mu(decompose\backslash^{\mathsf{R}}) = Id$, i.e. the condition (6.9). As mentioned already, the reductivity condition (6.10) is more suitable to show that a mapping is reductive than (6.9). To this end, we formulate the following axiom for the reductivity of *decompose*:

$$\{x \mid decompose\ x \in \mathsf{RS}(A)\} \subseteq A \Longrightarrow a \in A \text{ for any } A \text{ and } a \tag{7.1}$$

which is essentially the condition (6.10), adapted for the relational context. From this, the proposition $\mu(decompose\backslash^{\mathsf{R}}) = Id$ can be proved, as [Doornbos and Backhouse 1995] shows. In other words, we have replaced the axiom (6.9) for *decompose*, used in the Divide-and-Conquer formalisation on the abstract level of allegories, by the axiom (7.1).

The body part of the Divide-and-Conquer tactic. In the body part we instantiate the body part of the abstract *Divide_and_Conquer* and so the mapping *dac* from *Divide_and_Conquer* is obtained as the function
$dac :: (\gamma_1 ::_{sort} \{class_1\}, \gamma_2 ::_{sort} \{class_2\})\ Domain \Rightarrow (\gamma_1, \gamma_2)\ Range$
here, i.e. as the hylomorphism $\mathcal{H}^{\mathsf{R}}(decompose, compose)$, which is a function due to the functions *decompose* and *compose*.

Furthermore, the crucial results about *dac*

1. $dac \sqsubseteq dac\_spec$, and

2. $dac = compose \circ \mathsf{RF}\ (dac) \circ decompose$

are obtained by theorem translation.

7.2.2 Deriving QuickSort

Now, the Divide-and-Conquer tactic can be applied in order to derive polymorphic functions in correct-by-construction manner. As the first example, a derivation of QuickSort is considered.

Specification of sorting functions. First of all, we need to give a specification of sorting in form of a relation, since it should give us an instance of the formal parameter *dac_spec*. To this end, we take the permutation relation $\leadsto$ on lists, already defined in Isabelle/HOL. So, the type of the derived constant $\leadsto$ is $(\alpha\, list \times \alpha\, list)\; set$, and $xs \leadsto ys$ essentially holds iff the numbers of occurrences of any element in xs and ys coincide. Next, we 'wrap' the permutation relation by the derived constant *perm_rel*, which type matches the relational interpretation of the allegorical arrow type $\rightarrow$. More precisely, we declare
$perm\_rel :: \alpha\ list \Rightarrow (\alpha\ list)\ set$
and define it by
$perm\_rel\ xs \equiv \{ys \mid xs \leadsto ys\}$
In a similar manner we proceed with the predefined predicate
$sorted\ (\leq) :: (\alpha ::_{sort} \{linorder\}\ list)\ set$
for lists of any type of the sort $\{linorder\}$ which are sorted by an order $\leq$. That is, we define the relation
$sorted\_rel :: (\alpha ::_{sort} \{linorder\})\ list \Rightarrow (\alpha\ list)\ set$
using the embedding *rep_set* by
$sorted\_rel \equiv rep\_set(\{xs \mid sorted\ (\leq)\ xs\})$
and obtain the corresponding coreflexive relation.

Sorting functions can be now specified by the derived constant
$sorting\_spec :: (\alpha ::_{sort} \{linorder\})\ list \Rightarrow (\alpha\ list)\ set$
which is consequently defined by the composition of the previous two relations:
$sorting\_spec \equiv perm\_rel\ ;\ sorted\_rel$
That is, *sorting_spec* assigns to any list the (finite) set of its sorted permutations.

The intermediate data structure. The choice of an intermediate data structure is carried out by an instantiation of the formal relator parameter R. So, in the case of QuickSort we instantiate by the assignment

$$(\gamma_1, \gamma_2, \alpha)\ \mathsf{R} \mapsto \gamma_1 \times \alpha \times \alpha + \mathbf{1}$$

where $\mathbf{1}$ represents the singleton data type, containing only the element $\{*\}$. By this instantiation, only the first auxiliary parameter γ_1 of R will be used.

Intuitively, this assignment is motivated by the way in which the QuickSort algorithm works: it either terminates directly for primitive inputs which is represented by the $\mathbf{1}$ case, or proceeds otherwise with a 'pivot' element of auxiliary type γ_1, descending into two smaller partitions of the input each having the parameter type α, and thus represented by $\alpha \times \alpha$.

Instantiating ***Domain*** **and** ***Range*****.** Since we want to derive a sorting function which maps lists to lists, the formal parameters *Domain* and *Range* are instantiated by (γ_1, γ_2) *Domain* $\mapsto \gamma_1$ *list* and (γ_1, γ_2) *Range* $\mapsto \gamma_1$ *list*.

These instantiations impose that, e.g. *dac_spec* has to be instantiated by a constant having the type $(\alpha ::_{sort} \{?class_1\})\ list \Rightarrow (\alpha\ list)\ set$, where $\{?class_1\}$ should emphasise that the formal type class parameter $class_1$ has not been instantiated yet. Further, the definition of *sorting_spec* imposes the type class instantiation $class_1 \mapsto linorder$. Moreover, since γ_2 has not been used in all the instantiations above, an instantiation for $class_2$ can be omitted here.

The decomposition for QuickSort. From the previous instantiations follows that a constant, instantiating the formal parameter *decompose*, has the type
$(\gamma_1 ::_{sort} \{linorder\})\ list \Rightarrow (\gamma_1 \times (\gamma_1\ list) \times (\gamma_1\ list) + \mathbf{1})$
and the decomposition for QuickSort can be defined by the following two equations:

$$\begin{array}{lcl} decompose\ [\,] & = & Inr\ \{*\} \\ decompose\ (Cons\ x\ xs) & = & Inl\ (x, low\ x\ xs, high\ x\ xs) \end{array}$$

where *low* (*high*) is the well-known function, filtering those elements from xs which are less or equal (greater than) x.

Further, we also have to give some actual parameter for the specification of the decomposition. It is possible to supply *decompose* itself as the actual parameter for the formal parameter *dcmp_spec* as well, which also directly provides the condition $decompose \sqsubseteq dcmp\_spec$. On the other hand, such treatment would completely ignore the abstraction possibility, which is the main benefit of this Divide-and-Conquer tactic. More precisely, in the definition of the specification of the decomposition we should abstract as much as possible over implementation details. On the one hand this will make the proof of $decompose \sqsubseteq dcmp\_spec$ non-trivial, but on the other hand allows us to use only the essential properties of the decomposition function in the proof of the *SPRP* axiom, simplifying this core problem.

Following this principle, we define *dcmp_spec* by the equations

$$\begin{array}{lcl} dcmp\_spec\ [\,] & = & \{\ Inr\ \{*\}\ \} \\ dcmp\_spec\ (Cons\ x\ xs) & = & \{Inl\ (x, L, R) \mid (xs \rightsquigarrow L@R) \wedge \\ & & (\forall l\ mem\ L.\ l \leq x) \wedge (\forall r\ mem\ R.\ x < r)\} \end{array}$$

The composition for QuickSort. Similarly, a constant instantiating the formal parameter *compose* has the type

$(\gamma_1 \times (\gamma_1\ \mathit{list}) \times (\gamma_1\ \mathit{list}) + \mathbf{1}) \Rightarrow (\gamma_1 ::_{sort} \{\mathit{linorder}\})\ \mathit{list}$
and the composition for QuickSort can be defined by the following two equations:

$$\begin{array}{lcl} \mathit{compose}\ (\mathit{Inr}\ \{*\}) & = & [\,] \\ \mathit{compose}\ (\mathit{Inl}\ (x, L, R)) & = & L@[x]@R \end{array}$$

Since in this case the composition function does not make something apart of concatenation, it makes sense to use *compose* also as its specification, i.e. to instantiate the formal parameter *cmps_spec* by the function *compose*.

Proof obligations. Since all formal parameters of the Divide-and-Conquer tactic have been instantiated, we can consider the emerging proof obligations.

1. $\mathit{decompose} \sqsubseteq \mathit{dcmp\_spec}$, i.e. we have to prove that *decompose* satisfies the specification *dcmp_spec*. This follows immediately from the properties of *low* and *high*.

2. The reductivity property (7.1) for *decompose*, which we consider in more detail due to the involved relator RS on sets. In the previous section RS has been defined by $\mathit{abs\_set} \circ \mathsf{R} \circ \mathit{rep\_set}$. So, by the instantiation of R above, the term $\mathsf{RS}(A)$ is equivalent to the term

   $$\mathit{abs\_set}(\mathit{Id} \ast\!\ast\ \mathit{rep\_set}(A) \ast\!\ast\ \mathit{rep\_set}(A) +\!\!+\ \mathit{Id}).$$

   By the properties of *abs_set* and *rep_set*, this is in turn equivalent to $\mathit{UNIV} \times A \times A + \mathit{UNIV}$, where $\times$ denotes the Cartesian product and $+$ the disjoint sum on sets. That is, (7.1) reduces in this context to the condition

   $$\{x \mid \mathit{decompose}\ x \in \mathit{UNIV} \times A \times A + \mathit{UNIV}\} \subseteq A \Longrightarrow xs \in A$$

   which will be proved by the induction on the length of the list xs. This means, we have to show the same property, but under the additional assumption $\forall ys.\ \mathit{length}\ ys < \mathit{length}\ xs \longrightarrow ys \in A$. Then, we can show $xs \in A$ by showing $\mathit{decompose}\ xs \in \mathit{UNIV} \times A \times A + \mathit{UNIV}$. The proof proceeds further by cases on xs, such that the defining *decompose*-equations can be applied, and finally using the instances of the induction hypothesis by *low* and *high*.

3. The SPRP-property, which now comprises the condition

   $$\frac{b \in (\mathit{dcmp\_spec}\ ;\ (\mathit{Id} \ast\!\ast\ \mathit{sorting\_spec} \ast\!\ast\ \mathit{sorting\_spec} +\!\!+\ \mathit{Id})\ ;\ \mathit{compose})\ a}{b \in \mathit{sorting\_spec}\ a}$$

for any a, b of type $(\alpha ::_{sort} \{linorder\})\ list$ with a fixed α. As mentioned already, here we benefit of the abstraction of *decompose* by *dcmp_spec*, since we only consider what the decomposition function provides rather than how it has been implemented.

Obtaining a polymorphic sorting function. Now we can instantiate the body of the relational Divide-and-Conquer parametrisation. Provided by the simultaneous renaming of *dac* to *qsort* we obtain the function
$qsort :: (\gamma_1 ::_{sort} \{linorder\})\ list \Rightarrow \gamma_1\ list$
satisfying the following rules:

1. $qsort \sqsubseteq sorting\_spec$, i.e. a correct sorting function, and

2. $qsort = compose \circ (Id ** qsort ** qsort ++ Id) \circ decompose$, i.e. the recursive equation for *qsort*.

From the latter property together with the definitions of composition, decomposition, $\_ ++ \_$ and $\_ ** \_$, we can immediately derive the following two equations:

$$
\begin{array}{lcl}
qsort\ [\,] & = & [\,] \\
qsort\ (Cons\ x\ xs) & = & qsort(low\ x\ xs)@[x]@qsort(high\ x\ xs)
\end{array}
$$

which uniquely determine the *qsort*-function. But in these equations the decomposition and composition functions as well as the intermediate data structure are only implicitly present.

7.2.3 Updates on balanced binary search trees

In order to emphasise that the Divide-and-Conquer design tactic provides construction of solutions for a wide range of algorithmic problems, a function implementing updates on balanced binary search trees will be derived in the sequel.

Binary trees. A binary tree where each node contains a key (a term of type γ_1) and a value (a term of type γ_2) is specified in Isabelle/HOL by the following algebraic data type:

$$
\begin{array}{lcl}
\textbf{datatype}\ \ (\gamma_1, \gamma_2)\ Tree & = & Tip \\
 & | & Node\ ((\gamma_1, \gamma_2)\ Tree)\ \gamma_1\ \gamma_2\ ((\gamma_1, \gamma_2)\ Tree)
\end{array}
$$

Then the predicate $occurs :: \gamma_1 \Rightarrow (\gamma_1, \gamma_2)\ Tree \Rightarrow bool$, determining the set of keys occurring in a tree, can be recursively defined by the equations

$$
\begin{array}{lcl}
occurs\ k\ Tip & = & False \\
occurs\ k\ (Node\ L\ (k', v)\ R) & = & (k = k') \vee occurs\ k\ L \vee occurs\ k\ R
\end{array}
$$

Search trees. Further, the predicate
$search\_tree :: (\gamma_1 ::_{sort} \{linorder\}, \gamma_2)\ Tree \Rightarrow bool$
for search trees is defined only for trees with a linear order on keys:

$$\begin{array}{lcl} search\_tree\ Tip & = & True \\ search\_tree\ (Node\ L\ (k, v)\ R) & = & search\_tree\ L\ \wedge\ search\_tree\ R\ \wedge \\ & & (\forall k'.\ occurs\ k'\ L \longrightarrow k' < k)\ \wedge \\ & & (\forall k'.\ occurs\ k'\ R \longrightarrow k < k') \end{array}$$

An efficient look-up function
$lookup :: (\gamma_1 ::_{sort} \{linorder\}) \Rightarrow (\gamma_1, \gamma_2)\ Tree \Rightarrow \gamma_2\ option$
for search trees can be now defined in the standard way, where the data type $\gamma_2\ option$ is generated by the constructors $None :: \gamma_2\ option$ and $Some :: \gamma_2 \Rightarrow \gamma_2\ option$ and allows to represent partial maps, as in this case. So, the following equivalence between *occurs* and the keys, on which *lookup* is defined (i.e. the value is not *None*), is now derivable for any search tree T:

$$(occurs\ k\ T) = (\exists v.\ lookup\ k\ T = Some\ v)$$

Balanced trees. The predicate $balanced :: (\gamma_1, \gamma_2)\ Tree \Rightarrow bool$ for balanced trees can be defined recursively by the following two equations

$$\begin{array}{lcl} balanced\ Tip & = & True \\ balanced\ (Node\ L\ (k, v)\ R) & = & balanced\ L\ \wedge\ balanced\ R\ \wedge \\ & & (depth\ L = depth\ R\ \vee \\ & & \ depth\ L = depth\ R + 1\ \vee \\ & & \ depth\ L + 1 = depth\ R) \end{array}$$

Consequently, balanced binary search trees (AVL-trees) can be now specified by the predicate $avl :: (\gamma_1 ::_{sort} \{linorder\}, \gamma_2)\ Tree \Rightarrow bool$, simply defined by $avl\ T \equiv search\_tree\ T\ \wedge\ balanced\ T$.

Specifying update functions. Now, an update function should take a key-value tuple and an AVL-tree as arguments and return an AVL-tree, where the lookup on the given key returns the given value, while lookups on other keys stay unchanged. Moreover, that the difference between the depths of the output and input trees is 0 or 1, is additionally required. This specification of update can be formalised by the relation
$update\_spec :: (\gamma_1 ::_{sort} \{linorder\}) \times \gamma_2 \times (\gamma_1, \gamma_2)\ Tree \Rightarrow ((\gamma_1, \gamma_2)\ Tree)\ set$
defined by:
$update\_spec\ ((k, v), T) \equiv$

$$\{T' \mid avl\ T \longrightarrow (avl\ T' \ \wedge lookup\ k\ T' = Some\ v \wedge \\ \forall k' \neq k.\ lookup\ k'\ T' = lookup\ k\ T \wedge \\ depth\ T \leq depth\ T' \ \wedge \ depth\ T' \leq depth\ T + 1))\}$$

The domain, range and type class interpretations. The type of the specification *update_spec* above firstly imposes the following interpretations for the Divide-and-Conquer formal parameters for the domain and the range of the update function:

$(\gamma_1, \gamma_2)\ Domain \ \mapsto\ \gamma_1 \times \gamma_2 \times (\gamma_1, \gamma_2)\ Tree$

$(\gamma_1, \gamma_2)\ Range \ \mapsto\ (\gamma_1, \gamma_2)\ Tree$

Secondly, it imposes that the first type class parameter $class_1$ has to be instantiated by *linorder*, since it qualifies in this context the types of keys in a tree. Further, $class_2$ should be interpreted by *type*, i.e. by the most general type class in the Isabelle/HOL hierarchy, since a type for the values in a search tree should be arbitrary.

The intermediate data structure and decomposition. Now we turn towards the choice of an interpretation of the formal relator parameter R. This is done by a consideration of the functionality of the decomposition function, which basically determines the intermediate data type for updates.

So, the instantiations above yield that the function *decompose* has a type of the form:

$(\gamma_1 ::_{sort} \{linorder\} \times \gamma_2 \times (\gamma_1, \gamma_2)\ Tree) \Rightarrow (\gamma_1, \gamma_2, (\gamma_1 \times \gamma_2 \times (\gamma_1, \gamma_2)\ Tree))\ ?\mathsf{R}$

where ?R emphasises that the relator is still unknown. Let $((key, val), T)$ be an input of *decompose*. We consider the two cases for T:

1. $T = Tip$. Then the resulting tree is just $Node\ Tip\ (key, val)\ Tip$, i.e. we make an insertion, and also need to tag the output as constant, because no recursive calls are required.

2. $T = Node\ L\ (key', val')\ R$. Further, let $key < key'$. In this case we need to descend into the left sub-tree L with (key, val), but also need to tag (key', val') and R in such a way that the composition function knows that R is the right sub-tree of the input tree. The case $key' < key$ has to be treated symmetrically.

   Finally, if $key = key'$ then we make a constant step as in 1., but now updating the value, assigned to key, i.e. the result is the tree $Node\, L\, (key, val)\, R$. As in 1. we need to tag the output as constant.

As a result of this considerations we can specify the following auxiliary algebraic data type *Control*:

$$\textbf{datatype}\quad (\gamma_1,\gamma_2)\ \textit{Control} \;=\; \textit{Stop}\ ((\gamma_1,\gamma_2)\ \textit{Tree}) \mid \textit{Left}\ \gamma_1\ \gamma_2\ ((\gamma_1,\gamma_2)\ \textit{Tree}) \mid \textit{Right}\ \gamma_1\ \gamma_2\ ((\gamma_1,\gamma_2)\ \textit{Tree})$$

So, e.g. a term *Left k v T* captures the information that T is the left sub-tree of the input tree, and that we recursively descend into the right sub-tree.

Now an instantiation of the relator R can be made as follows:
$(\gamma_1,\gamma_2,\alpha)\ \mathsf{R} \mapsto (\gamma_1,\gamma_2)\ \textit{Control} \times \alpha\ +\ (\gamma_1,\gamma_2)\ \textit{Control}$
such that the decomposition function can be defined by the equations:

$$\begin{array}{l}
\textit{decompose}\ ((k,v),\textit{Tip}) \;=\; \textit{Node Tip}\ (k,v)\ \textit{Tip}\\
\textit{decompose}\ ((k,v),\textit{Node}\ L\ (k',v')\ R) \;=\\
\quad \textit{if}\ k < k'\ \textit{then Inl}\ (\textit{Right}\ k'\ v'\ R, ((k,v),L))\\
\quad \textit{else if}\ k' < k\ \textit{then Inl}(\textit{Left}\ k'\ v'\ L, ((k,v),R))\\
\qquad \textit{else Inr}(\textit{Stop}(\textit{Node}\ L\ (k,v)\ R))
\end{array}$$

This implementation of decomposition is also suitable to be used as the specification of decomposition. The reason for this decision is that the function *decompose* does not manipulate the sub-trees, such that this way we obtain the respective identities directly in the proof of *SPRP*.

Composition for updates. It remains to specify and to implement the composition function *compose* which consequently has the type
$((\gamma_1 ::_{sort} \{\textit{linorder}\},\gamma_2)\ \textit{Control} \times (\gamma_1,\gamma_2)\ \textit{Tree} + (\gamma_1,\gamma_2)\ \textit{Control}) \Rightarrow (\gamma_1,\gamma_2)\ \textit{Tree}$
and can be implemented by the following equations:

$$\begin{array}{l}
\textit{compose}\ (\textit{Inr}\ (\textit{Stop}\ T)) \;=\; T\\
\textit{compose}\ (\textit{Inl}\ (\textit{Left}\ k\ v\ L,\ T)) \;=\\
\quad \textit{if depth}\ L + 2 \leq T\ \textit{then rotateRight}(\textit{Node}\ L\ (k,v)\ T)\\
\quad \textit{else Node}\ L\ (k,v)\ T
\end{array}$$

where the function *rotateRight* re-balances the composed sub-tree, and the case *compose* (*Inl*(*Right k v R*, T)) is defined symmetrically.

In contrast to the decomposition, now it is worth to abstract the function above using the composition specification *cmps_spec*, because *compose* can manipulate (i.e. rotate) the tree, composed from the input sub-trees. This way we can abstract over the details of how the trees are manipulated, but just use the fact that the output is indeed a balanced search tree. Moreover, by an abstract specification we can explicitly formulate all further properties which are provided by composition, i.e. by rotations. However, finding a sufficient and compact abstraction is in general not straightforward. Also here, the specification *cmps_spec* can be defined by the following equations, but can be also possibly brought into a more compact and elegant form:

$$cmps\_spec(Inr\ (Stop\ T)) \ = \{\ T\ \}$$

$$\begin{aligned}
&cmps_spec\ (Inl\ (Left\ k\ v\ L,\ T)) \qquad = \\
&\{T' \mid search_tree(Node\ L\ (k,v)\ T)\ \wedge\ balanced\ L\ \wedge\ balanced\ T\ \wedge \\
&\qquad depth\ T \leq depth\ L + 2\ \wedge\ depth\ L \leq depth\ T + 1 \\
&\longrightarrow\ avl\ T'\ \wedge\ (\forall k'.\ lookup\ k'\ T' = lookup\ k'\ (Node\ L\ (k,v)\ T))\ \wedge \\
&\quad (depth\ T = depth\ L + 2 \longrightarrow \\
&\qquad depth\ (Node\ L\ (k,v)\ T) \leq depth\ T' + 1\ \wedge \\
&\qquad depth\ T' \leq depth\ (Node\ L\ (k,v)\ T))\ \wedge \\
&\quad (depth\ T < depth\ L + 2 \longrightarrow \\
&\qquad depth\ T' = depth\ (Node\ L\ (k,v)\ T))\ \}
\end{aligned}$$

$$\begin{aligned}
cmps_spec\ (Inl\ (Right\ k\ v\ R,\ T)) &= \text{symmetric} \\
cmps_spec\ (Inl\ _) &= UNIV \\
cmps_spec\ (Inr\ _) &= UNIV
\end{aligned}$$

The last two equations treat 'junk'-input, and are necessary in order to show that the implementation of composition meets the specification, i.e. that the condition $compose\ \sqsubseteq\ cmps\_spec$ holds, because the function *compose* is basically undefined for the cases like $Inr\ (Left\ k\ v\ L,\ T)$ or $Inl\ (Stop\ T)$.

Obtaining a polymorphic update function. The conditions, it remains to show, are the reductivity of *decompose* and the *SPRP*-axiom. That *decompose* is a reductive coalgebra is clear, since it descends into the input tree such that the depth of the output tree is strictly smaller than of the input. Further, as already mentioned above, in the proof of *SPRP* we benefit of the abstraction of the composition by *cmps_spec*. such that the proof becomes accordingly concise.

Finally, the Divide-and-Conquer tactic gives the polymorphic function $update :: (\gamma_1 ::_{sort} \{linorder\}) \times \gamma_2 \times (\gamma_1, \gamma_2)\ Tree \Rightarrow (\gamma_1, \gamma_2)\ Tree$ together with the following rules:

1. $update\ \sqsubseteq\ update\_spec$, and

2. $update = compose \circ (Id\ **\ update\ ++\ Id) \circ decompose$, i.e. the recursive equation for *update*.

7.3 Conclusion

The main topic of this chapter was the standard interpretation of relational algebra as a locally complete allegory, established via theory morphisms. Thus, we could reuse the entire development done on the abstract level in the two previous chapters. In particular, using theory instantiation we have obtained the Divide-and-Conquer design tactic for relations, allowing us to define polymorphic mappings in correct-by-construction manner in the relational context. This has been firstly applied in order to derive the well-known QuickSort algorithm, and secondly for a derivation of an update function for balanced binary search trees.

Chapter 8

Conclusion and Outlook

In this thesis the notions of signature and theory morphisms in an abstractly described logical framework have been introduced. The presented description of the framework is based on and inspired by the LCF-style theorem prover Isabelle. Several important aspects, regarding the implementation of the approach in Isabelle, have been highlighted as well.

The notions of morphisms provide that theories in the framework as objects and theory morphisms as arrows form a category. Moreover, the presented definitions of morphisms allow several further conclusions, as shown in Section 4.4. However, the central one is that any theory morphism $\tau : \mathcal{T}_1 \longrightarrow \mathcal{T}_2$ can be extended to the function $\overline{\tau}$, translating any proof term in $\mathcal{T}_1$ to a proof term in $\mathcal{T}_2$ in accordance with the mapping of propositions, induced by the underlying signature morphism σ, as shown in the following diagram:

$$\begin{array}{ccccc} AX(\mathcal{T}_1) & \stackrel{\imath}{\hookrightarrow} & \mathit{Proofs}^{Adm}(\mathcal{T}_1)^{\sharp} & \xrightarrow{\mathit{proof\text{-}of}_{\mathcal{T}_1}} & \mathit{Props}(\mathcal{T}_1) \\ & \searrow^{\xi^{\imath}_{\tau}} & \downarrow \overline{\tau} & & \downarrow \overline{\sigma}_{\tau} \\ & & \mathit{Proofs}^{Adm}(\mathcal{T}_2)^{\sharp} & \xrightarrow{\mathit{proof\text{-}of}_{\mathcal{T}_2}} & \mathit{Props}(\mathcal{T}_2) \end{array} \tag{8.1}$$

Here, $\imath$ is an injection of the axioms into the set of admissible proof terms, from which propositions in $\mathit{Props}(\mathcal{T}_1)$ are derivable. More precisely, if $a \in AX(\mathcal{T}_1)$

with $\mathit{typevars}(\mathit{thm\text{-}of}_{\mathcal{T}_1}(a)) = \{\alpha_1, \ldots, \alpha_n\}$ then $\imath(a) = a_{\overline{[\alpha_i := \kappa_i]}_n}$ with some $\kappa_1, \ldots, \kappa_n \in \mathit{Types}(\Sigma_{type}(\mathcal{T}_1))$. Further, $\xi^{\imath}_{\tau}$ is the mapping, induced by the theorem map ξ_{τ} and the given injection $\imath$ satisfying

$$\frac{\imath(a) = a_{\overline{[\alpha_i := \kappa_i]}_n}}{\xi^{\imath}_{\tau}(a) = \xi_{\tau}(a)_{\overline{[\theta(\beta_i) := \overline{\sigma}^{\sigma_{class}}_{type}(\kappa_i)]}_m}}$$

where $\mathit{typevars}(\overline{\sigma}(\mathit{thm\text{-}of}_{\mathcal{T}_1}(a))) = \{\beta_1, \ldots, \beta_m\} \subseteq \{\alpha_1, \ldots, \alpha_n\}$ and θ is the bijection obtained from $\mathit{thm\text{-}of}_{\mathcal{T}_2}(\xi_{\tau}(a)) \stackrel{\alpha}{=}_{\theta} \overline{\sigma}(\mathit{thm\text{-}of}_{\mathcal{T}_1}(a))$, i.e. the theorem map condition for a.

Using this property we can, e.g. reason about the consistency of axiomatic theories in the following manner. Let $\mathcal{T}_1$ and $\mathcal{T}_2$ be two, e.g. HOL-theories, and $\tau : \mathcal{T}_1 \longrightarrow \mathcal{T}_2$ a theory morphism. Then if $\mathcal{T}_2$ is consistent so is $\mathcal{T}_1$. For if $\mathcal{T}_1$ would be inconsistent, the HOL-proposition *False* would be derivable in $\mathcal{T}_1$ such that translation of the proof term of *False* by $\overline{\tau}$ would give the derivable proposition *False* in $\mathcal{T}_2$ in contradiction to the consistency of $\mathcal{T}_2$. Now, the significance and comprehension of conservative developments in Isabelle have been already emphasised and motivated in the introduction, such that in this context one can say that the theorem prover Isabelle provides a broad basis for reasoning about consistency of axiomatic developments using theory morphisms. More generally, theory morphisms structure the layer of axiomatic developments, and moreover bridge the gap between the axiomatic layer and the layer of conservative developments in an object logic of a logical framework.

On the other hand, the morphisms presented here are expressive in the sense that they allow us to relate a wide range of theories. So, class, type and operation signatures can be related by means of respective mappings. Especially the possibilities to relate type signatures have been augmented using type schemes, in comparison to the standard approach of rank-preserving type constructor mappings. This, in turn, augments the possibilities to relate operation signatures. This treatment of type mappings brings also some problems, discussed in this thesis. So, cases can occur where a base signature morphism is not normalisable, due to emerging loose type constructors which can make conservative extensions in the source to non-conservative extensions in the target. This also strongly complicates the problem of the construction of pushouts.

Case study. First of all, the presented case study emphasises the applicability of the approach. So, procedures and techniques for construction and transformation of morphisms and theories, presented in the first part of the thesis, have been successfully applied in the higher order logic of Isabelle.

Furthermore, the case study also comprise a structured abstract axiomatic theory development, formalising and connecting Kleene algebras, (locally complete) allegories, and V-valued relations. This development hints at the levels of abstraction which can be reached in such a powerful framework as Isabelle, and moreover how concepts, properties and even complete design tactics 'in-the-large' (like Divide-and-Conquer) derived on an abstract level can be specialised to more concrete contexts via theory morphisms and instantiation.

On the other hand, the case study has also sketched how we can proceed other way around, i.e. from the target to the source of an existing theory morphism. This has been done by the example in Section 6.1.2 of an abstraction of the Church-Rosser property from locally complete allegories to typed Kleene algebras having additionally an appropriate converse operator. Even though this is only a special example, it relies on general principles and leads to the general question of abstraction techniques in theorem provers.

Abstraction by proof term transformation. Formal meta-logical proof terms in a logical framework provide an approach for abstraction of derived propositions over a fixed type by transformation of the corresponding proof term, as proposed in [Johnsen and Lüth 2004]. This abstraction technique will be sketched below, and the connection to the presented approach will be indicated as well.

So, if $p \in \mathit{Der}(\mathcal{T})$ is a derivable proposition in $\mathcal{T}$ and $C \in \Sigma_{type}(\mathcal{T})$ a type constructor with $\mathit{rank}_{\mathcal{T}}(C) = 0$ the abstraction of p over C proceeds roughly as follows.

Since p is derivable, there is a proof term $\pi \in \mathit{Proofs}^{Adm}(\mathcal{T})$ with $\emptyset \vdash \pi\ \mathit{proof\text{-}of}_{\mathcal{T}}\ p$. Further, there is a finite set of constants $\{f_1, \ldots, f_m\} \subset \Sigma_{op}(\mathcal{T})$ which operate in some way on C. Now, let $d_{\overline{[\alpha_i := \kappa_i]}_n}$ be some sub-term of π such that there is some f_j with $1 \leq j \leq m$ which occurs in $\mathit{thm\text{-}of}_{\mathcal{T}}(d)$. This implicit context information can be made to an explicit assumption. More precisely, we can replace $d_{\overline{[\alpha_i := \kappa_i]}_n}$ by a fresh proof variable X_d in π, which gives the proof term π'. Further, let $p_d \stackrel{def}{=} \mathit{thm\text{-}of}_{\mathcal{T}}(d)\overline{[\alpha_i := \kappa_i]}_n$, and assume that p_d is a closed term. Then we define $\pi'' \stackrel{def}{=} \overline{\overline{\lambda}}(X_d ::_{prop} p_d).\pi'$, and thus obtain $\emptyset \vdash \pi''\ \mathit{proof\text{-}of}_{\mathcal{T}}\ p_d \Longrightarrow p$. This step is then iterated until all such implicit assumptions are made explicit, i.e. we obtain $\emptyset \vdash \pi_1\ \mathit{proof\text{-}of}_{\mathcal{T}}\ p_{d_1} \Longrightarrow \ldots \Longrightarrow p_{d_k} \Longrightarrow p$, such that all relevant properties of $\{f_1, \ldots, f_m\}$ are contained in $p_{d_1}, \ldots, p_{d_k}$.

In the second step, $f_1, \ldots, f_m$ are then replaced by fresh variables $x_1, \ldots, x_n$ in π_1, which gives the proof term π_1'. Additionally, $x_1, \ldots, x_n$ are bounded, i.e.

$\pi_2 \stackrel{def}{=} \overline{\lambda}(x_1 ::_{type} \kappa_1) \ldots (x_n ::_{type} \kappa_n).\pi'_1$. Consequently,

$$\emptyset \vdash \pi_2 \; \textit{proof-of}_{\mathcal{T}} \; \bigwedge(x_1 ::_{type} \kappa_1) \ldots (x_n ::_{type} \kappa_n).\, p'_{d_1} \Longrightarrow \ldots \Longrightarrow p'_{d_k} \Longrightarrow p'$$

holds, where p'_i is obtained from p_i, replacing the constants $f_1, \ldots, f_m$ by the variables $x_1, \ldots, x_n$. Finally, since all function symbols depending on C have been replaced by variables, the type C itself can be replaced by a fresh type variable, say α_C in π_2, which yields the proof term π_3, from which an abstract proposition

$$\bigwedge(x_1 ::_{type} \kappa'_1) \ldots (x_n ::_{type} \kappa'_n).\, p''_{d_1} \Longrightarrow \ldots \Longrightarrow p''_{d_k} \Longrightarrow p''$$

which does not depend on C is derivable. The type variable α_C in this proposition can be in particular instantiated again by C, such that the initial proposition p is then one of the instances of the resulting abstract proposition.

This abstraction approach has been also implemented in the Isabelle framework. One of the examples where it has been successfully applied was a derivation of the general connection between linear and tail recursion from a special case. The considered special case form two functions $\textit{sum}_1 :: \textit{nat list} \Rightarrow \textit{nat}$ and $\textit{sum}_2 :: \textit{nat list} \times \textit{nat} \Rightarrow \textit{nat}$, where the former sums a list of natural numbers by linear recursion and the latter by tail recursion. Thus, the proposition $\textit{sum}_1\; xs = \textit{sum}_2\; (xs, 0)$, stating the equivalence of this two functions can be proved in the standard way by structural induction on the list xs. Now, the proof term transformation technique can be applied to this derivation abstracting over the concrete type *nat*. This results in a generalised proposition:

$$\begin{array}{l} \bigwedge(f_1 :: \alpha\; \textit{list} \Rightarrow \alpha)\; (f_2 :: \alpha\; \textit{list} \times \alpha \Rightarrow \alpha)\; (\textit{plus} :: \alpha \Rightarrow \alpha \Rightarrow \alpha)\; (\textit{zero} :: \alpha). \\ \mathcal{A}_1 \Longrightarrow \ldots \Longrightarrow \mathcal{A}_k \Longrightarrow f_1\; xs = f_2\; (xs, \textit{zero}) \end{array} \tag{8.2}$$

Here, $\textit{sum}_1$ has been replaced by the variable f_1 and $\textit{sum}_2$ by f_2. Moreover, the addition on *nat* as well as 0 become explicit and replaced by the variables *plus* and *zero*, respectively. Further, the computed context $\mathcal{C} \stackrel{def}{=} \{\mathcal{A}_1, \ldots, \mathcal{A}_k\}$ contain not only the required properties of f_1 and f_2 but also the conditions that *plus* and *zero* form a monoid structure on α. The general theorem (8.2) can be now re-applied in a variety of settings. So, for example the well-known equality of the *foldl* and *foldr* functions for lists of elements of any type with a monoid structure is a simple instance of it.

Taking into account the concept of local proof contexts in Isabelle, sketched in the introduction, we can make the following observations

- for a given proposition p and a type constructor C with the rank 0, the abstraction technique essentially extracts the weakest local proof context $\mathcal{C}$, constituted by the considered proof term of p, and gives a derivation of the corresponding abstract proposition p'' in the context $\mathcal{C}$;

- the discussion in the introduction has shown that local proof contexts cannot have type constructors as parameters;

- hence, a possible extension of the abstraction technique to type constructors with the rank > 0 would in general result in a 'global' proof context, i.e. in a new theory $\mathcal{P}$, now introducing abstracted functions in form of logical constants as well as implicit assumptions, extracted from the proof, in form of axioms in $\mathcal{P}$, such that the derivation of the abstracted conclusion can be done in a theory $\mathcal{B}$ extending $\mathcal{P}$; in other words, such an abstraction produces a parameterised theory $\langle \mathcal{P}, \mathcal{B} \rangle$, which re-usability is provided by means of theory morphisms and instantiation of theories.

Finally, it should be mentioned that a general structure certainly cannot arise by transformation of a proof term which does not carry this structure in some special form. That is, some structure has to be recognised and used in a proof of some special problem in order to obtain a generalisation by abstraction of the proof. For instance, at the end of Section 7.2.2 the recursive equations

$$\begin{array}{lcl} qsort\ [\,] & = & [\,] \\ qsort\ (Cons\ x\ xs) & = & qsort(low\ x\ xs)@[x]@qsort(high\ x\ xs) \end{array}$$

have been easily derived from the hylomorphism equation for *qsort* having an explicit divide-and-conquer form: $qsort = compose \circ F\ qsort \circ decompose$. This step is easy, because it basically just 'unfolds' explicit structuring concepts as composition and decomposition functions. But the actual problem is to proceed other way around, because it requires some kind of an 'abstraction' decision, i.e. we need to recognise and to make the divide-and-conquer structure explicit in the two recursive equations of *qsort* above.

How far this approaches can be systematised and mechanised could be an interesting object of research.

Appendix A

Stack-Example Theory Sources Outline

Parametrisation

Parameter part

```
theory QuotientType_Parameters
imports Equiv_Relations Hilbert_Choice
begin
```

The type constructor *T* is the first formal parameter.

```
typedecl 'a T
```

The equivalence relation *eqR* on *'a T* is the second formal parameter.

```
consts
 eqR :: "'a T × 'a T ⇒ bool"

syntax
 eqR_ :: "'a T ⇒ 'a T ⇒ bool"   ("_/ ≃ _"  [50, 51] 50)
translations
"a ≃ b" == "(a, b) ∈ eqR"


axioms
```

```
 eqR_refl  : "a ≃ a"
 eqR_trans : "a ≃ b ⟹ b ≃ c ⟹ a ≃ c"
 eqR_sym   : "a ≃ b ⟹ b ≃ a"
```

end

Introducing the quotient type

theory *QuotientType*
imports *QuotientType_Parameters*
begin

typedef *'a TQ = "(UNIV // eqR) :: ('a T set) set"*
lemma *eqR_equiv : "equiv UNIV eqR"*

Class operations.

constdefs

```
 repr_of :: "'a TQ ⇒ 'a T"          ("^_" [99] 1000)
"repr_of q ≡ (ϵ a. a ∈ Rep_TQ q)"

 class_of :: "'a T ⇒ 'a TQ"         ("[_]" [99] 1000)
"class_of t ≡ Abs_TQ (eqR `` {t})"
```

lemma *class_of_eq : "(s ≃ t) = ([s] = [t])"*

lemma *class_of_repr : "[^q] = q"*

lemma *repr_of_class : "^[t] ≃ t"*

lemma *repr_inj : "^s = ^t ⟹ s = t"*

Congruences and extension.

constdefs

```
 congruence :: "('a T ⇒ 'b) ⇒ bool"
"congruence f ≡ (∀s t. s ≃ t ⟶ f s = f t)"

 qtype_ext :: "('a T ⇒ 'b) ⇒ ('a TQ ⇒ 'b)"  ("_#" [99] 1000)
"f# ≡ f ∘ repr_of"
```

---

lemma *congruence_comp :*
"congruence g ⟹ congruence (f ∘ g)"

lemma *ext1 :*
"congruence f ⟹ f# [s] = f s"

corollary *ext2 :*
"congruence (class_of ∘ f) ⟹ (class_of ∘ f)# [s] = [f s]"
by*(subst ext1, assumption, simp)*

lemma *congruence_req :*
"(∀s. f# [s] = f s) ⟹ congruence f"

lemma *ext_comp :*
"congruence (class_of ∘ f) ⟹ congruence g ⟹
g# ((class_of ∘ f)# q) = (g ∘ f)# q"

lemma *congruence_comp2 :*
"congruence (class_of ∘ g) ⟹ congruence (class_of ∘ f) ⟹
congruence (class_of ∘ f ∘ g)"

lemma *ext_property :*
"f s ≃ t ⟹ congruence (class_of ∘ f) ⟹ (class_of ∘ f)# [s] = [t]"
end

Specification of a stack

theory *StackSpec*
imports *HOL*
begin

typedecl *'a stack*

```
consts
 empty    :: "'a stack"
 is_empty :: "'a stack ⇒ bool"
 top      :: "'a stack ⇒ 'a"
 pop      :: "'a stack ⇒ 'a stack"
 push     :: "'a ⇒ 'a stack ⇒ 'a stack"

axioms
 S0 : "is_empty empty"
 S1 : "¬ is_empty(push x s)"
 S2 : "top(push x s) = x"
 S3 : "pop(push x s) = s"
 S4 : "¬ is_empty s ⟹ push (top s) (pop s) = s"
```

An immediate consequence:

```
lemma empty_neq_push : "empty ≠ push x s"
end
```

An implementation of stack

```
theory StackImpl
imports "$AWE_HOME/Extensions/AWE" Equiv_Relations Hilbert_Choice
begin
```

The well-known implementation of stacks by arrays with a pointer.

```
record 'a stack =
 size :: nat
 f    :: "nat ⇒ 'a"


constdefs
 empty :: "'a stack"
"empty ≡ (| size = 0, f = (λn. undefined) |)"

 is_empty :: "'a stack ⇒ bool"
"is_empty s ≡ (size s = 0)"

 top :: "'a stack ⇒ 'a"
"top s ≡ if size s = 0 then undefined else (f s) ((size s) - 1)"
```

```
 pop :: "'a stack ⇒ 'a stack"
"pop s ≡ (| size = ((size s) - 1), f = (f s) |)"

 push :: "'a ⇒ 'a stack ⇒ 'a stack"
"push x s ≡ (| size = ((size s) + 1), f = (f s)(size s := x) |)"
```

Defining an equivalence relation on concrete stacks.

```
constdefs
 eqR :: "'a stack × 'a stack ⇒ bool"
"eqR ≡ λ(s, t). (size s = size t) ∧ (∀n < (size s). (f s) n = (f t) n)"
```

Supplying the actual parameters *'a stack* and *eqR* to the parametrisation.

```
sigmorph parameters : QuotientType_Parameters ⟶ StackImpl
type_map: ["'a T" ↦ "'a stack"]
```

The operation assignment
QuotientType-Parameters.eqR ↦ *StackImpl.eqR*
is already uniquely determined by the given type assignment.

```
thymorph_goals parameters
```

Note: the infix annotation for the relation *eqR* has been imported from the theory QuotientType_Parameters by this *thymorph_goals* call, and is thus available here as well.

```
lemma eqR_sym:
"a ≃ b ⟹ b ≃ a"
lemma eqR_refl:
"a ≃ a"
lemma eqR_trans:
"a ≃ b ⟹ b ≃ c ⟹ a ≃ c"
```

Since the theory morphism goals have been proved, the signature morphism *parameters* can be now extended to a theory morphism as follows:

```
thymorph parameters by_sigmorph parameters
```

Obtaining the quotient type for *'a stack* by instantiation.

```
instantiate_theory QuotientType by_thymorph parameters
renames: [("QuotientType.TQ" ↦ stackQ)]
axiomatic
```

The instantiation has to be axiomatic only because of the type definition axiom introduced in QuotientType. Nevertheless, recent development of the AWE

Extensions provides a possibility to hide this,using the technique generating the corresponding type definition (in this case *stackQ*) in the target theory (for more details see the AWE Manual). However, this changes nothing on the axiomatic nature of the instantiation of type definitions in HOL.

This instantiation step results in the extended theory morphism *parameters'*, which is now renamed to *quotient*

rename_thymorph *parameters'* **to** *quotient*

Now, properties of congruences and extension can be translated from QuotientType, and will be used in the proofs of the stack specification below.

translate_thm *"QuotientType.congruence_comp"* **along** *quotient*
translate_thm *"QuotientType.ext1"* **along** *quotient*
translate_thm *"QuotientType.ext2"* **along** *quotient*
translate_thm *"QuotientType.ext_comp"* **along** *quotient*
translate_thm *"QuotientType.ext_property"* **along** *quotient*
translate_thm *"QuotientType.congruence_comp2"* **along** *quotient*

Extending the stack operations to *'a stackQ*.

constdefs

```
 qempty :: "'a stackQ"
"qempty ≡ class_of empty"

 is_qempty :: "'a stackQ ⇒ bool"
"is_qempty ≡ is_empty#"

 qtop :: "'a stackQ ⇒ 'a"
"qtop ≡ top#"

 qpop :: "'a stackQ ⇒ 'a stackQ"
"qpop ≡ (class_of ∘ pop)#"

 qpush :: "'a ⇒ 'a stackQ ⇒ 'a stackQ"
"qpush x ≡ (class_of ∘ push x)#"
```

Congruence properties of the stack operations.

lemma *is_empty_congruence :*
"congruence is_empty"
lemma *top_congruence :*
"congruence top"
lemma *pop_congruence :*
"congruence (class_of ∘ pop)"

---

lemma *push_congruence* :
"congruence (class_of ∘ push x)"

Proving the stack specification for the quotient type *'a stackQ* and the lifted stack-operations.

sigmorph *spec_interpretation : StackSpec* ⟶ *StackImpl*
type_map: ["'a stack" ↦ "'a stackQ"]

Switch-off the simplifier for AWE

ML {* AWE.proof_search_with_simp := false *}

thymorph_goals *spec_interpretation*

lemma *S0*: "is_qempty qempty"
lemma *S1*: "¬ is_qempty (qpush x s)"
lemma *S2*: "qtop (qpush x s) = x"

lemma *S3*: "qpop (qpush x s) = s"
lemma *S4*: "¬ is_qempty s ⟹ qpush (qtop s) (qpop s) = s"

The correctness of the implementation is finally established by the extension of the signature morphism to a theory morphism.

thymorph *spec_interpretation* **by_sigmorph** *spec_interpretation*

Now we can translate and combine general properties, derived in StackSpec and QuotientType.

translate_thm "StackSpec.empty_neq_push" **as** *qempty_neq_qpush*
along *spec_interpretation*

So we can obtain the inequality of any *push* with the empty stack.

corollary "¬ (push x s ≃ empty)"

end

Bibliography

[Baader and Nipkow 1998] BAADER, F. AND NIPKOW, T. 1998. *Term Rewriting and All That.* Cambridge University Press.

[Backhouse 2002] BACKHOUSE, R. 2002. Galois Connections and Fixed Point Calculus. *Algebraic and Coalgebraic Methods in the Mathematics of Program Construction*, Volume 2297 of *Lecture Notes in Computer Science*, pages 89–148. Springer.

[Ballarin 2009] BALLARIN, C. 2009. *Tutorial on Locales and Locale Interpretation.* Second revision for Isabelle 2009, available at Isabelle homepage `http://isabelle.in.tum.de`

[Berghofer and Nipkow 2000] BERGHOFER, S. AND NIPKOW, T. 2000. Proof terms for simply typed higher order logic. In Harrison, J., Aagaard, M., editors, *13th International Conference on Theorem Proving in Higher Order Logics (TPHOLs'00)*, Volume 1869 of *Lecture Notes in Computer Science*, pages 38–52. Springer.

[Bird and de Moor 1997] BIRD, R. AND DE MOOR, O. 1997. *Algebra of Programming.* Volume 100 of Prentice Hall Int. Series in Computer Science. Prentice Hall, London.

[Bortin, Johnsen and Lüth 2006] BORTIN, M., JOHNSEN, E. B., LÜTH, C. 2006. Structured Formal Development in Isabelle. In Volume 13 of *Nordic Journal of Computing*, pages 2–21.

[Burstall and Goguen 1980] BURSTALL, R. M., AND GOGUEN, J. A. 1980. The semantics of clear, a specification language. *Abstract Software Specifications*, Volume 86 of *Lecture Notes in Computer Science*, pages 292–332. Springer.

[Doornbos and Backhouse 1995] DOORNBOS, H. AND BACKHOUSE, R. 1995. Induction and Recursion on Datatypes. In Möller, B., editor, *Mathematics of Program Construction*, Volume 947 of *Lecture Notes in Computer Science*, pages 242–256. Springer.

[Ehrig and Kreowski 1999] EHRIG, H. AND KREOWSKI, H.-J. 1999. Refinement and Implementation. In E. Astesiano, H.-J. Kreowski, B. Krieg-Brückner, editors, *Algebraic Foundations of Systems Specification*, pages 201–242. Springer.

[Griffioen and Huisman 1998] GRIFFIOEN, D. AND HUISMAN, M. 1998. A comparison of PVS and Isabelle/HOL. *Theorem Proving in Higher-Order Logics (TPHOLs 1998)*, Volume 1479 of *Lecture Notes in Computer Science*, pages 123–142. Springer.

[Haftmann 2009] HAFTMANN, F. 2009. *Haskell-style type classes with Isabelle/Isar* available at Isabelle homepage `http://isabelle.in.tum.de`

[Harrison 2009] HARRISON, J. 2009. *Handbook of Practical Logic and Automated Reasoning.* Cambridge University Press.

[Johnsen and Lüth 2004] JOHNSEN, E. B., LÜTH, C. 2004. Theorem Reuse by Proof Term Transformation. In Konrad Slind and Annette Bunker and Ganesh Gopalakrishnan, editors, *International Conference on Theorem Proving in Higher-Order Logics (TPHOLs 2004)*, Volume 3223 of *Lecture Notes in Computer Science*, pages 152–167. Springer.

[Kammüller *et al.* 1999] KAMMÜLLER, F., WENZEL, M., AND PAULSON, L. C. 1999. Locales – A Sectioning Concept for Isabelle. In *Proc. 12th International Conference on Theorem Proving in Higher Order Logics (TPHOLs'99)*, Volume 1690 of *Lecture Notes in Computer Science*, pages 149–166. Springer.

[Kozen 1998] KOZEN, D. 1998. *Typed Kleene algebra.* Technical Report TR98-1669, Computer Science Department, 1998. Cornell University.

[Krieg-Brückner *et al.*1991] KRIEG-BRÜCKNER, B., KARLSEN, E. W., LIU, J., TRAYNOR, O. 1991. The PROSPECTRA Methodology and System: Uniform Transformational (Meta-) Development. In Prehn, S., Toetenel, W. J. (editors), *Formal Software Development Methods*,

Proc. 4th Intl. Symp. of VDM Europe, Volume 552 of *Lecture Notes in Computer Science*, pages 363–397. Springer.

[Mac Lane 1998] MAC LANE, S. 1998. *Categories for Working Mathematician.* Springer.

[Milner 1985] MILNER, R. 1985. The use of machines to assist a rigorous proof. In Hoare, C. A. R. and Shepherdson, J. C., editors, *Mathematical Logic and Programming Languages*, pages 77–88, Prentice Hall.

[Mosses 2004] MOSSES, P. D. Editor. 2004. *CASL – the common algebraic specification language*, Reference Manual. Volume 2960 of *Lecture Notes in Computer Science.* Springer.

[Nipkow *et al.* 2002] NIPKOW, T., PAULSON, L. C., AND WENZEL, M. 2002. Isabelle/HOL – a Proof Assistant for Higher Order Logic. Volume 2283 of *Lecture Notes in Computer Science.* Springer.

[Nipkow 1993] NIPKOW, T. 1993. Order-sorted polymorphism in Isabelle. In G. Huet and G. Plotkin, editors, *Logical Envinronments*, pages 164-188. Cambridge University Press.

[Owre at al. 1999] OWRE, S. AND SHANKAR, N. AND RUSHBY, J. M. AND STRINGER-CALVERT, D. W. J. 1999. PVS Language Reference. Computer Science Laboratory, SRI International, Menlo Park, CA.

[Owre and Shankar 2001] OWRE, S. AND SHANKAR, N. 2001. *Theory Interpretations in PVS.* Technical Report SRI-CSL-01-01, Computer Science Laboratory, SRI International, Menlo Park, CA.

[Paulson 1989] PAULSON, L. C. 1989. The foundation of a generic theorem prover. *Journal of Automated Reasoning*, Vol. 5, pages 363–397.

[Paulson 1990] PAULSON, L. C. 1990. Isabelle: the next 700 theorem provers. In Odifreddi, P., editor, *Logic and Computer Science*, pages 361–386, Academic Press.

[Peyton Jones 2003] PEYTON JONES, S., Editor. 2003. *Haskell 98 Language and Libraries. The Revised Report.* Cambridge University Press.

[Pfenning 2001] PFENNING, F. 2001. Logical Frameworks. In *Handbook of Automated Reasoning*, Robinson, A., Voronkov, A., editors, Elsevier, pages 1063–1147.

[Srinivas and Jullig 1995] SRINIVAS, Y. V. AND JULLIG, R. 1995. Specware: Formal Support for Composing Software. In *Proc. Conf. Mathematics of Program Construction*, Volume 947 of *Lecture Notes in Computer Science*. Springer.

[Wenzel 2001] WENZEL, M. 2001. *Isabelle/Isar – a versatile environment for human-readable formal proof documents.* PhD thesis. Technische Universität München.

Index